Gerald Farca
Playing Dystopia

Studies of Digital Media Culture | Volume 8

The series is edited by Gundolf S. Freyermuth and Lisa Gotto.

Gerald Farca, born in 1983, did his doctorate at the University of Augsburg (English Literature) and is a member of the Augsburg Cultural Ecology Research Group. In 2016, the digital culture and game studies scholar worked as a visiting researcher and lecturer at the Center for Computer Games Research of the IT University in Copenhagen.

Gerald Farca

Playing Dystopia

Nightmarish Worlds in Video Games and the Player's Aesthetic Response

[transcript]

Bibliographic information published by the Deutsche Nationalbibliothek
The Deutsche Nationalbibliothek lists this publication in the Deutsche Nationalbibliografie; detailed bibliographic data are available in the Internet at http://dnb.d-nb.de

Cover concept: Kordula Röckenhaus, Bielefeld
Cover illustration: Remember Me (Capcom, 2013)
Printed by Majuskel Medienproduktion GmbH, Wetzlar
Print-ISBN 978-3-8376-4597-2
PDF-ISBN 978-3-8394-4597-6
https://doi.org/10.14361/9783839445976

Contents

Acknowledgments

It has been quite a journey, and I want to thank a fair number of people in various countries for their support over these last few years.

I am greatly indebted to my supervisors, Prof. Dr. Martin Middeke and PD. Dr. Christoph Henke, for their invaluable help and inspiration during my Ph.D. studies and time at the University of Augsburg.

In addition, I want to express my gratitude to Espen Aarseth, Hans-Joachim Backe, and Susana Pajares Tosca for the fruitful conversations we had and their critical feedback on my work, and the entire team of the Center for Computer Games Research at the IT University of Copenhagen for involving me in a pleasant and creative ambiance.

Many thanks go as well to the Oberseminar of Augsburg University, including Alexander Lehner, Christian Attinger, David Kerler, Martin Riedelsheimer, Adina Sorian, Despina Repa, Bettina Auerswald, Korbinian Stöckel, Timo Müller, and many more.

In addition, I am particularly grateful to Gundolf S. Freyermuth, Björn Bartholdy, André Czauderna, and the team of the Cologne Game Lab.

Moreover, many thanks go to Sebastian Domsch, Jens Müller, Susanna Layh, Hubert Zapf, Günter Butzer, Michael Nitsche (from my hometown Augsburg), Miguel Sicart, Markus Schulzke, and to the many colleagues and friends I met at international conferences: Victor Navarro-Remesal, Lars de Wildt, Jan Wysocki, Ea Christina Old-Gregg Willumsen, Daniel Vella, Sebastian Möring, Oli Tapio Leino, Hugo Montembeault, Arno Görgen, Rudolf Inderst, Marc Bonner, Veli-Matti Karhulati, Souvik Mukherjee, Thomas Apperley, Tomasz Majkowski, Justyna Janik, Toh Weimin, Pawel Frelik, Feng Zhu, Astrid Ensslin, Joel Lehman, Anders Sundnes Løvlie, Ida Kathrine Hammeleff Jørgensen, Rune Kristian Lundedal Nielsen, Michael Debus, Sebastian Risi, Daniel Cermak-Sassenrath, Martin Pichlmaier, Anders Hartzen, Sebastian Zethraeus, Flaminia Grimaldi, Sigursteinn Gunnarsson, Alon Kaplan, Jennifer Killham, Kevin Lin, Julian Oberbeck,

Linda Kruse, Litos Sinmas, Mohamed Larbi Bouchoucha, Adam Russel, Juliette Lemaitre, Jackson Wood, Manas Gajare, Patrick Sobel, and Charlotte Ladevèze. I apologise to those I forgot to mention.

Further, I want to thank three foundations that helped me partially fund this book: FAZIT-Stiftung, Oskar-Karl-Forster Stiftung, Stiftung Digitale Spielekultur.

Lastly, and most importantly, I wish to thank my parents for their continuous support and love.

Introduction

In FALLOUT 3 (Bethesda Game Studios, 2008), the player embarks on a journey into the dystopia of the Capital Wasteland. The gameworld awaits her[1] in turmoil and in anarchical structures and factions that divide the wasteland. This survival of the fittest is experienced by the player in various forms and can be narrowed down to a chain of events that confronts her with a far-reaching *choice*—which is representative for my deliberations in this study.

Relatively early in the game, the player is sent on a quest to Tenpenny Tower, a (supposedly) utopian microcosm within the larger dystopia of the gameworld. Tenpenny Tower is fenced in by a wall to protect it from the surrounding areas and is the home of Allistair Tenpenny, a formerly renowned businessman who grants entrance only to the wealthy and noble. He is especially suspicious of a race called the ghouls. After the apocalypse, they were deformed by the radiation fallout and are now avoided by most wasteland inhabitants. Meanwhile, the word has spread about the sterile illusion of Tenpenny Tower, and the ghouls regularly attempt to gain residence in it. Based on this premise and the *perspectives* (positions) it holds, the player becomes involved in the events and is set in a precarious situation. In order to mediate between the positions, several possibilities are thinkable and potential for the player to enact: 1) eradicating the ghouls; 2) killing the inhabitants of Tenpenny Tower; 3) solving the issue through diplomacy (which might nonetheless end in catastrophe); 4) doing nothing. The prospect for an ethical solution, as Miguel Sicart claims, are thus slim[2]—but becoming creatively engaged in the situation holds *emancipatory potential* for the player.

1 Throughout this study, I will use the personal pronouns *she, he, her, him* interchangeably and in alternation with each new chapter. There is no agenda behind the constellations, the aim is for readability while upholding a gender-inclusive use of pronouns.

2 Miguel Sicart, *Beyond Choices: The Design of Ethical Gameplay* (Cambridge, Mass.: MIT Press, 2013), 96-98.

Consequently, to resolve the issue and have the situation exert its effect, the player becomes involved in a multi-faceted *form of play* that oscillates between her *literal* interaction with a set of rules and navigation of the gameworld and a *figurative* one that comprises her imaginative games within it.[3] In a mixture of *imaginative acts* and *ergodic actions,* therefore, the player approaches the issue and becomes *psychologically* and *aesthetically (emancipatedly) involved* in the action. To do so, she imaginatively closes the indeterminacies of the gameworld and sets the various elements and events in context. There are many blank spots and incomplete structures in FALLOUT 3—such as an inaccessible building or incomplete information about a character or a situation—and to fill in these particulars, the player draws from her real-world knowledge, that of other fictional worlds/games, and her experience so far in the game. This form of involvement is well-known in literature and film, but here it occurs simultaneously with the player's navigation of the gameworld and her performance in it. As Henry Jenkins holds:

> Narrative comprehension is an active process by which viewers assemble and make hypotheses about likely narrative developments on the basis of information drawn from textual cues and clues. … As they move through the film, spectators test and reformulate their mental maps of the narrative action and the story space. In games, players are forced to act upon those mental maps, to literally test them against the game world itself.[4]

Espen Aarseth has described this latter quality of video game play—which he calls *the ergodic*—as a sort of physical involvement between "*ergon* and *hodos*, meaning 'work' and path'," where the player exerts "nontrivial effort … to traverse the text [game]"[5] and fills in the gameworld indeterminacies by creating signifiers and personal perspectives on the game. This may occur when she decides to solve a quest in a particular manner, chooses from different dialogue options, discovers new parts of the gameworld, or may be as simple as the choice of certain equipment.

---

3 Marie-Laure Ryan, *Narrative as Virtual Reality: Immersion and Interactivity in Literature and Electronic Media* (Baltimore: John Hopkins UP, 2001), 16-17.

4 Henry Jenkins, "Game Design as Narrative Architecture," in *First Person: New Media as Story, Performance, and Game*, ed. Noah Wardrip-Fruin and Pat Harrigan (Cambridge, Mass.: MIT Press, 2004), 126.

5 Espen J. Aarseth, *Cybertext: Perspectives on Ergodic Literature* (Baltimore: John Hopkins UP, 1997), 1.

Ergodicity, therefore, enables the player to manoeuvre and act within a gameworld through some sort of input device (such as a controller)—but this also presupposes "*a disposition* and readiness to act" on the player's side, "not merely the action of pressing a button or pulling a joystick."[6] Gordon Calleja thus rightfully claims that the ergodic requires "the player's *cognitive* effort"[7] as she is planning her moves.[8] Still, this form of cognitive involvement is different from the imaginative involvement referred to above, as it is concerned with deliberations on ludic encounters and problem solving, whereas the latter also refers to the player's engagement with a storyworld. Naturally, both forms are often intertwined, but a separation seems beneficial, as there are different things going on in the player's mind in each situation of play.

Play, as such, involves the player in both imaginative and ergodic ways and draws on her real-world knowledge so that she can comprehend the gameworld and act within it. By doing so, play naturally interweaves gameworld considerations with those of the empirical world[9]—this is to say, while the player negotiates her options within the diegesis (ludic possibilities to exploration/interaction, gaining power/money by killing the ghouls, or solving issues in an ethical manner, and so on), her deliberations are informed by what she knows from the empirical world (familiar norms, conventions, processes, states of affairs, etiquettes, etc.). However, these elements are experienced differently in the dystopian gameworld, for the transfer to the fictional reality has distorted their appearance. By disrupting the vertical hierarchies of the empirical world elements, magnifying the differences, and reorganising them horizontally into gameworld perspectives/positions, they are partially freed from their usual interrelations. As such, the created reality confronts the player with a refracted mirror of what she knows and involves her in *games of estrangement* between what is familiar and unfamiliar. This state of matter may entice and bewilder the player at the same time, as it aggravates the referentiality between worlds, but, in doing so, it allows for the formation of hitherto unexpected connections between the dystopian gameworld and the opaque nature of social totality. Video game play, therefore, shows emancipatory potential so

6 Gordon Calleja, *In-Game: From Immersion to Incorporation* (Cambridge, Mass.: MIT Press, 2011), 41.

7 Ibid., 42.

8 Ibid., 41-42.

9 With the empirical world, I refer to the extratextual surroundings (outside of the gameworld) and to what is usually meant when saying the 'real world', the 'actual world', and so on.

fundamental to the dystopian genre and equals the appreciator's experience in non-ergodic art, which also "resists translation into referential meaning."[10]

Play, in this sense, can be "pleasurable but risky, and potentially harmful,"[11] as it "disruptively reveal[s] our conventions, assumptions, biases, and dislikes."[12] This quality of play is expressed in the resolution of the Tenpenny Tower quest in whatever form, when the player compares the enacted events (the perspectives she encounters and co-creates, which foreground certain empirical world elements) to what she knows from the real world and composes creative connections between both worlds. The image of Tenpenny Tower that emerges in the player's mind thereby allows for the establishment of links to and associations with the empirical world—whether these revolve around the policy of building a wall to protect the United States of America from Mexican immigration and to marginalise ethnic minorities, or the cultural swing to the right in parts of Europe, where countries such as Poland and Hungary refrain from accepting Syrian refugees for fear they would pollute their immaculate Utopias. Such free-floating implications stand not simply in metaphorical nor allegorical relation to the empirical world but result from an intricate experience in the gameworld and the negotiation of its contents. In addition, they hold the potential to influence the player's in-game actions, and through this test run in virtuality, she may explore solutions (or attenuations) for similar crises in the empirical world. Of course, imaginings and consequent actions differ from player to player and the cultural surroundings they are familiar with, but they are nonetheless outlined by a game's structure—that is, what the player encounters and how it is arranged.

It is this dialectic between *(dystopian) game, player, and culture (world)* that will be the object of scrutiny in this study and which is responsible for affecting the player in a lasting manner. Representational art—including literature, film, theatre, and games—involves the participant in creative *games of fictionality* in different ways. These are informed by the artwork's guiding structure and the participant's freedom of imagination and action, and require a specific approach from her. This is because she not only tries to make sense of the work at hand, which was encoded in its transition to the fictional realm, but also because in doing so, the participant composes creative connections to the empirical world in a process of decoding that is filtered through her self and set in motion by her attempts to *close the blanks* between the work's perspectives. Wolfgang Iser has described

10 Wolfgang Iser, *The Act of Reading: A Theory of Aesthetic Response*, trans. Wilhelm Fink (Baltimore: John Hopkins UP, 1978), 11; cf. 61, 99, 184.

11 Miguel Sicart, *Play Matters* (Cambridge, Mass.: MIT Press, 2014), 11.

12 Ibid., 15.

this process for the literary text as the reader's "attempt to ideate [*vorstellen*] that which one can never see as such" and that "the character of these images consists in the fact that they bring to light aspects which could not have emerged through direct perception of the object."[13] Therefore, playing dystopia (and video game narratives in general) comes close to—though it also differs in important respects—the reader's involvement in the literary text. For, in contrast to ideology, "heuristic fiction proposes *trial runs* for approaches to what is,"[14] and this allows "the reader to see everyday norms and conventions, social habits of thinking and feeling, in a different light ... [and] to explore, in a kind of *trial action in a virtual environment*, the consequences of breaking and transgressing norms without having to fear sanctions in real life."[15]

Such a playful engagement with fictional worlds is of great benefit to the philosophy of Utopia[16] and its manifestations in fictional narrative form: the genres of *utopia* and *dystopia*. These magnify the distance between empirical and fictional worlds even further as they involve the reader/appreciator in paradisiac or nightmarish worlds, respectively, and in an "exploration of alternatives in a way that supports or catalyzes social transformation."[17] In this study, I will focus on dystopian fiction, which is said to evoke a specific effect in the appreciator. By

13 Iser, *Act*, 137.

14 Wolfgang Iser, *The Fictive and the Imaginary: Charting Literary Anthropology*, trans. Frankfurt a.M.: Suhrkamp Verlag (Baltimore: John Hopkins UP, 1993), 143; emphasis added.

15 Ingo Berensmeyer, *Literary Theory: An Introduction to Approaches, Methods and Terms* (Stuttgart: Klett Lerntraining, 2009), 79; emphasis added.

16 Following Baccolini and Moylan, I use capitalisation to refer to the philosophy of Utopia (utopianism) and Anti-Utopia and to specific Utopias—a certain imaginary, virtual, or real place that fulfils the necessary aspects of a Utopia: for instance, the Utopia of Columbia in BIOSHOCK INFINITE (Irrational Games, 2013) or the island of Utopia in Thomas More's eponymous narrative (1516). Lowercase will be used for Utopia's manifestations, such as real-world practices or fictional experiments: the literary utopia/dystopia, those of film/theatre, and the video game utopia/dystopia. (Raffaella Baccolini and Tom Moylan, "Introduction. Dystopia and Histories," in *Dark Horizons: Science Fiction and the Dystopian Imagination*, ed. Raffaella Baccolini and Tom Moylan [New York: Routledge, 2003], 11.)

17 Ruth Levitas and Lucy Sargisson, "Utopia in Dark Times: Optimism/Pessimism and Utopia/Dystopia," in *Dark Horizons: Science Fiction and the Dystopian Imagination*, ed. Raffaella Baccolini and Tom Moylan (New York: Routledge, 2003), 13.

involving her in imaginative trial actions and letting her explore playfully a reduced and estranged version of social totality (in the fictional storyworld), the dystopian narrative functions as a *warning* to humankind. It reveals troublesome trends hidden within empirical reality while, at the same time, it investigates potential solutions to the nightmare[18]—and, thus, represents a "heuristic device … an epistemological and not ontological entity."[19] Consequently, and in a "*disruption (Beunruhigung)* of the present,"[20] Utopia evokes an *aesthetic response* in the appreciator that enlightens her through fictional trial actions and drives her to ethical action in the real world.

HYPOTHESIS: THE VIDEO GAME DYSTOPIA AS A NEW STRATEGIC ENTERPRISE OF UTOPIA AND PLAYFUL TRIAL ACTION

The main purpose of this study is to explain why these claims apply so well to the *video game dystopia* (VGD), and to explore the player's aesthetic response to nightmarish gameworlds. I will argue that the video game dystopia describes a new strategic enterprise of the utopian philosophy. By sending the player on a journey through hell but retaining a hopeful (utopian) core, it involves her in a *playful trial action (or test run)* in which she may test, track, and explore in detail an estranged gameworld and an alternative societal model through imaginative and ergodic means. This venture into the fictional reality of dystopia shows potential to warn the player about negative trends within empirical reality and to explore emancipatory routes that may transform the gameworld. It thus serves the player as a subversive example and inducement to effect social change and transformation in the empirical world.

This aesthetic response is meticulously outlined by dystopia's implied player, which can be described as the affordance and appeal structure of the game that

---

18 Tom Moylan, *Scraps of the Untainted Sky: Science Fiction, Utopia, Dystopia* (Boulder, Colo.: Westview Press, 2000), 96-97, 106-107; Fátima Vieira, "The Concept of Utopia," in *The Cambridge Companion to Utopian Literature,* ed. Gregory Claeys, (Cambridge: Cambridge UP, 2010), 6, 17; Hans Ulrich Seeber, *Die Selbstkritik der Utopie in der Angloamerikanischen Literatur* (Münster: LIT, 2003), 33, 45-46.

19 Darko Suvin, *Metamorphoses of Science Fiction: On the Poetics and History of a Literary* Genre (New Haven: Yale UP, 1979), 52.

20 Fredric Jameson, *Archaeologies of the Future: The Desire Called Utopia and Other Science Fictions* (London: Verso, 2005), 228.

holds all the predispositions necessary for the game to exercise its effect by endowing the player with a certain role (or roles). The effect I am referring to is well-known to the appreciator of representational art or the reader of literary texts and describes an *aesthetic effect* experienced in and through the act of play. Consequently, becoming involved in a dystopian game, the empirical player steps into a *dialectical communication* (a creative negotiation to discover the truth)[21] with the implied player, whose roles she accepts but, at the same time, exposes to scrutiny. These are embedded/virtualised in an entire system of perspectives (positions) that compose the game and its world and which the player co-creates through ergodic action. As the connections between the perspectives remain indeterminate (or are negated), the player is urged to close the blanks between them. She thus experiences art's aesthetic effect through acts of ideation and creates the revelatory connection to the empirical world in a continual revision of composed images and meanings.

This analysis of play's underlying structure shows the benefit of comprehending the player's involvement in dystopia—and video game narratives (VGN)—on a profound, structuralist level that anticipates participation. In other words, considering the specific mediality of the video game medium in relation to non-ergodic forms of representational art will help explain the structure that affords play in the first place and which drives the player to catharsis and aesthetic response. Such a player necessarily shows an open-minded attitude towards the work of art she is confronted with, and given the plethora of player types (that all savour play for different reasons: for ludic, narrative, or world pleasures, and so on), *the emancipated player* becomes a necessity for my deliberations. For this player type describes an empirical being who is willing to engage with the implied player on a complex level, to indulge in potentialities and imaginings that are evoked, while not blindly accepting any truths.[22]

What is more, by scrutinising the tripartite dialectic between *(dystopian) game, player, and culture (world)*, the study will illuminate the relationship between

21 The term dialectic as it is used here implies a negotiation/discourse between two or more parties (in this case: game, player, culture) that engage in a reasonable yet subversive argument to discover unspoken or general truths about life and the world. This use resembles the way Aristotle defines dialectic in *The Art of Rhetoric* (367-322BC), where it is described as similar though not equal to rhetoric. (Aristotle, *The Art of Rhetoric*, trans. W. Rhys Roberts [Megaphone eBooks, 2008], http://www.wendelberger.com/downloads/Aristotle_Rhetoric.pdf).

22 Henceforth, I will speak of the empirical player in terms of an emancipated player if not indicated otherwise.

gameworld and empirical world in terms of *fictionality* and from a *phenomenological* point of view. Fiction, as such, is construed as a semantic phenomenon and as a functional approach that involves the player in an aggravated communication with the dystopian game—whereby I primarily follow Walton, Doležel, Iser, and their counterparts from science fiction/dystopian studies, Suvin and Moylan.[23] Moreover, this conception of fiction does not contest the ontological dimension of the gameworld as a virtual artefact but organically integrates it.

STATE OF RESEARCH: THE ASSUMED POSITION WITHIN VIDEO GAME STUDIES

The benefits of these deliberations to video game studies involve first of all the establishment of a *framework* to categorise and describe the VGD as a genre. Such a work has not yet been done,[24] although there are several articles and conference proceedings that mostly focus on close playings of dystopian games[25] (often discussing the BIOSHOCK[26] series), while few attempt to categorise the VGD as a

23 Kendall L. Walton, *Mimesis as Make-Believe: On the Foundations of the Representational Arts* (Cambridge, Mass.: Harvard UP, 1990); Lubomír Doležel, *Heterocosmica: Fictional and Possible Worlds* (Baltimore: John Hopkins UP, 1998); Iser, *Act*; *Imaginary*; Suvin, *Metamorphoses*; Moylan, *Scraps*.

24 Frelik argues that "to date [2014] there has been little critical work on how video games in general and specific gaming genres in particular engage and transform traditional genre codes as derived from fiction and film." (Pawel Frelik, "Video Games," in *The Oxford Handbook of Science Fiction*, ed. Rob Latham [Oxford: Oxford UP, 2014], 229).

25 For example: William Gibbons, "Wrap Your Troubles in Dreams: Popular Music, Narrative, and Dystopia in *Bioshock*," *Game Studies* 11, no. 3 (December 2011) http://gamestudies.org/1103/articles/gibbons; Joseph Packer, "The Battle for Galt's Gulch: *Bioshock* as Critique of Objectivism," *Journal of Gaming and Virtual Worlds* 2, no. 3 (December 2010): 209-224, https://www.intellectbooks.co.uk/journals/view-Article,id=10275/

26 BIOSHOCK (2K Boston, 2007), BIOSHOCK 2 (2K Marin, 2010), BIOSHOCK INFINITE (Irrational Games, 2013), BURIAL AT SEA: EPISODE 1 and 2 (Irrational Games, 2013-2014).

narrative genre.[27] By doing so, the results of this study will be of additional importance to utopian studies in a transmedial environment, which invite the utopian inclined researcher to "stay open to new forms [of the genre] as they emerge" but that "will require new critical formulations."[28] What is more, the practical areas of game design and writing benefit from the established framework, as it guides the designer while creating a dystopian game. Finally, my claims on fictionality and the description of a certain subtype of games I will refer to as the video game narrative (which includes the VGD) address central concerns of video game studies, which I will outline in the following.

Throughout the brief but intense history of video game studies, a heated debate concerning the *ontological dimension* of the medium and its ways to *communicate meaning* has been conducted and is still ongoing (see also chapter IV). In this regard, three central issues have emerged that revolve around the following questions: 1) should video games be described in the tradition of non-digital games, or as forms of narrative, drama, or film, and so on;[29] 2) are gameworlds fictional or virtual realities; 3) how is meaning created: a) in the interaction of interrelated elements that necessitate the player's intervention or b) through game rules that afford different kinds of processes?

Whereas the first area of investigation has been partially answered in that scholars nowadays describe what is usually called a game[30] (or a certain, widespread subtype of games) as *hybrids* between traditional games and participatory

---

27 For example: Marcus Schulzke, "The Critical Power of Virtual Dystopias," *Games and Culture* 9, no. 5 (July 2014): 315-334, http://gac.sagepub.com/content/9/5/315; Sebastian Domsch, "Dystopian Video Games: Fallout in Utopia," in *Dystopia, Science Fiction, Post-Apocalypse: Classics – New Tendencies – Model Interpretations*, ed. Eckart Voigts and Alessandra Boller (Trier: Wissenschaftlicher Verlag Trier, 2015), 395-410.

28 Raffaella Baccolini and Tom Moylan, "Conclusion. Critical Dystopia and Possibilities," in *Dark Horizons: Science Fiction and the Dystopian Imagination*, ed. Raffaella Baccolini and Tom Moylan (New York: Routledge, 2003), 247.

29 For different positions on the game medium, see Noah Wardrip-Fruin and Pat Harrigan, eds. *First Person: New Media as Story, Performance and Game* (Cambridge, Mass.: MIT Press, 2004).

30 The term video game or game (which I use interchangeably) designates a vague umbrella term that includes different phenomena, ranging from games such as TETRIS (Nintendo, 1989) to GRAND THEFT AUTO V (Rockstar North, 2013). It can therefore be misleading and should rather be used in a metonymic way. (Calleja, *In-Game*, 3; Espen

narratives,[31] the nature of this new genre and its specific modes of discourse remain unclarified.

Issues two and three, however, evoke more controversial arguments. This discrepancy is based on two positions that aim to describe the ontology of games from either a "ludo-fictionalist" perspective, "inspired by Kendell Walton's radical and influential *Mimesis as Make-Believe* (1990)"[32] that regards the gameworld and its objects as props in imaginative games of make-believe[33] or from a "ludo-realist" point of view.[34] This latter perspective perceives the gameworld as a virtual artefact and as closer to the empirical world.[35] Consequently, the potential for confusion here is substantial when not clarifying one's terms in a precise manner and pinpointing the concept of fiction/virtuality one is employing.

In addition, the innumerable attempts to determine the meaning-creation in games have proceeded similarly. An important school in this respect is the proceduralist approach inspired by Ian Bogost's influential theory on procedural rhetoric, which argues that the emergence of meaning in games resides in formal structures and rules that give rise to interrelated processes—in other words: "The art of persuasion through rule-based representations and interactions rather than the spoken word, writing, images, or moving pictures."[36] This approach is challenged by

J. Aarseth, "A Narrative Theory of Games," *Proceedings of the International Conference on the Foundations of Digital Games* (2012), 130, 133, http://dl.acm.org/citation.cfm?id=2282365).

31 For example: Aarseth, "Narrative"; Calleja, *In-Game*; Marie-Laure Ryan, *Avatars of Story* (Minneapolis: University of Minnesota Press, 2006); Sebastian Domsch, *Storyplaying: Agency and Narrative in Video Games* (Berlin: De Gruyter, 2013).

32 Espen J. Aarseth, "Ontology," in *The Routledge Companion to Video Game Studies*, ed. Mark J. P. Wolf and Bernard Perron (New York: Routledge, 2014), 491.

33 For example: Ryan, *Narrative*; *Avatars*; Domsch, *Storyplaying*; Grant Tavinor, "Art and Aesthetics," in *The Routledge Companion to Video Game Studies*, ed. Mark J. P. Wolf and Bernard Perron (New York: Routledge, 2014), 59-66; "Fiction," in *The Routledge Companion to Video Game Studies*, ed. Mark J. P. Wolf and Bernard Perron (New York: Routledge, 2014), 434-421; *The Art of Videogames* (Oxford: Wiley-Blackwell, 2009).

34 Aarseth, "Ontology," 491.

35 For example: Calleja, *In-Game*; Espen J. Aarseth, "Doors and Perception: Fiction vs Simulation in Games," *Escritas Mutanes*, 2006, http://www.luisfilipeteixeira.com/fileManager/file/fiction_Aarseth_jan2006.pdf

36 Ian Bogost, *Persuasive Games: The Expressive Power of Videogames* (Cambridge, Mass.: MIT Press, 2010), ix.

scholars who draw on the traditions of arts and aesthetics and describe the meaning-making process in terms of a dialectic between game, player, and culture. Thereby, a game's layers (system of rules and semiotic world elements) are engaged by a player in different modes of involvement.[37] In this conception, "procedurality" is seen as "necessary but not sufficient" to describe the act of play.[38] "The message is not in the mechanic or the fiction but in how both are put together—in the dissonance between action and meaning. The mechanic is not the message. The gameworld is not the message. Play is the message."[39]

It is needless to say that such an approach leaves the realms of ontology (at least partially) and enters a discussion of *phenomenology* and *cognitive sciences*. Of specific interest here is Gordon Calleja's influential work on player involvement that describes the relation between player and game as a form of incorporation that sets the player in a limbo state between virtual and empirical world. Calleja thus builds on the long discussion of immersion and presence in video game studies[40] and describes the player's multifarious involvement in a game through kinesthetic, ludic, narrative, affective, shared, and spatial participation.[41] Various studies have continued in this phenomenological direction[42]—not to men-

---

37 For example: Aarseth, "Ontology"; *Cybertext*; Calleja, *In-Game*; Sicart, *Beyond*; *Play*; Domsch, *Storyplaying*; Susana P. Tosca, "*Amnesia: The Dark Descent:* The Player's Very Own Purgatory," in *Analyzing Digital Fiction*, ed. Alice Bell, Astrid Ensslin, and Hans K. Rustad (London: Routledge, 2014), 109-123.

38 Sicart, *Beyond*, 38.

39 Ibid., 96.

40 For example: Ryan, *Narrative*; Janet H. Murray, *Hamlet on The Holodeck: The Future of Narrative in Cyberspace* (Cambridge, Mass.: MIT Press, 1998); Carl Therrien, "Immersion," in *The Routledge Companion to Video Game Studies*, ed. Mark J. P. Wolf and Bernard Perron (New York: Routledge, 2014), 451-458.

41 Calleja, *In-Game*.

42 For example: Rune Klevjer, "What is the Avatar? Fiction and Embodiment in Avatar-Based Singleplayer Computer Games" (PhD diss., University of Bergen, 2006), https://pdfs.sematicscholar.org/b666/b2b75bb7923607a2cb202ff6c222707d3703.pdf; Daniel Vella, "The Ludic Subject and the Ludic Self: Analyzing the 'I-in-the-Gameworld'" (PhD diss., IT University of Copenhagen, 2015), https://en.itu.dk/~/media/en/research/phd-programme/phd-defences/2015/daniel-vella---the-ludic-subject-and-the-ludic-self-final-print-pdf.pdf?la=en; Brendon Keogh, "A Play of Bodies: A Phenomenology of Videogame Experience" (PhD diss., RMIT University, 2015), https://researchbank.rmit.edu.au/eserv/rmit:161442/Keogh.pdf

tion those on affect and the psychology of playing games. They address the relation of the player's body to both the player-character (PC)/avatar and the gameworld, while regarding this interrelation as one that oscillates between proximity and distance.[43]

Given these diverse strands in video game studies, I will describe the act of playing dystopia from a *phenomenological point of view* that is, nonetheless, anchored in *structuralism* and *narratology*. For as Iser argues: whereas "a theory of the aesthetics of reception (*Rezeptionstheorie*) … deals with existing readers, whose reactions testify to certain historically conditioned experiences of literature," "a theory of aesthetic response (*Wirkungstheorie*) … has its roots in the text."[44] However, such a theory never loses sight of the empirical reader's importance to this process and analyses this interaction "in terms of a dialectical relationship between text, reader, and their interaction."[45] As such, a theory of aesthetic response necessarily involves a discussion of *fictionality,* which I will describe as upholding a specific, aggravated relation (referentiality) between the fictional and empirical world. I am thus influenced by the previously discussed positions of video game studies in different ways, but I strongly tend towards a conception of games as multi-faceted phenomena and as hybrids between many things, the predominant form of which intermingles the qualities of traditional games and participatory narratives to involve the player in diverse ways.[46] Such a conception of games and play construes the meaning-making process as an interaction between different elements (culture, player, game), while rejecting linear approaches to the issue.

---

43 My notion of player-character is similar to Daniel Vella's "playable figure." (Vella, "Ludic Subject,"10). The playable figure combines the qualities of the terms avatar and player-character in that "it encapsulates both the fact that the entity is taken on and 'played out' by the player (in the sense that the player might say she is 'playing Bilbo' in The Hobbit), but also the fact that it remains a figure in its own right." (Ibid., 10). However, as I see no reason why the term player-character does not exhibit this meaning as well—and because of its acceptance in video game studies and design—I will stick to it, but in the mode of the playable figure.

44 Iser, *Act*, x.

45 Ibid., x.

46 Aarseth, "Narrative"; Sicart, *Beyond*; Calleja, *In-Game*; Tosca, "Amnesia"; Domsch, *Storyplaying*; Tavinor, "Art," *Art of*; Klevjer, "Avatar"; Ryan, *Narrative*; *Avatars*; Michael Nitsche, *Video Game Spaces: Image, Play, and Structure in 3D Game Worlds* (Cambridge Mass.: MIT Press, 2008).

Moreover, I will circumvent the ontological discussion of fiction and virtuality by describing the former phenomenon from a functional and phenomenological point of view. From this perspective, fiction respects the ontological integrity of the gameworld as a virtual artefact but endows it with a specific quality that sensualises the abstract gamespace and involves the player in a meticulous dialectic between gameworld and empirical world through acts of ideation. All in all, these are necessary steps to describe the nature of dystopia's aesthetic effect on the player and the meaning-making process in VGDs, which is the focus of this study.

METHOD: A CREATIVE ENCOUNTER BETWEEN GUIDING STRUCTURE AND PLAYFUL EXPRESSION

It has been clarified that the VGD involves the player in a struggle for Utopia that is both similar to and different from the reader's/appreciator's participation in non-ergodic storyworlds. The similarity rests in the gameworld's imaginative evocative qualities, while the difference concerns the player's navigation of and interaction with this world, which is extended to the ergodic and situates the player as an active "agent … capable of effecting real transformation of the global social and economic system."[47] These facets partially continue the tradition of non-ergodic dystopias in that the VGD relies "on the same utopian and dystopian tropes"[48] but extends them by altering the former's plot structure. Tom Moylan and Raffaella Baccolini have described this framework as the "clash of the official narrative," the description of the dystopian society, "and the oppositional counter-narrative,"[49] the rebellion against the hegemonic order conducted by one more dissident thinkers.[50]

This framework is dynamically widened in the VGD by both the *systemic nature* of a game and the player's *agency*. For the video game as a procedural system offers the player an insight into the "underlying logic of how these [dystopian]

47 Levitas and Sargisson, "Utopia in Dark Times," 16; cf. Alexander C. H. Hall, "'I am Trying to Believe': Dystopia as Utopia in the *Year Zero* Alternative Reality Game," *Eludamos. Journal for Computer Game Culture* 3, no. 1 (2009), 71-72, http://www.eludamos.org/index.php/eludamos/article/view/vol3no1-8/112

48 Elizabeth Nyman and Ryan L. Teten, "Lost and Found and Lost Again: Island Utopias and Dystopia in the *BioShock* Series," *Games and Culture* 13, no. 4 (November 2015): 6, http://gac.sagepub.com/content/early/2015/11/19/1555412015616510.abstract

49 Moylan, *Scraps*, 152.

50 Ibid., xiii; Baccolini and Moylan, "Introduction," 5.

worlds are created and sustained"[51] while sending her on a quest for hopeful possibilities and the "agency" capable "to influence and change the system."[52] This increased involvement in dystopia reconfigures the old "conventions" by the "additional facet of … [the] ludic (i.e. play) dimension"[53] and creates an experience "that focuses anger"[54] through *agency*, whose lessons may "carry over into reality, giving gamers a better sense of how they can make a difference in life."[55]

Doing justice to such a complex involvement in dystopia is the main task of this study. In disclosing the *preconditions of the player's aesthetic response* and describing the *underlying structure* that affords play, I wish to arrive at *a unified theory of aesthetic response* for VGNs. Such an investigation is lacking in video game studies and represents a fundamental gap to be filled. Furthermore, this perspective on games is beneficial in that it

> never allow[s] players to be insulated from gameplay. That means never forgetting, while observing game dynamics, that gameplay isn't solely about what games make the player do, but about how and why he does it, what it does to him, and what he makes of it retrospectively.[56]

Consequently, and to avoid the mistakes of linear approaches to video games, my deliberations are heavily informed by theories on representational art and the appreciator's involvement in it. Of specific interest here is Kendall Walton's seminal work on mimesis and make-believe that describes the creative involvement of appreciators in the guiding structures of representations (which are equal to fictions in the specific Waltonian sense).[57] I will then refine these deliberations with

51 Schulzke, "Virtual," 10.

52 Domsch, "Dystopian," 401.

53 Rowan Tulloch, "Ludic Dystopias: Power, Politics, and Play," *Proceedings of the Sixth Australasian Conference on Interactive Entertainment*, no. 13 (2009), http://dl.acm.org/citation.cfm?id=1746063

54 Jane Donawerth, "Genre Blending and the Critical Dystopia," in *Dark Horizons: Science Fiction and the Dystopian Imagination*, ed. Raffaella Baccolini and Tom Moylan (New York: Routledge, 2003), 30.

55 Hall, "Trying to Believe," 80.

56 Samuel Archibald, "Literary Theory (Appendix)," in *The Video Game Theory Reader 2*, ed. Bernard Perron and Mark J. P. Wolf (London: Routledge, 2009), 362.

57 Walton, *Mimesis*.

Jacques Rancière's thoughts on an emancipated spectator who imaginatively engages with theatre plays,[58] Lubomír Doležel's contemplations on fiction as a semantic phenomenon that involves the reader in a multifarious communication with both the empirical and literary world,[59] and, primarily, Wolfgang Iser's influential theory on aesthetic response.[60] Here, the reader steps in an imaginative dialectic with the structural concept of the implied reader, which is composed of a system of perspectives. These outline the empirical reader's imaginative involvement in the fictional text through structural indeterminacies that open up between the perspectives and evoke acts of ideation in the reader—that is, the creation and continuous revision of images by using her real-world knowledge.

The truth of these observations and the creative encounter between *formal structure* and *playful expression* applies as much to the engagement with non-ergodic representations as it does to video game play or realms beyond. It can be found when a band composes a piece of music—orienting themselves in the elegance of structure and breathing in life through creativity and passion—in performing a play, dance, art, life, and the universe itself. For the empirical player this means stepping into a creative dialectic with dystopia's implied player and in an emancipatory encounter between the poles of orientation and free interplay, of conforming to the rules of a specific role and scrutinising one's performance—which sets the player in a liminal position between gameworld and empirical reality.

Of great importance to my theory is therefore the concept of the *implied player*, which I borrow from Iser's original take on the implied reader and Aarseth's use of the implied player in game studies.[61] This construct will be refined and, as briefly outlined above, is by no means to be confused with the empirical player. Instead, *it* can be described in a non-personified form as both the *structure of the game* and *a specific role (or roles) ascribed to the player.* In other words, *the implied player represents the affordance and appeal structure of the (dystopian) game that drives the player to aesthetic response. To do so, it confronts her with a vast system of perspectives and structural indeterminacies that require the player to complete them through ergodic and imaginative action.* These perspectives offer windows into the estranged reality of the gameworld and

58 Jacques Rancière, *The Emancipated Spectator* (London: Verso, 2009).

59 Doležel, *Heterocosmica.*

60 Iser, *Act*; *Imaginary.*

61 Iser, *Act*; Espen J. Aarseth, "I Fought the Law: Transgressive Play and the Implied Player," *Proceedings of the 2007 DiGRA International Conference: Situated Play* 4 (2007), http://www.digra.org/wp-content/uploads/digital-library/07313.03489.pdf

outline the player's involvement in and comprehension of it. Through their combination in acts of ideation, the player experiences a game's aesthetic effect and a cathartic insight into the empirical world. Thereby, the perspectives oscillate in function between virtual *game objects* with which the player can interact and Waltonian *props* that evoke specific imaginings. They comprise:

1. **Sensorial perspective:** the sensorial filter (visual, auditory haptic) through which the player is granted access to the gameworld.
2. **World perspective:** the (dystopian) gameworld including its settings, objects, architecture, and topological/labyrinthine structures; the sounds and music of this world; and the characters who inhabit it.
3. **Plot perspective:** the plot developments that are outlined according to dystopia's narrative framework: the official narrative and counter-narrative.
4. **System perspective:** processes, playing styles, and player actions that are outlined by the game's dynamic system and rules.

To explain the phenomenon of the implied player bears a couple of considerations, and the discussion of the appreciator's/reader's involvement in representational art/literature will be complemented by an investigation into *narratology* and the *indeterminacy/virtuality* of gameworlds. As such, I will describe the perspectival system of the (dystopian) game by resorting to *classical narratology* and intertwine these deliberations with relevant work from video *game studies* and *game writing*. Moreover, my claim that the gameworld formulates an incomplete work world in the Waltonian sense, and so requires the player's participation, will be fortified by insights on 1) *game fiction*; 2) *the structural peculiarities of gameworlds*; and 3) *environmental storytelling techniques* that describe the gameworld as an imaginatively incomplete space that extends into the past and future.

The most interesting of these indeterminacies is the *blank*, which evokes the player's urge for combination and the negotiation between the different perspectives she encounters and co-creates. As such, the blank describes a vacancy in the overall system of the gameworld and will be distinguished from what can be called a *gap*. This latter form of indeterminacy rather addresses the player's desire for completion and the filling in of imaginative inconsistencies (such as an inaccessible house) or virtualised potentialities that can be actualised through ergodic action, which results in the creation of perspectives. Nonetheless, both forms of indeterminacy are very much intertwined and may occur in parallel.

Having clarified this structural framework of dystopia's implied player, I will continue to explain its dialectical communication with the empirical player. For this purpose, I will turn to the estranged nature of dystopian science fiction, which further aggravates the player's comprehension of the gameworld and her own agency in relation to the empirical world.[62] These insights will be approximated to the video game by intertwining Iser's notion of imaginative play as the interaction between *the fictive, the imaginary, and the real*[63] with Roger Caillois' concept of *ludus and paidia*[64] and Sicart's conception of *subversive play* (what I will call *precarious play*) that oscillates between guiding structure and playful expression.[65]

Through these interrelated steps, I will arrive at a discussion of the player's formation of *images* during the act of play—based on Iser's deliberations on the act of reading[66]—and it will become clear that in *playing dystopia* the experience of meaning and the aesthetic effect do not reside in isolated elements such as the system of rules or semiotic layer of a game. Instead, they result from the complex interaction between the guiding structure of the implied player (which is permeated by the *fictive*) and the player's imaginative and ergodic interactions (that are propelled by the *imaginary*). Meaning is thus experienced in and through the act of play and through the player's acts of ideation that are informed by empirical reality (the *real*) but restructure the player's perception of it. I will demonstrate the viability of these theoretical manoeuvres in the analysis of various examples and close playings of dystopian games.

STRUCTURAL OVERVIEW OF THE ARGUMENT

To give my argument a clear structure this study is divided into three interrelated parts:

'Part I: Towards the Video Game Dystopia' builds a solid foundation in discussing the concept of Utopia and its fictional narratives genres: utopia and dystopia. It thereby creates a framework to classify dystopian games and their plot structures according to several subgenres and introduces the game scholar/designer to the thematic.

62 Suvin, *Metamorphoses*; Moylan, *Scraps*.

63 Iser, *Imaginary*.

64 Roger Caillois, *Man, Play and Games*, trans. Meyer Brash (Urbana: University of Illinois Press, 2001).

65 Sicart, *Beyond*; *Play*.

66 Iser, *Act*; *Imaginary*.

Chapter I will describe Utopia as a philosophy of hope that is deeply anchored in the human psyche.[67] This utopian impulse finds expression in a variety of artwork in different forms and manifests itself in video game fiction. To illustrate this facet, I will compare Thomas More's *Utopia* (1516) to Irrational Games' BIOSHOCK INFINITE (2013), which was released roughly 500 years after its predecessor. Although separated in time, BIOSHOCK INFINITE continues the Morean tradition in that it employs utopia's plot framework of a stranger coming to an exhilarating Other. The chapter will end with a discussion of definitions of the literary utopia to situate the dystopian genre within a utopian framework and draw inspiration for a definition of the VGD.

Chapter II builds on these observations and describes the fictional dystopia as a transmedial genre that functions as a strategy of Utopia. By retaining a hopeful core within its nightmarish worlds,[68] dystopia is opposed to anti-utopia—two genres that are often regarded as inseparable.[69] Building on a differentiation of the fictional dystopia into various subgenres, I will finally establish a typology of dystopian genres in games. These include:

1. **The VGD as anti-utopia:** a deceptive strategy of the status quo and negation of utopian thought.
2. **The VGD as classical dystopia:** an efficient type in which the player relentlessly, though unsuccessfully, struggles against a merciless hegemonic order.
3. **The VGD as critical dystopia of variant I:** a frequent type of dystopia that often discloses the history of the dystopian society and ascribes the role of the catalyst to the player in having her actualise one or more utopian enclaves leading out of the dystopian confinement.

---

67 Primarily following Jameson, *Archaeologies*; Vieira, "Concept"; ed., *Dystopia(n) Matters: On the Page, on Screen, on Stage* (Newcastle-upon-Tyne: Cambridge Scholar Publishing, 2013); Chris Ferns, *Narrating Utopia: Ideology, Gender, Form in Utopian Literature* (Liverpool: Liverpool UP, 1999); Lyman T. Sargent, "The Three Faces of Utopianism Revisited," *Utopian Studies* 5, no. 1, (1994), http://www.jstor.org/discover/10.2307/20719246?uid=3737864&uid=2129&uid=2&uid=70&uid=4&sid=21101560295713

68 For example: Moylan, *Scraps*; Ferns, *Narrating Utopia;* Vieira, *Dystopia(n) Matters*; "Concept"; Raffaella Baccolini and Tom Moylan, eds., *Dark Horizons: Science Fiction and the Dystopian Imagination* (New York: Routledge, 2003).

69 For example: Krishan Kumar, *Utopia and Anti-Utopia in Modern Times* (Oxford: Basil Blackwell, 1987); M. Keith Booker, *The Dystopian Impulse in Modern Literature: Fiction as Social Criticism* (Westport Conn.: Greenwood Press, 1994).

4. **The VGD as critical dystopia of variant II:** a most promising variant of the critical dystopia that makes full use of the game medium's possibilities. By confronting the player with both the possibility of attaining a better future and its loss, the critical dystopia of variant II makes the player choose whether she wants to become a catalyst for change.

The chapter comes to a close by illuminating dystopia's plot structure of official narrative and counter-narrative. This framework formulates an integral part of dystopia's implied player, and its application to video game fiction will be scrutinised through a comparison between George Orwell's *Nineteen Eighty-Four* (1949) and Valve's HALF-LIFE 2 (2004).

Chapter III pursues the question whether the VGD is effective in conveying a warning. For this purpose, I will first discuss the nature of the dystopian warning to then transpose these deliberations to the VGD. In doing so, I will formulate an initial hypothesis that the premise for effective VGDs rests in the diversity of their perspectival network and consequent aesthetic complexity (by which I mean the potential range of playthroughs, imaginings, and interpretations). Consequently, whereas effective dystopias intermingle several points of view on game events (through characters, world issues, ludic interactions) and constitute multi-layered artefacts that are both pleasurable and subversive, ineffective dystopias focus on the pleasures of combat without giving at least a believable justification for the spectacle. The chapter ends with an overview of real-world targets common to the VGD and how these employ dystopia's framework to establish perspectival diversity. Thereby, the targets can be grouped into three overarching categories that centre on the general point of attack of (human) nature: the threat of oppressive regimes, the excesses of capitalism, and the dangers of science and technology.

'Part II: Playful Trial Actions in Estranged Gameworlds' composes a unified theory of aesthetic response to playing dystopia in particular and VGNs in general. It links a discussion of representational art, fiction, and reader-response theories to suitable research from video game studies and narrows down the established framework to the VGD.

Chapter IV describes the intersubjective framework of the implied player as the affordance and appeal structure of the game and as a network of perspectives that outlines the player's involvement. To do so, it introduces central concerns of video game studies that revolve around questions of narrative and fiction. These will be approached from a phenomenological perspective that views the appreciator's interaction with them in terms of a mutual dialectic. By describing two perspectives on the game medium—games as objects (framework) and processes (gameplay)—the discussion will be expanded to games, and I will describe the implied player as an incomplete construct and a potentially multi-layered artefact that structures a work world in fundamental ways and implies different types of players. To fully actualise this virtualised potential, however, requires both active and reflective participation, and I will hint at the importance of the blank in the player's acts of ideation in an analysis of Thatgamecompany's JOURNEY (2012, 2015). The chapter will end on a detailed description of the VGN's perspectival system, which affords these blanks and involves the player in games of fictionality.

Chapter V will close several strands the previous chapter opened up and discuss the empirical player's interaction with the implied player in terms of a playful trial action that guides the former to aesthetic response. It will lay the focus on the importance of indeterminacies in the act of play which permeate the gameworld and can be grouped in two primary types: gaps (that fuel the player's urge to completion) and blanks (as overall vacancies in the system that drive the player to combination). The discussion will be geared towards dystopia here and complemented by theories of estrangement, since the gameworld awaits the player in modes of defamiliarisation that not only aggravate her understanding of this world but also her agency within it. The strings converge in a close playing of 4A Games' METRO 2033 (2010), where I will transpose Iser's deliberations on the act of play and describe the player's attempt to make sense of the gameworld and its relation to empirical reality by composing two interlinked *gestalts*: the first to understand the gameworld, its characters, and plot, and a second that allows the player to weave connections to the empirical world, thus decoding the games of fictionality. Play, in this sense, is seen as *precarious* and as a *regenerative* force that comes to the fore in the interaction between the fictive, the imaginary, and the real.

'Part III: Playing Dystopia' puts my theoretical manoeuvres on trial by discussing two VGDs to the satisfaction of the established framework. It will complement the briefer examples from Part I and II—and it should be noted that my focus in this study lies on single-player games, for multi-player games and online worlds require an analytical framework of their own.

Chapter VI conducts a close playing of Irrational Games' BIOSHOCK INFINITE (2013), a critical dystopia of variant I that sends the player to the floating city of Columbia. The game targets a variety of issues, including theocratic regimes, a rapturous form of capitalism leading to a culture of Disneyfication, and humankind's unrelenting desire for power. To warn the player of such potentialities, BIOSHOCK INFINITE entices her with a quest to save a mysterious woman by the name of Elizabeth as she comes to uncover society's true nature. To do so in an efficient manner, the game involves the player in games of estrangement and in a struggle for forgiveness that is both personal and universalised. Whereas on a basic level, this struggle revolves around Booker DeWitt's guilt (the PC) for having committed racist atrocities and losing his daughter Anna in a gambling debt, this guilt is transposed to the player's current objective of saving Elizabeth. Here, the player faces the loss of Utopia by participating in hectic combat and an estranged form of capitalism. This gamist playstyle (which is focused on spectacle and gathering points) is foregrounded by the game and aligned with consumer conduct in the empirical world (buying goods one is not in need of, while ignoring the atrocities of the capitalist production machinery). In an intricate story of multiverses and negative potentialities, BIOSHOCK INFINITE'S ambiguous ending reunites the thematic strands of the BIOSHOCK games and fortifies their attempt to denounce human nature. Yet it also expresses the possibility to break free from this vicious circle in the protagonists' struggle for forgiveness through self-sacrificial acts.

Chapter VII explores the post-apocalypse of Naughty Dog's THE LAST OF US (Remastered Edition, 2014), a critical dystopia of variant I in which humankind has been decimated by the Cordyceps Brain Infection and where nature has made a majestic return. THE LAST OF US warns the player about the ramifications of an uncontrolled capitalism that extends into most ecospheres and disrupts nature's intimate balance. Thereby, it involves the player in a dialectical opposition between confining city spaces and liberating outdoor spaces. Whereas the former remind her of the shortcomings of bureaucratic consumer capitalism and the scientific hubris of man, the latter offer a safe haven from the intense gameplay of the city and suburban areas. Although seemingly a didactic opposition, the journey is one of gradual realisation, where Ellie aids both Joel (the PC) and the player in savouring aspects of the natural world they might otherwise have forgotten. Given these juxtapositions, THE LAST OF US makes creative use of dystopia's plot framework of official narrative and counter-narrative.

Finally, chapter VIII summarises my findings in this study and dares to look into the future of the VGD in anticipation of games that involve the player in new forms of creativity and social subversiveness.

Part I:
Towards the Video Game Dystopia

Preface to Part I

The dream is to Utopia as the starship is to the traveller in his desire to find his inmost self. It is a vessel to explore the unknown, to seek out possibilities beyond our current imagination, and to uncover and follow one's hidden wishes. The road leading there may be bumpy and plastered with hurdles, tough. One may get lost from time to time, and even hurt, since the dream is reluctant to give clear directions. Instead, its appearance is distorted in all senses and confronts the utopian traveller with a hardly comprehensible but fascinating realm.

However, despite all these challenges, the route of the traveller is one well worthy to pursue. For the dream hides a secret within all its misconfigured appearance and unconsciously guides the traveller in desperate times. The pathway towards this secret is thus a cathartic, emancipatory one. It not only grants the traveller a more thorough understanding of the own self—by letting him explore the deepest regions of his unconscious and having him confront his inner demons—but it also enables him to strife for something far greater. It is this universal desire for a better future that is valid for all of humankind and comes in the form of a regenerative wish for a Utopia.

Utopia, one may say, lets dreams become a reality, whereas dreams guide the utopian traveller towards Utopia's realisation. To fulfil such a wish in the empirical world is an entirely different matter, however, and may run into severe problems. But dreaming and the playful exploration of hidden possibilities are the first step.

Part I of this study introduces the reader to the philosophy of Utopia (or utopianism) and its manifestations in fictional narrative form. As philosophy of hope, Utopia can be found in many different real world attempts to create a better society but also in a more or less disguised form in representational artwork. Whether these fictional explorations come in the form of dreams of a future Eden or visions of nightmare, they all hold dear Utopia's desire and hope for the betterment of society as a whole and embark the participant on a journey towards emancipation, to seeing the necessity for Utopia.

The genre of the video game dystopia is no different in this respect. Although such games involve players in nightmarish gameworlds, which most often seem beyond the possibility for redemption, they retain the notion of hope. Consequently, the VGD fuels players' creative faculties and involves them in a trial action and a struggle for Utopia in both the gameworld and the empirical world.

1 Utopia and the Dream of a Better World

> Utopia is forward-looking, yes. Always just around the corner, always on the other side of the horizon, Utopia is 'not yet', elusive, glimpsed but never grasped. That's one of the things I love about Utopia. And yet, like you [Ruth Levitas], I want the world to be very different from the way it is now. I want to ride the wave of utopian impulse toward a new now.
>
> *Levitas and Sargisson, "Utopia in Dark Times," 20.*

The philosophy of Utopia, as Lucy Sargisson refers to it, fundamentally revolves around the principle of hope and the dream of a better world. It begins with the *act of imagination* itself, when the utopian thinker explores real or fictional alternatives to the contemporary present and, in doing so, induces a "transformative process"[1] that will grasp the recipient's attention.[2] This process begins "in the now,"[3] as the Utopian scrutinises empirical reality for potential ills and composes a vision of the future in which these issues are either solved or have turned into a nightmare.[4] Notwithstanding the direction of the utopian impulse, whether it results in a *utopia* or *dystopia*, "the desire called Utopia"[5] is deeply anchored in the human psyche and looks forward. However, there is a pitfall here, as Slavoj Žižek

1 Levitas and Sargisson, "Utopia in Dark Times," 16.

2 Ibid., 13, 16; Seeber, *Selbstkritik der Utopie*, 58.

3 Levitas and Sargisson, "Utopia in Dark Times," 17.

4 Booker, *Impulse*, 15; Jameson, *Archaeologies*, 11ff.

5 Ibid., 84.

warns: if desire is tipped out of balance, it may indeed be the greatest enemy to human happiness.[6]

Utopians have become aware of this peril, and the concept underwent critical changes mainly because of Utopia's abuse which occurred with the totalitarian regimes of the 20th century, including communist dictatorships.[7] This chapter will explore the route Utopia has taken to reach its *modern function* and will create a strong foundation for the discussion of dystopia in chapter II, which at its core similarly revolves around the wish for the betterment of society. For this purpose, I will make use of both older and contemporary deliberations on Utopia and its manifestations in fictional narrative form. I will describe the philosophy as moving away from abstract ideals and near-perfectionism. Utopia in the 21st century, then, evokes a *cautious desire* and indicates "a direction for man to follow, but never a point to be reached."[8] It thus takes on "the shape of a process" and "a programme for change and for a gradual betterment of the present."[9]

Defining the utopian terrain in such a way will be beneficial to describing Utopia's *function* and how the philosophy continues to exert influence on modern works of art such as the video game. Thereby, I will follow Fredric Jameson's overarching claim that most new utopias—whether they appear as real or fictional experiments in political thought, intentional communities, literature, film, theatre, or video games—function as "wish-fulfillments ... and hallucinatory visions in desperate times"[10] but, in addition, conform to the postmodern need for "pluralism" and "the fact that all knowledge and all aspirations emanate from specific standpoints, that others will see differently, that negotiation is necessary."[11]

Such an understanding of Utopia (or utopianism) lays the focus on a *transformative discourse* and a *creative dialectic* between various parties, while it upholds Utopia's primary function as *warning*. In this sense, Utopia is driven by a relentless but cautious pursuit of an unfulfilled future and is best described as "a strategy"[12] and "a critical and diagnostic instrument"[13] to remind humankind that what is at stake is the future itself. To put it in the words of Jameson: Utopia shows

---

6 Slavoj Žižek, *Welcome to the Desert of the Real! Five Essays on September 11 and Related Dates* (London: Verso, 2002), 58-59.

7 Levitas and Sargisson, "Utopia in Dark Times," 25.

8 Vieira, "Concept," 22.

9 Ibid., 23; cf. 22-23.

10 Jameson, *Archaeologies*, 233.

11 Levitas and Sargisson, "Utopia in Dark Times," 15.

12 Vieira, "Concept," 23.

13 Jameson, *Archaeologies*, 148.

us "the future as *disruption* (*Beunruhigung*) of the present, and as a radical and systematic break with even that predicted and colonized future which is simply a prolongation of our capitalist present."[14]

In order to approach this *disruptive effect* on the reader/player (which will be the main focus of Part II), I will begin with an investigation of the philosophy of Utopia and humankind's striving towards it, then move on to Utopia's fictional narrative form: the literary utopia. There, I will argue that Thomas More's legacy with *Utopia* (1516) continues to exert a strong influence on contemporary artwork. By comparing its plot structure to Irrational Games' BIOSHOCK INFINITE (2013), which was released roughly 500 years later, I will additionally argue that both works centre around a wish deeply ingrained in the human psyche. This wish for Utopia finds expression in fictional artwork in a disguised form and needs the reader/player to uncover it by comparing the fictional world to their empirical surroundings. Finally, the chapter will come to a close by detecting commonalities in definitions of the literary utopia, which will later inform my hypotheses on the video game dystopia.[15]

---

14 Ibid., 228.

15 A question I do not wish to answer in this study (for it goes beyond its scope) is whether there is a video game utopia. Such a potential genre, as Oliver Hall claims, springs from a deep dissatisfaction with reality and the urge to seek alternatives. (Hall, "Trying to Believe," 74). If this is so, what games would be classified as such? Can one include Maxis' THE SIMS 4 (2014) or Related Design's ANNO 2070 (2011), arguing that they project virtual spaces that allow for the exploration of utopian (and dystopian) possibilities? Could games like Blizzard's WORLD OF WARCRAFT (2004) be classified in this way as well? Needless to say, massive multiplayer online role-playing games enable the creation of online communities where players come together to experience adventures in a make-believe world. (Rudolf Inderst, *Vergemeinschaftung in MMORPGs* [Boizenburg: Werner Hülsbusch, 2009]). But do such communities create spaces that allow for the negotiation of potential alternatives to the empirical present? Do they enable explorations of better ways of social life that critically comment on real world issues—and, thus, as Patrick Jagoda claims, enable experimentation "with utopian societies and collective forms of organization?" (Patrick Jagoda, "Digital Games and Science Fiction," in *The Cambridge Companion to American Science Fiction*, ed. Eric C. Link and Gerry Canavan [New York: Cambridge UP, 2015], 146).

1.1 UTOPIANISM

In order to uncover the utopian impulse at the core of artworks, I wish to begin by referring to BIOSHOCK INFINITE, which thematises not only humankind's guilt for having lost Utopia but also the desire to regain it (see chapter VI for an in-depth analysis). The plot of BIOSHOCK INFINITE revolves around the floating city of Columbia (which took inspiration from the great world fairs of the late 18th and early 19th century in the U.S.). New discoveries in quantum mechanics have enabled scientists to build a city that hovers in the air, towering above the land of the United States of America. Columbia is an emblem of American exceptionalism, science, and religious beliefs, and was built following a dream of the self-proclaimed prophet Father Zachary Hale Comstock. However, the city is much more than a flagship of scientific excellence. It functions as a metaphor for a basic human trait: the dream of a better life. It is such a dream—of a utopia in the sky—that fuels humankind's utmost desires and gives people hope where none can be found. This is at the core of utopianism itself. As Henry David Thoreau once proclaimed: "If you have built castles in the air, your work need not be lost; that is where they should be. Now put the foundations under them."[16]

Utopianism can be considered "a philosophy of hope,"[17] a hope that, according to Lyman Tower Sargent, is imperative to any endeavour to improve upon society[18]—and, obviously, there is good reason for mankind to try. For in living memory, people have always been frustrated and discontented with their own lives. This situation provides the main wellspring of utopianism itself, a deep-seated dissatisfaction with life and the urge to change things for the better.[19]

Utopianism, then, is of essential importance to humankind. For the German philosopher Ernst Bloch, Utopia even lingers in every aspect of the human experience. He opens the discussion in his magnum opus *The Principle of Hope* (1955–59) with the idea that people daydream with the "explicit wish for something we lack."[20] Even though not all of these dreams are utopian per se (for instance, dreams about private concerns, sex, or food), many of them are (dreams about peace, equality, freedom, or a sufficient food supply for all of mankind). To bring

---

16 Henry D. Thoreau, *Walden and Civil Disobedience* (New York: Barnes & Noble Classics, 2003), 292.

17 Lyman T. Sargent, *Utopianism: A Very Short Introduction* (Oxford: Oxford UP, 2010), 8.

18 Ibid., 8.

19 Ibid., 4, 102.

20 Ibid., 109.

about real change, however, the utopian wish must not remain a daydream but has to take on the form of a forward vision. With this in mind, Bloch's famous dictum of *not-yet* becomes of paramount importance, implying a utopia that "express[es] possibility,"[21] a dream that can be realised sometime in the future.[22]

It is for these reasons that utopianism has often been considered a "common human phenomenon"[23] which "has existed in every cultural tradition."[24] This "anthropological dimension", "the wishing nature of man" and his "very disposition ... towards utopia,"[25] represents an indestructible force that may drive people forward in times of discontent.[26] Scholars have defined the phenomenon of utopianism (or the utopian impulse) as "social dreaming,"[27] as "the basic human yearning for a better world" and "the urge to dream or speculate about a more perfect social order,"[28] or as "obscure yet omnipresent ... finding its way to the surface in a variety of covert expressions and practices."[29] Each time it is the *universal wish* hidden within the dream, the hope for the betterment of society as a whole, that plays a vital role. When Ruth Levitas speaks about her hopes for a different world—"I wanted, and still want, the world to be changed"[30]—it becomes clear what utopianism is about.

This universality of Utopia can, however, quickly become an issue, and differentiation becomes imperative—a pitfall that Sargent explicitly stresses. He therefore underlines the necessity to distinguish between the concept of utopianism, on the one hand, and the literary genre of utopia on the other, and further subdivides the thematic into what he calls "the three faces of utopianism:"[31] 1) the literary utopia; 2) utopian practice; and 3) utopian social theory. Whereas the literary utopia is comprised of narrative texts that can be categorised as fictional

---

21 Ibid., 111.

22 Ibid., 109 ff.; Ernst Bloch, *The Principle of Hope: Volume One* (Cambridge Mass.: MIT Press, 1986), 1ff.

23 Sargent, "Three," 19.

24 Sargent, *Utopianism*, 126.

25 Vieira, "Concept," 20.

26 Ibid., 20.

27 Sargent, "Three," 3.

28 Ferns, *Narrating Utopia*, 32.

29 Jameson, *Archaeologies*, 3.

30 Levitas and Sargisson, "Utopia in Dark Times," 13.

31 Sargent, *Utopianism*, 5.

"thought experiments"[32] in which the utopian society is only imagined, the intentional communities of utopian practice rather serve as actual experiments.[33] In these, people tinker "with their own lives"[34] and organise their community according to what they believe is a more perfect principle.[35] Finally, there is utopian social theory, which, according to Sargent, deals with a plethora of phenomena: "utopia as a method of analysis", "the relationship between utopia and ideology," utopianism as an explanation of "social change," and "the role of utopianism in religion."[36]

Yet Sargent is not the only researcher to place importance on such distinctions. The Marxist and literary theorist Fredric Jameson suggests a similar segmentation of the utopian thematic when he distinguishes between "utopian form" (written texts) and "utopian wish" (the general utopian impulse he detects in everyday life). As a third category, Jameson adds "political practice," to which he assigns social movements that have actually attempted to achieve Utopia, such as intentional communities. More interestingly, though, is that he discerns an "indispensable starting point"[37] in his attempt to define the utopian thematic, which is, of course, Sir Thomas More's *Utopia* (1516). From there, Jameson posits "two distinct lines of descendency:" a *utopian programme* and a more general *utopian impulse*. Under the first line he subsumes the literary genre of utopia as well as revolutionary praxis, the space of the city, and self-conscious attempts to achieve Utopia. The second line, on the other hand, "is more obscure and more various"[38] and revolves around the unconscious utopian impulse universal to human beings. Here, Jameson includes political social theory and reform, the individual building, and the four levels of utopian allegory: the collective, temporality, the body, and utopian investment.[39]

For the literary scholar, this bewildering variety of the utopian thematic suggests one thing above all: to limit one's interest to literary aspects and focus on how utopian ideals are represented in fictional narrative form. Naturally, the crossing of boundaries in an interdisciplinary field, such as Utopia, cannot be neglected—in particular utopia's complex relation to the empirical present and to a

32 Baccolini and Moylan, "Introduction," 11.

33 Sargent, *Utopianism*, 5 ff.

34 Ibid., 7.

35 Levitas and Sargisson, "Utopia in Dark Times," 19 ff.

36 Sargent, *Utopianism*, 7.

37 Jameson, *Archaeologies*, 1.

38 Ibid., 3.

39 Ibid., 3-9.

specific historical time period.[40] The literary utopia is only one of utopianism's potential manifestations, yet a very important one.[41]

1.2 THE LITERARY UTOPIA OF SIR THOMAS MORE AND *Bioshock Infinite*'s COLUMBIA

The founding text of the literary utopia and that which gave the utopian tradition its name is Sir Thomas More's *Utopia*, which was first published in 1516. To name both his book and the imaginary island he describes, More created the neologism *utopia* from two Greek morphemes: the prefix *ouk* (later reduced to *u*, which means *no* or *not*) and the lexeme *topos* (which means *place*). By adding the suffix *ia*, the term *utopia* was born: the place that does not exist.[42] If not confusing enough, More then created a further neologism, and with *eutopia*—the good place—he slightly yet decisively changed the meaning of the previous term.[43] In essence, then, the island of Utopia can be described as both "desirable, yet unattainable"[44] and "[m]ore perfect than the real world, yet non-existent."[45] This "duality of meaning"[46] and its inherent ambiguity have accompanied the literary utopia and its manifestations ever since.[47]

With *Utopia*, More gave rise not only to a new literary genre but also established a blueprint that many authors have followed since.[48] Principally, any utopia (or dystopia) discusses the relationship between two societies: the author's empirical society and its fictional counterpart, the Utopia. Accordingly, More's story revolves around the travels of Raphael Hythloday, who recounts his trip to the remote island of Utopia, where he discovers an astonishing society. The book is divided into two sections: the first part portrays More's empirical present in a negative way (Europe and England in the early 16th century), and the second part details an imaginary society in which the issues of More's time have been overcome.

40 Ferns, *Narrating Utopia*, 50; Seeber, *Selbstkritik der Utopie*, 60.

41 Ibid., 11; Vieira, "Concept," 7.

42 Ibid., 3-4; Sargent, *Utopianism*, 2.

43 Vieira, "Concept," 4-5.

44 Ferns, *Narrating Utopia*, 39.

45 Ibid., 3.

46 Vieira, "Concept," 5.

47 Ibid., 5.

48 M. Keith Booker, *Dystopian Literature: A Theory and Research Guide* (Westport, Conn.: Greenwood Press, 1994), 53.

In Utopia, private property has been abolished, which has led to a quasi-socialist society that has done away with capitalist greed and its resultant hierarchical structures.[49] This "dual emphasis" on the empirical and fictional world has led to *primarily*[50] two readings of *Utopia*: "a clever parody of the social system in More's England" and "*Utopia* as a serious effort to envision an ideal society."[51]

As a literary genre, then, utopia enjoys the special status of a "hybrid genre."[52] It is, according to Chris Ferns, "both political … and aesthetic."[53] But this need not be a handicap and can be quite rewarding. In understanding utopia as a hybrid genre, the boundaries to neighbouring fields such as political and philosophical discussion open up, which enables a fruitful dialectic.[54] Was it not More himself who once claimed of his *Utopia*: "the truth, as if smeared with honey, might a little more pleasantly slide into men's minds."[55]

More's merit in *Utopia* is undeniable, and although he might not have invented utopianism itself, "he considerably changed the way this desire was to be expressed"[56] and gave it a "new distinctive form."[57] For Krishan Kumar, the modern utopia is a "product of a new age,"[58] the result of both the Renaissance and the Reformation.[59] It therefore represents "an attribute of modern thought"[60] based on the "humanist logic"[61] that reason and volition could actually transform the future—as for the first time in history, man's destiny was laid into his own hands.[62]

49 Ibid., 53-54.

50 Primarily, because it is difficult to establish a clear separation for More's narrative, since the book's second part may be seen as a satire with ambiguous intentions. Ferns traces back this "ambivalence" to "the complex and sometimes contradictory character of the author himself," who, as a follower of the Roman Catholic Church, composed a fictional society devoid of religion. (Ferns, *Narrating Utopia*, 36; cf. 36-37).

51 Booker, *Theory*, 53-54.

52 Ferns, *Narrating Utopia*, 11.

53 Ibid., 5.

54 Seeber, *Selbstkritik der Utopie*, 11 ff.

55 Thomas More, "Letter 15: To Peter Giles [41A]," in *For All Seasons – Selected Letters of Thomas More*, ed. Stephen Smith (New York: Sceptre Publishing, 2012).

56 Vieira, "Concept," 6.

57 Ferns, *Narrating Utopia*, 32.

58 Kumar, *Modern*, 21.

59 Ibid., 22.

60 Vieira, "Concept," 6.

61 Ibid., 4.

62 Ibid., 4-7; Ferns, *Narrating Utopia*, 32.

As a result, *Utopia* is "not an abstract ideal,"[63] as Kumar observes with Plato's *Republic*, but "a concrete fictional portrayal"[64] "of a more perfect society in operation"[65] "in which we are invited vicariously to participate."[66]

Over the years, More's *Utopia* has inspired many followers and an entire tradition of utopian and dystopian narratives—with one of its most recent being Irrational Games' BIOSHOCK INFINITE. Released in March 2013, the game stands as a reminder that *Utopia*'s narrative blueprint still exerts a strong influence on contemporary artwork. The plot in BIOSHOCK INFINITE follows the journey of private detective Booker DeWitt to a wondrous city in the sky that was named after the fictive Archangel Columbia. In order to repay his gambling debts, DeWitt was assigned the task of freeing a mysterious woman. In the best of utopian tradition, he discovers and steadily explores the floating city of Columbia, which represents exactly such a society as described above—and one that shares many similarities with More's *Utopia*. When Father Zachary Hale Comstock first envisioned the city, he had a definite goal in mind: the dream of a society in which life would be better. But this time, it is not a distant reader who travels to Utopia in his imagination, but the player himself who embarks on this trip—not in person, of course, but in the form of a player-character (PC) he controls.

Assuming the role of DeWitt, the player experiences first-hand the journey to and tour around Utopia. Columbia presents itself as majestic, a new Eden on earth, floating many miles above the United States. The moment the player enters the city, he witnesses an extraordinary society rife with scientific marvels and technological brilliance. Most importantly, though, he sees a vibrant society that, as Kumar and others have ascertained for More's *Utopia*, is in "full operation."[67] Everywhere around the player neoclassical buildings built on hovering islands tranquilly bob up and down, while steampunk-style zeppelins, marvels of engineering, punctuate the summer sky. Meanwhile, the city's inhabitants go about their daily routines. One can hear their chatter and see their children play cheerfully in the streets. Everyone seems happy and content. Walking down the alley, a gentle breeze blows. The alley leads to a gigantic statue of Father Comstock, the city's prophet, which overlooks the square. Today is the day of the annual Columbian Raffle and Fair, and people all around the city have gathered to celebrate Columbia's magnificence. Obviously, I stress this vividness to underline a certain

63 Kumar, *Modern*, 25.

64 Ferns, *Narrating Utopia*, 34.

65 Ibid., 32.

66 Kumar, *Modern*, 25.

67 Ibid.

point. Columbia is an *archetype utopia*, a good place that only exists virtually. Once there, the player encounters an active society that is more perfect than his contemporary one—at least, at first glance[68]—in which he is encouraged to participate in a way that is not possible in non-ergodic media.

Figure 1: The city of Columbia and Father Comstock's statue.

BIOSHOCK INFINITE (Irrational Games, 2013), ch. Welcome Center.

The similarities between More's *Utopia* and BIOSHOCK INFINITE do not end here. As touched on above, More's blueprint of the journey to Utopia and the tour around it are present as well—on this occasion, though, enacted by the player. To explain it more clearly, let me briefly illustrate the origins of this narrative device. For all its variety, the literary utopia shows an invariable fixed point that is still present 500 years after *Utopia* was published (referring to BIOSHOCK INFINITE's release in March 2013): the genre's "framing narrative."[69] The typical utopian structure, Ferns argues, clarifies how the main character (and narrator) "reaches

68 Grzegorz Maziarczyk, "Playable Dystopia? Interactivity and Narrativity in *BioShock* and *BioShock Infinite*," in *(Im)perfection Subverted, Reloaded and Networked: Utopian Discourse across Media*, Peter Lang Edition, ed. Barbara Klonowska, Zofia Kolbuszewska, and Grzegorz Maziarczyk (Frankfurt a.M.: Internationaler Verlag der Wissenschaften, 2015), 241-242.

69 Ferns, *Narrating Utopia*, ix.

the more perfect society and obtains the opportunity to witness its distinctive excellences."[70] He bases his claims on Angelika Bammer's observations about utopia's basic plot and Northrop Frye's reflections on the nature of "a Socratic dialogue between guide and narrator, in which the narrator asks questions or thinks up objections and the guide answers them."[71] Ferns comes to the conclusion that the utopian narrative constitutes a "curious hybrid of classical dialogue and traveller's tale."[72] As a result of these observations, the basic utopian plot might be described as follows (describing Raphael Hythloday's voyage): the narrative structure of *Utopia* outlines the journey of a random traveller to the remote island of Utopia, which lies somewhere in the southern hemisphere. Utopia is isolated from the familiar world and its exact location is never revealed (which implies the sheer difficulty of attaining Utopia). Once there, the utopian traveller is given a guided tour of this different society and is told why Utopia is superior to other places. The journey comes to a close when the utopian traveller returns home, where he will be able to share his newly found insights with his contemporaries.[73]

The reason for such a structure can easily be discerned, for utopias are always "constructs made up of bits and pieces of the here and now,"[74] and this *now* for More was Europe in the 15th and 16th century. For Kumar there is no doubt here: the emergence of utopia shows "a direct causal connection" to "the European voyages of exploration and discovery."[75] In those times, Amerigo Vespucci and his followers "were literally discovering a New World, which was bound to stir the utopian imagination."[76] When Columbus discovered the American continent, he was convinced that he had found the lost Eden. Indeed, what the explorers had found was an unknown, exotic world inhabited by an indigenous culture that showed a principal similarity to More's Utopia: the absence of private property.[77] To put it short, "America, now newly found, was Utopia,"[78] yet as Kumar adds: a

70 Ibid.

71 Northrop Frye, "Varieties of Literary Utopias," in *Utopias and Utopian Thought*, ed. Frank E. Manuel (Boston: Beacon, 1967), 326. Quoted in Ferns, *Narrating Utopia*, 13.

72 Ferns, *Narrating Utopia*, 13.

73 Ibid., 38-40; Vieira, "Concept," 7.

74 Jameson, *Archaeologies*, xiii.

75 Kumar, *Modern*, 23.

76 Ibid.

77 Ibid., 23, 70-71.

78 Ibid., 70.

utopia "to be *made*."[79] Disappointed with the primitiveness of the indigenous people, the discoverers nevertheless aimed at the realisation of a better world.[80] "America had a *space* where utopia could be constructed"[81] and, therefore, this was "a God-given opportunity – or duty – to create a new Garden Eden, a new earthly paradise."[82]

Since Hythloday's sea voyage, then, the journey to Utopia has adopted many forms. There are parallel universes or travels through space to faraway planets, a sleeper awakes in a distant future, or a traveller ventures through time.[83] It is exactly this allure of the unknown and mysterious Other that has constituted a huge part of utopia's fascinating spell ever since.[84]

BIOSHOCK INFINITE continues the Morean tradition and lets the player set out on his personal journey to Utopia as "an outsider"[85] who will encounter the exotic place and enigmatic Other that is Columbia.[86] Every journey has a starting point, and this one begins with a lighthouse. The year is 1912, and Booker DeWitt is escorted by Rosalind and Robert Lutece to a small island off the coast of Maine. They travel in a rowing boat and it is pouring with rain. Once on the island, the player enters the lighthouse. Its floorboards creak, and the place is littered with religious writings that mention a "New Eden."[87] The player reaches the observation deck, where he gains access to the lantern room upon entering a secret code. The sky now flashes red, and horns sound penetratingly as Columbia and the lighthouse prepare for the pilgrim's ascension. DeWitt is strapped to a chair and finds himself locked within a space pod. With a bang, he is rocketed up into the sky on a marvellous journey to Utopia. When the pod breaches the clouds, DeWitt and the player are left breathless at the view of the extraordinary society of Columbia.

Shortly upon arrival, the player's tour around Utopia begins. In contrast to More's story, however, BIOSHOCK INFINITE uses a more implicit variant of the Socratic dialogue. When the player reaches Columbia, there is no guide to show him around, no one to tell him about the city's distinctive excellences. Yet, this is not necessary, for he experiences Columbia directly and absorbs its ambiance. Not

79 Ibid., 73.

80 Ibid., 71-78.

81 Ibid., 81.

82 Ibid., 73.

83 Ibid., 25; Ferns, *Narrating Utopia*, 99.

84 Seeber, *Selbstkritik der Utopie*, 9.

85 Maziarczyk, "Playable Dystopia," 239.

86 Ibid.

87 BIOSHOCK INFINITE (Irrational Games, 2013), ch. Coast of Maine.

only does the player see the city through his own eyes (in a first-person perspective mediated through the PC's point of view) but he may move around it, enter its shops, read its billboards, and listen to the Utopians' conversations. Most importantly, though, the player witnesses the city's events and enjoys engaging in some of its activities.

Figure 2: The scientifically advanced island of Laputa in Swift's Gulliver's Travels. *The city hovers above the realms of Balnibarbi and is ruled by an uncompromising king.*

Jonathan Swift. *Gulliver's Travels*, 1992.

Finally, there is another aspect implied by the player's journey to the sky, which again links BIOSHOCK INFINITE to *Utopia.* Like More's island, Columbia still belongs to this world, but its location has become remote.[88] After the events of the Boxer Rebellion, in which Columbia was revealed to be a battleship in disguise, there was a disagreement between Father Comstock and the U.S. government which led to a rupture. Out of this disappointment, Columbia vanished into the skies and could never again be found. Similar to Utopia—whose remote location and defensive structures provide shelter from the outer world[89]—Columbia is fortified to protect itself against unbelievers. It assumes a superior position over them by hovering above the land below and through the use of sophisticated technology—a fact that closely links Columbia to Jonathan Swift's Laputa in *Gulliver's*

88 Elizabeth Nyman and Ryan Teten link this aspect to the general conception of gameworlds as separate Otherworlds or islands. They argue that both gameworlds and island are similar in structure, for they "are by their very nature isolated and bounded," and because of this isolation and artificiality, they serve well "to mimic the unnaturalness of their society" necessary to utopian and dystopian games. (Nyman and Teten, "Lost and Found," 6; cf. 1-6). For these reasons, they come to the conclusion that video game utopias and dystopias (in their nature as islands) stand in clear connection to the tradition first introduced in More's Utopia. (Ibid. 5).

89 Ferns, *Narrating Utopia*, 39.

Travels (1726). This is so because Laputa's bottom plate is constructed out of a robust substance called adamant. Consequently, whenever the king is in disagreement with the population below, he uses the flying city to block rain or sunlight. In extreme circumstances, he lowers the island onto the lands beneath to crush its cities.

As in the two above-mentioned fictions, the negative aspects of Columbia are hidden at first, and when the player is granted admittance to the city, he encounters a world of many wonders. He has reached *heaven*, or as one inhabitant of Columbia formulates: "Heaven, friend. Or as close as we'll see till Judgment Day."[90] There is no doubt here, the city of Columbia is a religiously inspired Utopia, though one that lies in this world. As such, it cannot be declared a metaphysical figment of the imagination but rather a dream that could be realised through human effort.

The issue at hand brings the investigation to the relationship between Utopia and religion, a topic that has given rise to much controversy. Generally speaking, Utopia belongs to this world, while religious projections of a better world belong to the next life. But this may not be entirely accurate.[91] Christianity's most famous visions of the afterlife are certainly heaven and hell, which have often been linked to utopia and dystopia, respectively.[92] For Sargent, hell represents an "archetypal dystopia."[93] Such religious visions may, however, not exclusively refer to the next world but also to this one: for example, the Garden of Eden, the millennium, and the Second Coming of Christ.[94] Eden, as they say, once "was eutopia,"[95] an earthly paradise where human beings (Adam and Eve) lived together in harmony with nature. Yet they were expelled from it when Eve tasted the forbidden fruit.[96] For humankind, the Fall marked a turning point that led away from closeness to God towards a life of despair. For all its elusiveness, though, the dream of Eden was never lost, and "actual and fictional expeditions to discover"[97] its whereabouts were undertaken.[98]

90 BIOSHOCK INFINITE (Irrational Games, 2013), ch. Welcome Center.

91 Kumar, *Modern*, 10.

92 Sargent, *Utopianism*, 96-97.

93 Sargent, "Three," 13.

94 Sargent, *Utopianism*, 86.

95 Sargent, "Three," 20.

96 Ibid., 20.

97 Sargent, *Utopianism*, 88.

98 Ibid., 87-88.

In addition, there are the myths of the millennium, the apocalypse, and the Second Coming of Christ, which occupy a similar limbo between metaphysical and this-worldliness. For Kumar, the concept of the millennium is "most replete with utopianism."[99] The idea illustrates "an imminent cataclysm in which God would destroy the wicked and raise the righteous for a life in a messianic kingdom."[100] The Christian Apocalypse especially, Kumar argues, is closely linked to Utopia. In order to prepare for Christ's Second Coming, which "would spell the end of all earthly history, and the beginning of eternal life in heaven,"[101] life in the City of God would offer perfect preparation. Of particular interest here are Kumar's observations of a specific "understating of Augustine's description of the City of God."[102]

> Regarding the City of God as a state of earthly preparation for the Second Coming, this could lead to the setting up of communities of perfected men and women who set themselves apart from the unregenerated society around them, and who saw themselves as owing no allegiance to the laws and customs of the secular state.[103]

In BIOSHOCK INFINITE, Columbia exactly fulfils this purpose. It is a city of more perfect men and women, prepared for Judgment Day, who show no allegiance to the lands below. The apocalypse in this case, however, is laid into the hands of Father Comstock (not to mention, the destructive force of Columbia) and, specifically, into his daughter's: Elizabeth.

Considering the previously mentioned facts, then, Kumar comes to the conclusion that

> the concept of the millennium [offers] an intermediate term between the purely earthly existence of fallen man and the purely heavenly existence of man redeemed. It is a transcendental concept, truly, but a transcendence that links earth and heaven rather than irreconcilably separating them.[104]

Without being able to solve the complex issue of Utopia and religion (which has only been touched on here), it still seems appropriate to distinguish between these

99 Kumar, *Modern*, 13.

100 Sargent, *Utopianism*, 92.

101 Kumar, *Modern*, 15.

102 Ibid.

103 Ibid.

104 Ibid., 17.

terms. Certainly there are convincing arguments that favour their close relation: for instance, that Sargent discerns a "close connection between Christianity and utopianism,"[105] and that Kumar declares religion to be "the unconscious of utopia,"[106] while underlining that "without the hope that religion ultimately offers, without specifically the paradisiac and millennial expectations that Christianity inspires, it may be that utopia becomes a lifeless shell … [and may not be] capable of arousing a significant and heartfelt response."[107] These arguments notwithstanding, it is important to mention that "Utopia is not religion"[108] and that the idea of the millennium may be "as close to it as Christianity ever got."[109]

1.3 THE SECRET WISH HIDDEN WITHIN THE DREAM

In order to define the genre of utopia, there remains an important question to be answered, namely that of its direction. Utopia, as previously clarified, occupies a universal dimension and formulates an inherent part of the human psyche. If that is the case, *the nature of the dream and the wish* itself become of central concern. The question to be asked, then, is: are utopias *progressive* or *regressive*, are they directed towards the future of humankind—"toward the new now"[110]—or are they "nostalgic in that they look back to an idealized past which is then moved into the future?"[111] Probably utopias are both. But the answer to the question is more intricate.

In *The Interpretation of Dreams*, Sigmund Freud argues that "[*t*]*he dream is the* (*disguised*) *fulfilment of a* (*suppressed, repressed*) *wish.*"[112] "These wishes," he continues, "existing in repression are themselves of infantile origin."[113] Based on this premise, Ferns describes the utopias of More and his successors as expressions of the "unconscious and unavowable desires"[114] of men and their longing for

105 Sargent, *Utopianism*, 101.

106 Kumar, *Modern*, 421.

107 Ibid., 420-421.

108 Ibid. 421.

109 Ibid., 17.

110 Levitas and Sargisson, "Utopia in Dark Times," 20.

111 Sargent, *Utopianism*, 21.

112 Sigmund Freud, *The Interpretation of Dreams*, trans. A. A. Brill (Hertfordshire: Wendsworth, 1997), 68.

113 Ibid., 392.

114 Ferns, *Narrating Utopia*, 40.

"maternal security through the institution of a masculine order."[115] That is to say, these early utopias depict metaphorical journeys to a lost "prenatal security"[116] and stand for "a *return* to the womb."[117] Especially indicative of this claim is Utopia's inherent symbolism, which becomes apparent in its shape and form. On closer inspection, More's island resembles a crescent in the form of the maternal womb, at the bottom of which a "phallic tower"[118] penetrates its entrance.[119] Ferns further justifies his assertion with the fact that More's Utopia is governed "by a conquering king [who is] remaking the world to his own image."[120] King Utopus thus has a strong similarity to European explorers such as Christopher Columbus.[121] However, it is not "the colonization of *terra incognita*"[122] per se that links these early utopias to prenatal wish-fulfilment, but more importantly, that the land to be conquered is often gendered as female.[123] Consequently, Ferns comes to the conclusion that "the utopian dream – at least as first formulated – is rather one of recreating … [the womb's] security by distinctive male means."[124] In this sense, the Renaissance utopias are partially regressive in that they depict a return to the safety, security, and well-being of a past lost long ago.

Figure 3: Cover of More's Utopia*. The island resembles a crescent in the form of the maternal womb.*

Thomas More, *Utopia*, 2009.

115 Ibid., 104.

116 Ibid., 5.

117 Ibid., 41.

118 Ibid., 47.

119 Ibid., 41, 47.

120 Ibid., 49.

121 Ibid.

122 Ibid. 48.

123 Ibid., 45.

124 Ibid., 47.

Similar to the Renaissance utopias, BIOSHOCK INFINITE can be partially described as regressive, and as a return to the lost Eden. Still, such a description is incomplete, and one cannot help but stress the game's progressive vision, which reveals itself in the wish-fulfilment of a particular character: Booker DeWitt and his attempt to regain his daughter Anna (or Elizabeth). In the fashion of the Renaissance utopias, the city of Columbia is a society dominated by men, led by the father-figure Comstock. In one of his many incarnations, DeWitt (who at the same time *is* Comstock in another of the many worlds of the BIOSHOCK multiverse) travels to the city of Columbia to regain his lost daughter. With the trip to Utopia (Columbia), the regressive aspects mentioned earlier become apparent: the sexual connotations of the womb, the phallic tower, and male dominance. It is a lighthouse where DeWitt's journey to the sky begins, a phallic symbol par excellence. Furthermore, on top of the lighthouse is a space pod which ejects from the lighthouse's peak with DeWitt inside and rockets up into the sky, where in the moment it bursts through the white of the voluptuous clouds, the city of Columbia appears in all its glory. Such a chain of events naturally evokes an image of fertilisation in the player's mind.

Yet dreams, according to Freud, distort the original wish beyond recognition, to ensure the sanity of the dreamer. This wish, in its pure and undisguised form, may be found deeply suppressed within the unconscious.[125] It is for this reason that DeWitt, at least initially, is not aware that Elizabeth is his own flesh and blood and only unconsciously tries to fulfil his wish to retrieve her. Indeed, the entire game can be interpreted as only occurring in the mind of DeWitt, while he is dreaming in his office (having lost Anna in a gambling debt).[126] In this scenario, the game events represent a Freudian anxiety dream of loss and illustrate the unconscious battle against a dark part of the self (Comstock, but also Booker himself, having taken part in the Wounded Knee Massacre), which raises the additional question of *forgiveness*. Going one step further, DeWitt's dream may again be interpreted as a distorted anxiety dream that, on a more general level, depicts humankind's innate fear of losing something of value, like the future itself. In this

125 Freud, *Dreams*, 161, 382.

126 However, the game leaves open whether DeWitt really lost Anna, because she might still be lying in her cradle in the adjoining room after DeWitt awakes from his nightmare. In the case of loss, the dream functions as mental trauma processing and harbours the wish to regain Anna. In the case of her safety, conversely, the dream depicts anxiety about losing something of value, in this case, DeWitt's daughter. In both cases the wish remains the same: the safety of something precious.

regard, Elizabeth functions as a symbol of hope that should be preserved and safeguarded at all costs, and the true meaning of DeWitt's dream may revolve around *the collective fear of losing the prospect of Utopia*. Consequently, and in line with these facts, the Utopia of Columbia includes both regressive and progressive elements, and a strict delineation seems hard to establish.[127]

I will explain this matter in more detail in chapter VI, but what is of importance for the moment is the fact that, for all their emphasis on "stability, security, freedom from hunger, from endless toil, from war,"[128] the Renaissance utopias can be classified as "dreams of order" (29), which include: Thomas More's *Utopia* (1516), Tommaso Campanella's *The City of the Sun* (1623), Francis Bacon's *New*

---

127 Viewed from a psychoanalytical perspective, it is interesting that the VGD comes first and foremost out of the pen of men—like the early Renaissance utopias, the later ones from Bellamy to Wells, and the first wave of classical dystopias. Ferns *Narrating Utopia*, 27. Consequently, gender-specific urges hidden within the unconscious may have ramifications for the content and form of the utopian dream—and thus on games such as BIOSHOCK INFINITE. In this respect, Keith Stuart comes to the conclusion that the VGD is dominated by "violence" and represents "masculinised post-apocalyptic fantasies." This argument is based on the fact that the game development community largely consists of men, and this is reflected by the content of the dystopian games. In these, the player takes control of vicious male protagonists who assume the role of father figures to the secondary and female cast. As a result of the player's involvement, he thus accepts the fact "that the future will be ruled by men of violence and fervour, and that we have to become them in order to survive." While basically true, Stuart's arguments oversimplify the act of playing dystopia. This is because he first of all neglects the games' endings and themes (BIOSHOCK INFINITE, THE LAST OF US), which are about forgiveness and a return to a peaceful society in balance with nature. Second, he ignores (or is unaware) that violent processes/actions in dystopian games seldom mean what they seem, because of their fictional nature. They rather refer to the social wrongs of the player's empirical present, which is dominated by the carnivorous mechanisms of capitalism and global powers (see chapter V). However, there is one aspect in which I agree with Stuart. This point revolves around the question of when the VGD will follow in the steps of the science fiction of the 1970s and 80s, for during this period the genre experienced a resurgence dominated by female writers. As such, a second (female) wave of VGDs is highly desired and anticipated. (Keith Stuart, "*The Last of Us*, *BioShock: Infinite* and why all Video Game Dystopias Work the Same," *Guardian*, July 1, 2013, https://www.theguardian.com/technology/gamesblog/2013/jul/01/last-of-us-bioshock-infinite-male-view).

128 Ferns, *Narrating Utopia*, 14.

Atlantis (1627), and the later socialist Utopias of Edward Bellamy, *Looking Backward* (1887), and H. G. Wells, *A Modern Utopia* (1905).[129] This "traditional utopia" (14) is "primarily centralist and authoritarian" and illustrates a safe haven for people "in a world of disorder"[130]—a fact that also applies to BIOSHOCK INFINITE. As a consequence, these utopias have reached "a state of perfection or near-perfection"[131] and have often been described as "static, ahistorical utopias" that "offer a frozen image of the present, and eliminate the idea of a future from their horizon."[132] The logic behind such a conception is simple: "there is no progress after the ideal society has been established,"[133] after "the end of history"[134] is reached.

With the dawn of the Age of Enlightenment, however, and the increasing faith in human reason and science, a major shift within the utopian genre occurred. Although Renaissance utopias offered visions of a better future, their realisation in the empirical world was to be doubted—as is implied by Utopia's remote location, which is nowhere to be found.[135] In the 18th century, then, a change occurred concerning "the nature of the *connection* between the writer's society and the utopian alternatives proposed."[136] Eutopia became *euchronia*: the good place that does not exist turned into the good place "elsewhere in time."[137] Heading towards the future of mankind, Utopia became progressive and was now seen as a true possibility, rather than mere speculative thinking.[138] Of course, this change of direction correlates with the mood of the time period and, as Fátima Vieira observes, is due to "a change of mentality presided over by the optimist worldview that prevailed in Europe in the Enlightenment."[139] Having in mind the significant breakthroughs of this period—in science, new technologies brought about by the Industrial Revolution, and Darwin's theory of evolution—the belief in progress as a constant variable in human history had reached unprecedented magnitude.[140] It was this "new

129 Ibid., 14-15, 74, 97.

130 Ibid., 14.

131 Ibid., 189.

132 Vieira, "Concept," 9.

133 Ibid., 9.

134 Ferns, *Narrating Utopia*, 59.

135 Ibid., 38-40.

136 Ibid., 67-68.

137 Ibid., 83.

138 Ibid., 68; Kumar, *Modern*, 38, 39, 45; Vieira, "Concept," 9.

139 Ibid.

140 Ibid., 10; Kumar, *Modern*, 44-45.

zest for the future"[141]—or, as Vieira observes (referring to Anne-Robert Turgot and the Marquis de Condorcet), the strong belief in "the inevitability of progress" associated "with the idea of infinite human perfectibility" that led humankind to believe in "the splendour that would await,"[142] if they only worked hard enough to attain it. Hope now glimmered at the edge of the horizon, and mankind was eager to grasp it.[143]

Underlining Utopia's newly found forward character, which began with the euchronia, Jameson declares that "[t]he desire called Utopia must be concrete and ongoing,"[144] with the literary utopia being a manifestation "of historical and collective wish-fulfillment."[145] In order to justify his thesis, Jameson refers to Samuel Taylor Coleridge's categories of *Imagination* and *Fancy*, which he describes as follows:

> On some metaphysical level, Imagination is a theoretical concept, designating the primal creative force of God: which an aesthetic context reduces to the shaping power appreciated in the architecture of monumental literary plots (the so-called primary and secondary imaginations).[146]

He then links the two categories to Immanuel Kant's distinctions between *the Sublime* and *the Beautiful* and to Sigmund Freud's *dream-thoughts* (the undistorted dream harbouring the original wish) and *dream-content* (the manifest dream as experienced by the dreamer—that is, after the dream work has completed its work and distorted the original dream wish by means of condensation and displacement).[147] What is of interest now is that Jameson discerns a similar relationship between "the primal architecture of the *Utopian Imagination*"[148] and *expressions of utopian thought*—that is "Utopian fancies"[149] such as the literary utopia.[150]

Again, he turns to Freud and his observations on daydreams and art—the work of art represents the artist's disguised and distorted wish-fulfilment of which he is

141 Ibid., 39.

142 Vieira, "Concept," 10.

143 Ibid., 9-10.

144 Jameson, *Archaeologies*, 84.

145 Ibid., xiii; cf. 84.

146 Ibid., 44.

147 Freud, *Dreams*, 169ff., 190ff.; Jameson, *Archaeologies*, 44-45.

148 Ibid., 53; emphasis added.

149 Ibid., 53.

150 Ibid., 45, 53-54.

secretly ashamed[151]—to distinguish two types of wish-fulfilment: "a repellent purely personal or individual 'egoistic' type, and a disguised version which has somehow been universalized and made interesting, indeed often gripping and insistent, for other people."[152] What this means is that a personal, egotistical wish—that of the artist or that of a fictional character when dreaming—might become universalised through "aesthetic or artistic decoration"[153] to capture the mind of the public. But this wish has never been purely personal in the first place, which would reduce Utopia to a decorative farce of egotistical daydreams, yet it harbours a more profound signification. As Jameson declares: "Utopian fancies [are] … placeholders and symptoms of *a more fundamental repression*, of the coming up short of the Utopian imagination against taboos that prevent any wholesale redesigning of the social order as such", and these taboos Jameson identifies "as the baleful effect and influence" of Anti-Utopia.[154]

Building on these observations, I can confirm an initial hypothesis: Utopia is irrevocably ingrained in the human psyche and can be connected to the well-being and the protective shell of the mother's womb—which shall not be seen as regressive but as progressively heading towards a better future. This universal striving for Utopia is inherent to all of utopian and dystopian fiction, and to other fictional works out there. For the appreciator this means being confronted with the wish in its distorted form, and in order to uncover it, he has to exert interpretive effort, thus exorcising his inner demons. In other words, Booker DeWitt's unequivocal wish to retrieve his daughter Anna (Elizabeth) signifies on a further level—and for the player—the progressive wish to strive for Utopia and the fight against the delusive forces of Anti-Utopia. Utopia, therefore, assumes the form of *a collective wish-fulfilment* that is deeply ingrained in the human psyche and inscribed in a more or less disguised form in representational artwork. Such a statement, as my further investigation will show, holds true for the video game medium as well—and, in particular, for the VGD, which harbours "utopian energy"[155] that expresses itself in the act of play.[156]

---

151 Sigmund Freud, "Creative Writers and Day-Dreaming," in *The Standard Edition of the Complete Psychological Works of Sigmund Freud Volume IX*, trans. The Institute of Psycho-Analysis (London: Vintage, 2001), 141-154.

152 Jameson, *Archaeologies*, 46; cf. 45ff.

153 Ibid., 47.

154 Ibid., 53; emphasis added.

155 Hall, "Trying to Believe," 70.

156 Ibid., 70, 72, 74, 80.

1.4 DEFINING THE LITERARY UTOPIA

Utopia can assume a very comprehensive form, and there exists a general consensus that defining the concept presents the researcher with severe problems, even if he confines himself to the literary utopia.[157] Nonetheless, a precise definition that delimits the subject and establishes boundaries seems inevitable. Otherwise, as Sargent confirms, "we [would] have no subject."[158] In the following, I will therefore undertake an attempt to define utopia in its fictional narrative form (the literary utopia) and establish a preliminary hypothesis on the video game utopia and dystopia based on these observations.

For such an enterprise, the definition(s) of Lyman Tower Sargent appear particularly useful. He defines *u/eutopia* as follows:

> A non-existent society described in considerable detail and normally located in time and space. In standard usage utopia is used both as defined here and as an equivalent for eutopia or a non-existent society described in considerable detail and normally located in time and space that the author intended a contemporaneous reader to view as considerably better than the society in which that reader lived.[159]

Probably most important to Sargent's definition is utopia's need to describe *a society*, which he already underlined in a previous essay: "But it must be a society – a condition in which there is human (or some equivalent) interaction."[160] However, this society does not refer directly to any empirical one, but is *non-existent*—a fact that underscores utopia's imaginary and fictional nature. In the case of utopia, though, the fictional place shows a specific quality: "All fiction describes a no place," and "utopian fiction generally describes good or bad no places."[161] Consequently, a certain utopia *describes in detail* (owing this part to its verbal form) an imaginary society that *a contemporaneous reader* shall view as *considerably better* than his empirical one. Through this juxtaposition of worlds—and the dialectic quality it implies—utopia holds a critical attitude, which arises out of the connection between a specific historical moment (that is to be considered negative) and an imagined solution in fictional narrative form.

157 Ferns, *Narrating Utopia*, 10.

158 Sargent, "Three," 12; cf. 5ff., 12.

159 Sargent, *Utopianism*, 6.

160 Sargent, "Three," 7.

161 Ibid., 5.

A similar definition is Darko Suvin's, who regards utopia as primarily "a literary genre or fiction."[162]

> Utopia is the verbal construction of a particular quasi-human community where sociopolitical institutions, norms, and individual relationships are organized according to a more perfect principle than in the author's community, this construction being based on estrangement arising out of an alternative historical hypothesis.[163]

Like Sargent, who speaks of a society that is considerably better than the present one, there is no hint of *perfection* in Suvin's definition either, and he prefers the term *more perfect*. Both scholars thus explicitly stress that the literary utopia, although sometimes near to it, does not aim for perfection. It is something "not inherent in the genre"[164] and has primarily been used by opponents of Utopia (the so-called anti-utopians) as ammunition for their attacks.[165] The second aspect worth mentioning is Suvin's notion of *cognitive estrangement* (which he derives from Bertolt Brecht's *Verfremdungseffekt*)[166] that arises out of an *alternative historical hypothesis*. Such a quality, he continues, fundamentally links the SF genre to utopia,[167] as both start with "a cognitive hypothesis"[168] that is *extrapolated* into the future "with totalizing ('scientific') rigor"[169] and in "a wide-eyed glance from here to there."[170] Through this creative approach, a fictional alternate world is created—both familiar and unfamiliar to the appreciator—that works around the logic

---

162 Suvin, *Metamorphoses*, 46.

163 Ibid., 49.

164 Ibid., 45.

165 Ibid., 45; Sargent, "Three," 9-10.

166 Bertolt Brecht, *Bertolt Brecht Poems 1913-1956*, ed. John Willet and Ralph Manheim (New York: Routledge, 1987).

167 Having defined science fiction "as the literature of cognitive estrangement," Suvin subordinates utopia to the former and declares it to be a "sociopolitical subgenre." (Suvin, *Metamorphoses*, 4, 61). Yet there are other opinions, like that of Sargent, who asserts that "utopia is the well-spring." (Sargent, "Three," 11). For the purpose of this study, I do not dwell on the issue of genre but appreciate that both SF and utopia share fundamental similarities.

168 Suvin, *Metamorphoses*, 75.

169 Ibid., 6.

170 Ibid., 37.

of "*a fictional 'novum' (novelty, innovation) validated by cognitive logic*."[171] During this process, called estrangement, the reader shall confront "a set normative system – a Ptolemaic-type closed world picture – with a point of view or look implying a new set of norms,"[172] and he thus gains a fresh perspective on the empirical world.[173] From this view, "utopia is a possible impossible,"[174] a fictional alternate world that nevertheless retains a strong connection to the empirical present from which it originates.[175]

It is this fundamental characterisation of utopia that Hans Ulrich Seeber stresses as well, who deems *the imagination of alternatives* to be the genre's thematic core.[176] His definition of utopia thus includes:

1) The outline of an alternative order in which human beings live together in a mostly closed social system, which in comparison to the particular empirical society is either better (positive utopia) or worse (negative utopia), but at least different and hypothetically possible.
2) The alternative order (explicitly or implicitly) criticises the social ills of the author's contemporary present (utopian intention).

171 Ibid., 63.

172 Ibid., 6.

173 Ibid., 6, 70, 71, 75.

174 Ibid., 43.

175 It is here that one should separate the genres of utopia, dystopia, and SF from those of the fantasy, the fairy-tale, and the folk-tale. Following Suvin, I will not discuss these as instances of dystopia, because they lack hypothetical possibility. This Suvin explains in that the folktale is "indifferent to cognitive possibilities" and that "SF retrogressing into fairy tale … is committing creative suicide." (Suvin, *Metamorphoses*, 8). "Even less congenial to SF is the fantasy … a genre committed to the interposition of anti-cognitive laws into the empirical environment." (Ibid., 8). One has to confess that Suvin's claims are bold, but it is true that whereas fantasy, the folk-tale, and the fairy tale lie in "the realm of the impossible," SF, utopia, and dystopia must remain in the "domain … [of] the possible." (H. Bruce Franklin, "What is Science Fiction – and How it Grew," in *Reading Science Fiction*, ed. James Gunn, Marleen S. Barr, and Matthew Candelaria (Basingstoke: Palgrave MacMillan, 2009), 23). This is a feature that fictions with dragons, magic, or dwarfs do not normally possess.

176 Seeber, *Selbstkritik der Utopie*, 11.

3) The rhetoric of fiction, which by means of its ability to sensualise the abstract (description, narration, dialogue) may give the utopian outline the illusion of the real and the probable. From the depicted, the utopia emits a persuasive intent.[177]

Taking a closer look at Seeber's argumentation, four observations seem particularly interesting: 1) the emphasis on utopia's *function* as an instrument of social criticism; 2) the potential of utopia as fiction to evoke a certain reaction in its audience; 3) the negative possibility of utopia—*dystopia*—as a manifestation of the genre; 4) the emphasis on the genre's believability and *hypothetical possibility*. Consequently, and having in mind the aspects of the two previous definitions, a first exploration into the genre of the video game utopia becomes possible, thereby gathering insights that will be of use for a definition of the video game dystopia.

1.5 TOWARDS THE VIDEO GAME DYSTOPIA

The preceding chapter has shown that *hope* has always been Utopia's dear companion and that the call for a continuous struggle towards it represents its essence. This dream of a better world, of a Utopia within reach, is universal to every cultural tradition and designates something deeply ingrained in the human psyche. Jameson has called the phenomenon historical and collective wish-fulfilment, and it is because of this quality that manifestations of utopianism can be found in a plethora of phenomena. One of these is the literary utopia: a representational work of art in which the reader embarks on an imaginative journey to a fictional and estranged place. The literary utopia thereby harbours a certain wish, slumbering in all of humankind, for the safety and the well-being of living in Utopia. In confronting the reader with distorted and condensed visions of his unconscious yearnings, utopia holds all the preconditions necessary to exorcise his inner demons.

Over the course of the last 500 years, the literary utopia and its plot structure have shown a strong endurance and continue to exert great influence on contemporary artwork. I have discussed one of these in the comparison of Thomas More's *Utopia* with Irrational Games' BIOSHOCK INFINITE. The game begins with Booker DeWitt's—and, vicariously, the player's—journey to a virtual society where they discover a foreign and estranged city that goes by the name of Columbia. At least in its opening hour, BIOSHOCK INFINITE represents an archetypal utopia (although already showing hints of dystopia) and mainly follows the traditional utopian plot: the journey to a remote location, the tour around Utopia, and the awareness that

177 Ibid., 70; my trans.

the society at hand shows considerable improvements over the player's empirical one. However, BIOSHOCK INFINITE is no utopia but rather resembles a malignant anxiety dream that in a Freudian sense harbours the wish for Utopia within all its terror. This insight leads to the more general conclusion (which I will explicate in the following chapter): that utopia is not opposed to dystopia, and that both genres essentially hold the same function—for at their core they are wish-fulfilment, revolving around the dream for a better future.

Moreover, the chapter has formulated an indispensable starting point for the venture into the VGD. It has detailed the similarities between utopian texts and games, and for this reason, I wish to end the chapter with a first hypothesis on the video game utopia—which, although not the focus of my argument, will be invaluable for the later discussion of the VGD. However, it would be a mistake to apply literary utopian theory to an ergodic medium without having in mind the differences between the media types in question (specifically pertaining to the player's involvement in a game). I will anticipate these and continue to elaborate on them in Part II, where precise advancements of these hypotheses shall be conducted. Consequently:

1. *The video game utopia represents a virtual society outlined in minute detail that the game designer(s) intended contemporary players to view as considerably improved over their empirical present.*
2. *This society is marked by estrangement and stands in direct or analogic extrapolation to the designers' empirical world, thus retaining the notion of a hypothetical possibility. As a consequence, the gameworld shows a specific referentiality to the empirical world—which is that of fictionality—and aims to criticise the latter's social ills.*
3. *Such an aesthetic effect and persuasive attempt is facilitated through the sensualising abilities of fiction that endow the abstract gamespace with life and creativity. They thus involve the player in a creative dialectic with Utopia in which she or he may participate ergodically, imaginatively, psychologically.*
4. *This playful trial action in an estranged gameworld offers the potential for utopian explorations (at times, laying the prospect of Utopia directly into the player's hands) and harbours at its core the desire to live in a better world.*

2 Dystopia: Nightmarish Worlds as Distorted Anxiety Dreams

The previous chapter went into detail about the philosophy of Utopia and its manifestations in fictional narrative form. At their core was the drive towards a better future and the striving for the continual betterment of society as a whole. This undercurrent in the human psyche, called the utopian impulse, led to the formulation of Utopia's strategy, which extends to its fictional forms and revolves around enticing the reader to view empirical reality through a critical lens and to attempt to change reality. A further manifestation of Utopia's strategy can now be found in the *fictional dystopia*, a transmedial genre that encompasses literary works, poetry, painting, and film, as well as plays and video games. Though seemingly pessimistic, the dystopia similarly works towards achieving the utopian dream, though with an adjusted strategy.

Such a conclusion is not shared by every scholar, and there are tendencies to view the dystopian genre as the opposite of Utopia—and, specifically, the literary utopia. As Arthur Blaim claims: "The distinction between *dystopia* and *anti-utopia* has always been rather vague and confusing, to the extent that there is an evident tendency to use the terms … interchangeably."[1] Kumar traces back this *conservative thinking* and *antagonism* to the traditions of the Augustinians and the Pelagians. While the Augustinian (the utopian, such as Wells and Bellamy) is driven by a positive image of humankind and denies the concept of the original sin, the Pelagian (like Orwell or Huxley) is convinced that human beings are weak

1 Arthur Blaim, "Hell upon a Hill: Reflections of Anti-Utopia and Dystopia," in *Dystopia(n) Matters: On the Page, on Screen, on Stage*, ed. Fátima Vieira (Newcastle-upon-Tyne: Cambridge Scholar Publishing, 2013), 80; emphasis added.

and regularly succumb to "the sins of pride, avarice and ambition, however favourable the circumstances."[2] From this, Kumar concludes that "[t]he anti-utopian['s] … pessimistic and deterministic view of human nature leads him to the conviction that all attempts to create the good society on earth are bound to be futile."[3] In other words, "[t]here are those who fear … that utopia *can* be attained, and that it will be a nightmare."[4]

Figure 4: The fictional dystopia is a transmedial genre that immerses the reader/player in nightmarish worlds. In this picture of Irrational Games' BIOSHOCK INFINITE, *the destructive force of Columbia—the supposed Utopia and castle in the sky—can be discerned.*

BIOSHOCK INFINITE (Irrational Games, 2013), ch. The Atrium.

Contrary to these descriptions is a more modern consensus amongst scholars who opt for a differentiation between the philosophy of Utopia and what is called *Anti-Utopia*.[5] In their view, the latter tradition rather serves to undermine the utopian dream and to consolidate dominant ideology, thereby contributing to the downfall

2 Kumar, *Modern*, 100.

3 Ibid.

4 Ibid., 102; cf. 100-102.

5 Vieira, "Concept"; *Matters*; Baccolini and Moylan, *Horizons*; Moylan, *Scraps*; Sargent, "Three," *Utopianism*; Seeber, *Selbstkritik der Utopie*; Levitas and Sargisson, "Utopia in Dark Times."

of Utopia by locking it up "in the iron cage of Anti-Utopia."[6] Notwithstanding this distinction, these scholars remain aware of the perils of a misuse of utopian images.[7] Such an idea, of "utopia gone wrong,"[8] can be found in many fictions where promised castles in the sky turn out to be battleships in disguise—destructive forces of evil like Jonathan Swift's hovering island Laputa or BIOSHOCK INFINITE'S Columbia (Irrational Games, 2013). They remind the reader/player that the dream of Utopia, however noble it initially was, is fragile and may easily turn into a vicious nightmare if not treated with caution.[9]

Having this possibility in mind, the dystopian genre should *not* be regarded as an enemy of the utopian enterprise. I will thus follow Vieira et al.'s line of thinking in a recent anthology that "there are more affinities binding utopia and dystopia together, with regard to their aims and objectives, than differences setting them apart."[10] In fact, both genres are "almost correlative in their function."[11] Consequently, and to avoid definitional misunderstandings, this chapter will clearly distinguish between *dystopia* and *anti-utopia*,[12] which brings me to the following statement.

Triggered by trends hidden within the author's empirical present and fuelled by the fear of losing something of value—that is to say, the future itself—the *fictional dystopia* may be described as a vicious nightmare, *an anxiety dream in the*

6 Moylan, *Scraps*, 154; cf. 104; Levitas and Sargisson, "Utopia in Dark Times," 15.

7 Sargent, *Utopianism*, 9.

8 Vieira, "Concept," 16.

9 At this point, one has to underline that "one man's utopia" may easily be "another man's dystopia." (Booker, *Impulse*, 15). And that in order to declare a particular society—be it fictional or real—utopian, dystopian, or something in between largely depends on the personal disposition of the reader/observer—and one could extend this to intentional communities or other forms of belief systems. (Kumar, *Modern* 105, 125; Seeber, *Selbstkritik der Utopie*, 27; Lyman T. Sargent, "Do Dystopias Matter?" in *Dystopia(n) Matters: On the Page, on Screen, on Stage*, ed. Fátima Vieira [Newcastle-upon-Tyne: Cambridge Scholar Publishing, 2013], 11). In addition, there is the issue of the historical time period, since the perception of a certain utopia may change according to the norms and perceptions of a better life that prevail in a specific historical moment. (Seeber, *Selbstkritik der Utopie*, 235; Sargent, *Utopianism*, 23).

10 Fátima Vieira, "Introduction," in *Dystopia(n) Matters: On the Page, on Screen, on Stage*, ed. Fátima Vieira (Newcastle-upon-Tyne: Cambridge Scholar Publishing, 2013), 1.

11 Ibid.

12 Baccolini and Moylan, "Possibilities," 240-241; Moylan, *Scraps*, xiii.

Freudian sense, which has distorted[13] and disguised its original meaning: *the wish for Utopia*. As such, dystopias across the media critically investigate the possibility of a worse future in order to *warn* their readers, viewers, appreciators, or players about potential trends apparent or present in discernible forms in their specific empirical present. As a strategy of Utopia, then, the fictional dystopia retains the strong urge to change the future for the better, and thus the element of hope. Consequently, it scrutinises utopian proposals (such as Columbia and Rapture—the gameworld of 2K Boston's 2007 BIOSHOCK) in a constructive manner, while the *anti-utopia* shall be confined to such fiction or nonfiction that rejects utopianism or a particular utopia—suggesting that the possibility for a better future is non-existent. To put it simply, *there can be no hope in anti-utopia.*[14]

The reason for such a description of dystopia and the elaboration of the thematic that follows is not so much about recreating the academic argument that preceded my claims (for there has been a fair amount of discussion in this area)[15] but to establish *clear lines of orientation* for an investigation of the video game dystopia as a genre. The chapter will thereby answer two main questions: 1) what subgenres can be found in the fictional dystopia, and are these potentially useful for *a typology of the VGD*? 2) Is dystopia's *traditional plot structure*—the clash between official narrative and counter-narrative—to be found in the VGD, and if so, how do the ergodicity and spatial capabilities of the medium influence this framework? To give my discussion an illustrative example, I will conduct an in-depth comparison between Valve's HALF-LIFE 2 (2004) and George Orwell's *Nineteen Eighty-Four* (1949) and describe the former's methods of involving the player. Finally, I will come to a first conclusion about the VGD, postulating the genre *as a new strategic form of the utopian enterprise*.

---

13 Kumar poses the question whether dystopias "contain, in however distorted and diminished a form, a utopian impulse." (Kumar, *Modern*, 286).

14 There is a slight problem with such a definition, because fictional narratives are ambiguous by design, and the appreciator might find glimmers of hope in the most pessimistic instances. However, in anti-utopia a strong trend towards the Anti-Utopian pole and a rejection of Utopia is discernible.

15 See Blaim, "Hell upon a Hill," 80-95 and Moylan, *Scraps*, 124-145 for a detailed discussion of the subject.

2.1 ANTI-UTOPIA OR DYSTOPIA?

Since the advent of the literary utopia, historical responses in the form of so-called *anti-utopias* were not long in coming.[16] Considered by Kumar "as the greatest of the older anti-utopias,"[17] Swift's *Gulliver's Travels* (1726) presented not only a ferocious attack (or even an outright rejection) on the society it sprang from (18th century England) but also chose its prey in Bacon's scientific wonderland *New Atlantis*.[18] Viciously haunting utopian aspirations, "[a]s a nightmare to its dream, like a malevolent and grimacing *doppelgänger*, anti-utopia has stalked utopia from the very beginning."[19] But quite unlike utopia, which pictures an improved society where many of the problems of the empirical present are overcome, the anti-utopia takes its readers "on a journey through hell, in all its vivid particulars. It makes us live utopia, as an experience so painful and nightmarish that we lose all desire for it."[20]

Naturally, there will always be those inclined to view the future with either hope or alarm.[21] But the anti-utopian persuasion goes further, mostly stemming from a deep frustration with utopianism itself.[22] In the 20th century, the "the two great utopian experiments [Russia and the USA]"[23] had failed, and the promised futures of Bellamy and Wells, their utopian thought experiments on what a socialist and scientific world-state could look like, could now be found as real manifestations in the perverted realities of the 20th century.[24] Yet narrative fiction had already warned what could come of these utopian visions. Viewed by Kumar as attacks on the Wellsian Utopia, authors such as E. M. Forster ("The Machine Stops", 1909), Yevgeny Zamyatin (*We*, 1920), Aldous Huxley (*Brave New World*, 1932), and, most famously, George Orwell (*Nineteen Eighty-Four*, 1949), explicated what the utopias of Bellamy and Wells implied.[25] They "simply [took utopia] one step further"[26] and showed the Wellsian world-state as the nightmare they felt

---

16 Kumar, *Modern*, 126.

17 Ibid., 105.

18 Ibid., 105-106.

19 Ibid., 98.

20 Ibid. 103.

21 Sargent, *Utopianism*, 8.

22 Kumar, *Modern*, 104.

23 Ibid., 381.

24 Ibid., 133ff., 194ff., 206-207.

25 Ibid., 205-206.

26 Ferns, *Narrating Utopia*, 113.

it to be.[27] In this sense, the fictions above could be defined according to Sargent's definition of the anti-utopia as

> a non-existent society described in considerable detail and normally located in time and space that the author intended a contemporaneous reader to view as a criticism of utopianism or some particular eutopia.[28]

Still, the definition at hand would do these literary works no justice, as their potential targets are not limited to the attack on Wells or utopianism in general. Indeed, utopianism might not be their target at all—as will be made clear later on.

"Can there be anything more commonplace than the pronouncement that, in the twentieth century, utopia is dead – and beyond any hope of resurrection?"[29] With hindsight it is clear that the 20th century became mankind's darkest hour so far. Events such as the rise of totalitarian regimes (communism, fascism, and national socialism), two World Wars, the atomic bombing of Hiroshima and Nagasaki, the Holocaust, and the Cold War smothered the belief in Utopia and left the world in a state of "cosmic pessimism."[30] Moreover, the rise of capitalism and modern consumer society led to the devaluation of the individual, reducing him to a cog in the mechanisms of corporate orchestras.[31]

It comes as no surprise, then, that narrative fiction should respond. Out of the horrors of the 20th century, dystopian fiction arose, producing minute "[f]orecasts" and "anxious anticipations"[32] of even darker times.[33] Consequently, the dystopias of Zamyatin, Huxley, and Orwell (Forster's can be included as well), represent fictional narratives that focus on "social and political critique."[34] All three "great defining texts of the genre of dystopian fiction"[35] qualify in that their targets are

27 Kumar, *Modern*, 205ff.

28 Sargent, "Three," 9.

29 Kumar, *Modern*, 380.

30 Ibid., 381.

31 Kumar, *Modern*, 380-381; Vieira, "Concept," 18-19; M. Keith Booker, "On Dystopia," in *Critical Insights: Dystopia*, ed. M. Keith Booker (Ipswich, Mass.: Salem Press, 2013), 2-3.

32 Ian F. Clarke, "The Pattern of Prediction 1763-1973: Methods of Prediction 1918-1939," *Futures 2*, no. 4 (December 1970): 376, 379, https://www.sciencedirect.com/journal/futures/vol/2/issue/4

33 Moylan, *Scraps*, xi.

34 Booker, *Impulse*, 19.

35 Ibid., 20.

to be found in *a specific historical moment*, which is their author's empirical present. Whereas *We* comments on dangerous trends Zamyatin spotted in post-revolutionary Soviet Russia, *Brave New World* illuminates the concerns Huxley had about Western consumer capitalism. Finally, *Nineteen Eighty-Four* targets a variety of issues, including socialism, capitalism, and totalitarianism.[36] As instances of "social criticism,"[37] then, dystopian fictions primarily revolve around "'real-word' societies and issues"[38] extrapolated into the future in order to show them in their terrifying magnitude—showing the reader not only *what could be* but, in a way, *what already is*.[39] The function of such a strategy is simple. In setting the reader "within the defamiliarizing context of an extreme fictional society,"[40] she is made vulnerable to her own critical judgment.[41]

As a result, dystopia elicits a specific *aesthetic (ethical) response*, which is that of a *warning*. Although a warning underlines the potentiality of the depicted, the reader is nonetheless reminded that the narrative she has read remains fictional, and that it might not be too late[42] to avoid the nightmare in the empirical world.[43] According to this analysis, then, the works of Forster, Zamyatin, Huxley, and Orwell could be classified as instances of dystopia, which Sargent has defined as

36 Booker, *Impulse*, 19-20; "On Dystopia," 8; Kumar, *Modern*, 266ff., 293ff.

37 Booker, *Impulse*, 18.

38 Ibid., 19.

39 Ferns, *Narrating Utopia*, 15, 107; Lucy Sargisson, "Dystopias Do Matter," in *Dystopia(n) Matters: On the Page, on Screen, on Stage*, ed. Fátima Vieira (Newcastle-upon-Tyne: Cambridge Scholar Publishing, 2013), 40.

40 M. Keith Booker and Anne-Marie Thomas. *The Science Fiction Handbook* (Hoboken: Wiley-Blackwell, 2009), 65.

41 Suvin, *Metamorphoses*, 6, 71.

42 In this sense, Sargent compares the dystopia to the *jeremiads* of ancient times. These criticise people for their wrongdoings and detail the ways in which God will punish the sinners. But jeremiads also hint at the possibility of redemption, suggesting that if people are open to change, God may reward them instead of punishing them. Similarly, the dystopia illustrates the results of human wrongdoings (without the religious connotation) but includes the notion that it need not be that way. (Sargent, "Matter," 12).

43 Ferns, *Narrating Utopia*, 107; Vieira, "Concept," 17; M. Keith Booker, "About this Volume," in *Critical Insights: Dystopia*, ed. M. Keith Booker (Ipswich, Mass.: Salem Press, 2013), vii.

a non-existent society described in considerable detail and normally located in time and space that the author intended a contemporaneous reader to view as considerably worse than the society in which that reader lived.[44]

It seems that these two perspectives on the same literary works have brought the investigation even further away from the attempt to distinguish between anti-utopia and dystopia. It is probably for these reasons that many authors conflate the terms dystopia and anti-utopia—and to a certain extent they seem to be right. Ferns, for instance, categorises literary works in which the former dream of Utopia has turned into a nightmare as examples of "dystopia – or anti-utopia – which both parodies and subverts the traditional utopian model as a means of satirizing and warning against some of the more alarming trends in contemporary society."[45] He thus describes dystopia to hold *a double function*: 1) parodying the traditional utopia and 2) warning against some contemporary trend the author deemed dangerous.[46] Similarly, Booker describes dystopian literature "not so much a specific genre [but rather] as a particular kind of oppositional and critical energy or spirit."[47] To him,

dystopian literature is specifically that literature which situates itself *in direct opposition to utopian thought*, warning against the potential negative consequences of arrant utopianism. At the same time, dystopian literature generally also constitutes a critique of existing social conditions or political systems, either through the critical examination of the *utopian premises* upon which those conditions and systems are based or through the imaginative extension of those conditions and systems into different contexts that more clearly reveal their flaws and contradictions.[48]

Although there is much truth to this claim, Booker might be mistaken in two points: first, not all trends in contemporary society qualify as utopian proposals.[49] Second, if dystopian fiction serves as a warning of a nightmarish future to come, by definition it cannot oppose Utopia. For as Sargent correctly notices: "[a] warning implies that choice, and therefore hope, are still possible."[50] As critique does

44 Sargent, "Three," 9.

45 Ferns, *Narrating Utopia*, 15.

46 Ibid., 15, 109.

47 Booker, *Theory*, 3.

48 Ibid., 3; emphasis added.

49 Seeber, *Selbstkritik der Utopie*, 233-234.

50 Sargent, "Three," 26.

not entail rejection, it follows that *dystopia does not stand in opposition to Utopia* but rather functions as *a strategy* of the latter—or as Moylan puts it: "places the dystopia at least potentially in the camp of critical utopianism."[51] Dystopia, then, still remains a nightmare, but one that retains the wish for a better future in the midst of its despair—and it is exactly this notion of hope which the term anti-utopia does not imply.

2.2 ANTI-UTOPIA AS A REJECTION OF UTOPIANISM

> Throughout modernity ... the anti-utopian persuasion has systematically worked to silence and destroy Utopia, but Utopia ... has always offered a way to work against and beyond these attacks.[52]

Actually, the problem may lie with the term anti-utopia itself and with its denouncing connotation. Not only does the term imply a direct attack on a particular utopia, but it also evokes the disqualifying connotation of a rejection of utopian thought.[53] There is, however, a reason for such a denial, albeit a questionable one. In the opinion of so-called *anti-utopians*, "utopia is defined rigidly in terms of the quest for a state of impossible perfection, thus giving rise to a dystopian parody that satirically illustrates the futility and folly of all utopian aspiration."[54] This perfection can, according to the anti-utopian, only be reached through brutal force, by transforming the former utopian ideal into a totalitarian regime.[55] Looking below the surface of such an assertion exposes a particular strategy or agenda, however. In order to justify their ideologically induced claim that the empirical present "is indeed the best of all possible worlds,"[56] the anti-utopians resort to "the label *perfect* as a political weapon to justify their opposition."[57] Anti-Utopia, in this sense, functions as a strategy of dominant ideology. It postulates the end of history and discredits Utopia to praise and safeguard "the status quo and the satisfactions that

51 Moylan, *Scraps*, 136; cf. 105.

52 Ibid., 104.

53 Seeber, *Selbstkritik der Utopie*, 223-226.

54 Laurence Davis, "Dystopia, Utopia and Sancho Panza," in *Dystopia(n) Matters: On the Page, on Screen, on Stage*, ed. Fátima Vieira (Newcastle-upon-Tyne: Cambridge Scholar Publishing, 2013), 23.

55 Sargent, "Three," 9, 23.

56 Fredric Jameson, *The Seeds of Time* (New York: Columbia UP, 1994), 8.

57 Sargent, "Three," 9.

it delivers to its beneficiaries."[58] But perfection, as pointed out in the previous chapter, has never been on the agenda of Utopia in the first place. Indeed, "[w]ithout the use of the word *perfect*, part of the logic of the anti-utopian argument disappears."[59]

It is best, then, to differentiate between dystopia and anti-utopia in relation to *the question of hope*. Anti-utopias, in this sense, refuse "all utopian hope and effort."[60] They ridicule the utopian enterprise and are characterised by an utter disbelief in any utopian proposal.[61] Yet a slight problem emerges with such a strict classification, for absolute categories are difficult to uphold, above all when it comes to narrative fiction or art's inherent ambiguity. Consider an appreciator's response, and it might be difficult for a work of fiction to bear the impress of outright hopelessness, for the depiction may still evoke the desire to counteract dystopia in the real world.[62] In this line of thinking, Moylan's suggestion to situate Utopia's literary manifestations on "a complex continuum that stretches between"[63] "the historical antimonies of Utopian [sic] and Anti-Utopia"[64] is very beneficial.

2.3 Variants of Dystopia: The Classical and Critical Dystopia

In *Scraps of the Untainted Sky: Science Fiction, Utopia, Dystopia*, Moylan maintains that in order to critically investigate dystopian narratives, it is first necessary to establish both "the formal differentiation of the dystopian from the anti-utopian text; and [also] ... the political differentiation between dystopian texts that work from a position of utopian pessimism and those that are fully anti-utopian."[65] Following this premise, he locates the literary forms of *u/eutopia*, *dystopia* and *pseudo-dystopia*, *anti-utopia* on the established continuum, claiming that "[d]ystopias negotiate the social terrain of Utopia and Anti-Utopia in a less stable

58 Moylan, *Scraps*, 131; cf. 129, 157.

59 Sargent, "Three," 10.

60 Moylan, *Scraps*, 139.

61 Vieira, "Concept," 16.

62 Moylan, *Scraps*, 157-158.

63 Ibid., 122.

64 Ibid., 157.

65 Ibid., 155.

and more contentious fashion than many of their eutopian and anti-utopian counterparts."[66] In other words, the tension between *hope* and *pessimism* and how this is negotiated in a dystopian narrative (and by the reader) becomes of vital importance.[67] To further differentiate dystopias on this point, Moylan deems it necessary to distinguish between those that maintain a "horizon of hope (or at least invite readings that do)" and so-called *pseudo-dystopias*, which "only appear to be dystopian allies of Utopia, as they retain an anti-utopian disposition that forecloses all transformative possibility."[68] As such, the (hopeful) dystopia is characterised by a "militant pessimism" and "works with an open, epical strategy [that] maintains a possibility for change or identifies a site for an alternative position in some enclave or other marker of difference."[69] On the other hand, the pseudo-dystopia (or "the anti-utopia-as-dystopia") is marked by a "resigned pessimism" and "recycles a closed, mythic strategy."[70] In doing so, the pseudo-dystopia "fails (or chooses not) to challenge the ideological and epistemological limits of the actually existing society"[71] and retains a close connection to the "cynicism" and "despair" of the anti-utopia.[72]

By means of this strategy, it becomes possible to differentiate dystopias according to how they negotiate the possibility of hope. The primary locus for this negotiation is dystopia's traditional plot structure: "the clash of *the official narrative* and the oppositional *counter-narrative*"[73] that emerges in the course of the story.[74] While the official narrative describes the dystopian society in detail and lays emphasis on its confinements, the counter-narrative deals with the life or lives of dissidents who struggle with their surroundings. They only slowly come to recognise the situation for what it is and eventually start to revolt against the ruling regime.[75] It is by means of this oppositional structure and by laying an emphasis on the negotiation of hope that a classification between different dystopian subgenres becomes possible.

---

66 Ibid., 147; cf. 157.

67 Ibid., 155-156.

68 Ibid., 147.

69 Ibid., 157.

70 Ibid.

71 Ibid., 156.

72 Ibid., 157; cf.156-157.

73 Ibid., 152; emphasis added.

74 The official narrative is also called "the narrative of the hegemonic order." (Baccolini and Moylan, "Introduction," 5).

75 Ibid.; Moylan, *Scraps* xiii, 152.

Let me illustrate this point with a few examples. While Zamyatin's *We* ends in the defeat of its protagonist D-503 and the obliteration of I-330, the attempted resistance (the counter-narrative) stands as a reminder that transformation towards Utopia remains an option (specifically if viewed in the larger context of the Mephi resistance and utopian enclaves located outside the walls of the One State). The defeat may only be temporary, and, as such, *We* can be classified as a dystopia that leaves a possibility for hope beyond its final pages.[76] Orwell's *Nineteen Eighty-Four*, conversely, pictures a more pessimistic scenario. Indeed, Winston Smith's outright defeat (the fact that he betrays Julia and comes to love Big Brother) and the brainwashing he endures in Room 101 locate the novel close to the anti-utopian end of the continuum. With the view that no utopian enclaves are discernible—"[f]rom the proletarians nothing is to be feared"[77]—the counter-narrative results in utter failure.[78] However, *Nineteen Eighty-Four* was written in "a very specific context"[79] and cannot be declared a conservative manifesto like those that Ayn Rand or Karl Popper have produced.[80] Even Orwell maintains that his prime intent with the novel is that of a warning:

> I think that, allowing for the book being after all a parody, something like NINETEEN EIGHTY-FOUR *could* happen. This is the direction in which the world is going at the present time, and the trend lies deep in the political, social and economic foundations of the contemporary world situation. … The moral to be drawn from this dangerous nightmare situation is a simple one: *Don't let it happen. It depends on you.*[81]

In these lines, Orwell makes something very clear: for all its despairing trains of thought and the terrors its diegetic characters are subjected to (above all Winston and Julia), *Nineteen Eighty-Four* may still elicit a *positive response* in its readers. It may give them the incentive to act in the real world before it is too late.[82] As an extreme form of the dystopia, then, Orwell's nightmare vision may be situated in the realms of the latter, yet it inevitably evokes the notion of anti-utopia.[83]

---

76 Ibid., 139, 160-161.

77 George Orwell, *Nineteen Eighty-Four* (London: Penguin Group, 1987), 219.

78 Moylan, *Scraps*, 139, 161-162.

79 Ibid., 161.

80 Ibid., 162.

81 Bernard Crick. Introduction. Nineteen Eighty-Four, by George Orwell, (New York: Clarendon Press, 1984), 152-153, quoted in Kumar, *Modern*, 291.

82 Vieira, "Concept," 17.

83 Moylan, *Scraps*, 162-163.

In view of these findings, the narrative genre of dystopia can be defined as a work of fiction depicting "a social 'elsewhere' that appears to be far worse than any in the 'real' world."[84] It reminds the appreciator of man's folly for always choosing the "longer and more tortuous"[85] route and foregrounds the need for a cleansing disaster necessary for mankind to learn its lesson. As a fictional thought experiment, dystopia fulfils said role and functions as "a dire warning" and a cathartic wake-up call "from our nightmare to realise that we still have time."[86] As such, it becomes a crucial step towards achieving Utopia and cannot be regarded as a far-fetched vision of a future not to be feared.[87] On the contrary, dystopia brutally recalls that the depicted "in many ways is already present and in the making."[88] Still, not everything is lost, and hope, in the sense of a "gloomy optimism, fearing for the worst, hoping for the best,"[89] may remain an option (at least for the appreciator). As a result, dystopias do not "temporarily refuse the possibility of radical social change [like the anti-utopia does]; rather they look quizzically, sceptically, critically not only at the present society but also at the means needed to transform it."[90] Consequently, "the depiction works *not* to undermine Utopia but rather to make room for its reconsideration and refunctioning in even the worst times."[91]

The message of dystopia, then, becomes very clear: even though we may live in precarious times, hope has not vanished entirely. Glimmers of utopian expectations notwithstanding, the dystopia in its traditional form remains a very pessimistic genre, which shatters the hopes and beliefs of its diegetic characters. It is for this reason that the classical dystopias already mentioned (*We, Nineteen Eighty-Four, Brave New World*, and "The Machine Stops") have been attested to locate

84 Moylan, *Scraps*, xiii.

85 Lucian M. Ashworth, "Dystopia and Global Utopias: A Necessary Step Towards a Better World," in *Dystopia(n) Matters: On the Page, on Screen, on Stage*, ed. Fátima Vieira (Newcastle-upon-Tyne: Cambridge Scholar Publishing, 2013), 68.

86 Ibid., 71.

87 Ibid., 69ff.; Levitas and Sargisson, "Utopia in Dark Times," 16-17.

88 Aline Ferreira, "Biodystopias Matter: Signposts of Future Evolution," in *Dystopia(n) Matters: On the Page, on Screen, on Stage*, ed. Fátima Vieira (Newcastle-upon-Tyne: Cambridge Scholar Publishing, 2013), 52.

89 Vincent Geoghegan, "Darkness and Light," in *Dystopia(n) Matters: On the Page, on Screen, on Stage*, ed. Fátima Vieira (Newcastle-upon-Tyne: Cambridge Scholar Publishing, 2013), 48.

90 Moylan, *Scraps*, 133.

91 Ibid.

a horizon of hope only outside their storyworlds, that is, with the reader. They thus paved the way for film adaptions such as Stanley Kubrick's A CLOCKWORK ORANGE (1971) and François Truffaut's FAHRENHEIT 451 (1966), and films such as Terry Gilliam's BRAZIL (1985), Richard Fleischer's SOYLENT GREEN (1973), and Ridley Scott's BLADE RUNNER (1982)—which are all caught up in pessimism about the future.

2.3.1 The Classical Dystopia and the Player's Enclosure Within its System

The question to be asked is whether video games follow this tradition. Given that the prevailing mood of the 1980s and 1990s was negative, it comes as no surprise that the *classical dystopia* can be found in game fiction as well, where it created "pessimistic dystopias" in which "the system has found stability, and it therefore negates human agency."[92] Like their forbearers in non-ergodic forms of art, such VGDs address the current state of the empirical world by virtualising imaginations of the future that are outright destructive.

In contrast to the classical dystopias of literature and film, however, the VGDs make use of the medium's ergodic capacities to involve the player and convey their message. They thus "emphasize the irreversibility of dystopia"[93] by making players "active participants in creating or perpetuating the problems that make game worlds dystopian."[94] In other words, the player and the diegetic characters alike are confined in a system of rules from which there is no escape, and therefore these games deny the player the possibility of redeeming these worlds: that is to say—she *cannot win the game* from both a story and (often) ludic point of view. Consequently, as Schulzke argues, such dystopias implicitly warn the player not to let the fictional nightmare become a reality, and invite her to work towards a solution in the empirical world, where these issues are still manageable.[95]

The 1980s and 1990s gave rise to a fair number of games with these attributes: for example, Vid Kidz's arcade game ROBOTRON: 2084 (1982), where the player struggles against never diminishing waves of enemies, or Evryware's MANHUNTER: NEW YORK (1988), which ends with destruction and mass murder. In addition, there is Bullfrog Productions' SYNDICATE (1993), where it is revealed that the player is in charge of a corporation in a ruthless struggle for dominance,

92 Domsch, "Dystopian," 401.

93 Schulzke, "Virtual," 330.

94 Ibid., 316.

95 Ibid., 330-331.

Beam Software's SHADOWRUN (1993), Infocom's A MIND FOREVER VOYAGING (1985), and I HAVE NO MOUTH, AND I MUST SCREAM (The Dreamers Guild, 1995). The latter game is an adaption of Harlan Ellison's short story of the same name (1967) and involves the player in a journey through infernal punishment. Although such classical dystopian games are not the majority—probably because some of them contradict the prevalent gaming trope of having the player *win* a game—the subgenre should not be underestimated. Domsch, for instance, negates the existence of pessimistic dystopias entirely[96]—which fails to acknowledge the diversity of the dystopian genre in games.

Besides these older classical dystopias, there are some recent games—such as EVERY DAY THE SAME DREAM (Molleindustria, 2009) and THE STANLEY PARABLE (Galactic Cafe, 2013)—that continue this legacy. The latter game especially is a suitable example to illustrate the VGD's unique features in this respect. THE STANLEY PARABLE is a classical dystopia that involves the player in the routine of the bureaucratic labourer Stanley and has her experience the confinements of an office job. During the game, Stanley eventually reaches a room with *two open doors*, which confronts him with a profound choice. This choice is not only representative for branching narrative structures in games but also addresses the player's curiosity about exploring additional parts of the gameworld. The narrator adds further zest when he exclaims: "When Stanley came to a set of two open doors, he entered the door on his left."[97]

Given the constellation of these perspectives, THE STANLEY PARABLE can be described as holding a dual meaning. It first of all addresses the player's *lack of agency* in VGNs on a meta-level. In this interpretation, the narrator assumes the role of a game designer who meticulously plans a story to drive a player to catharsis.[98] On a second level, the game makes accessible the confines of a bureaucratic consumer capitalism crammed into the microcosm of an office building and again foregrounds the player's (and Stanley's) lack of agency—reducing them to faceless workers stuck in servile routines.

THE STANLEY PARABLE thus offers the player an experience that viewers of films like Terry Gilliam's BRAZIL (1985) or Alex Proyas' DARK CITY (1998) are

---

96 Domsch, "Dystopian," 402.

97 THE STANLEY PARABLE (Galactic Cafe, 2013).

98 Michael J. Heron and Pauline H. Belford, "All of Your Co-Workers are Gone: Story, Substance, and the Emphatic Puzzler," *Journal of Games Criticism* 2, no. 1 (January 2015), http://gamescriticism.org/articles/heronbelford-2-1; Lars de Wildt, "Precarious Play. To Be or Not to Be Stanley," *Press Start* 1, no. 1 (November 2014): 3-6, http://press-start.gla.ac.uk/index.php/press-start/article/view/10/4

familiar with. Yet it grants her a more involved role and shows overlap with the experience of participants in the real-world Milgram experiment. Conducted by Stanley Milgram in 1961 and described in *Obedience to Authority: An Experimental View* (1974),[99] the Milgram experiment tested the willingness of participants to obey authority figures without questioning their motives. Within the experiment, the participant assumed the role of a teacher, with the task of asking questions to a student she could not see (who was played by an actor). Whenever the student failed to give a correct answer, the participant was ordered to give him an electric shock by pushing a button. These increased in strength, but, encouraged by an authority figure, most participants obeyed even though they could hear the student's shouts of pain.

Such a connection to the empirical world is deeply ingrained in THE STANLEY PARABLE's structure and plot framework of *official narrative* and *counter-narrative*. Thereby, the official narrative involves the player in the oppressive structures of corporate powers and video game design alike and makes use of various gaming tropes to do so. This can be discerned in affordances that invite the player to push buttons with the expectation of consequences to their action (when, in fact, nothing happens), or doors closing behind the player's back once she has stepped through them (which stand as a reminder of spatial confinement in video games).

What follows from these observations is that player agency in THE STANLEY PARABLE is confined to moving within the limits of the gamespace and a few decisions that feign agency within "an artificial and fragile network of choices."[100] The narrator guides the player through the branching plot alongside the pre-established path. Following his orders, Stanley comes to experience a heroic story in which he discovers the machinations of the system—by encountering a Mind Control Facility and a Panopticon-style[101] observation room that monitors the workers—and eventually comes to destroy it. As a reward, Stanley is led to a gate that

99 Stanley Milgram, *Obedience to Authority: An Experimental View* (New York: Perennial Classics, 2004).

100 Heron and Belford, "Co-Workers;" cf. de Wildt, "Precarious Play," 9.

101 The Panopticon is a late 18th century experimental prison design by Jeremy Bentham, who can be considered the first representative of Utilitarianism—a branch of ethical theory that revolves around the maximisation of utility for the general public through action. Consequently, the Panopticon is constructed in such a way that all inmates can be simultaneously and constantly observed by a single watchman from the tower located in the middle of the circular structure of cells. As such, it reflects a more general societal tendency in which authorities constantly need to control data streams and the

opens on to a lush environment. His dreams of freedom are about to come true. But happiness comes at a price, and the player who wishes to explore her newly found freedom is frustrated when the game triggers a cutscene that strips her of control.

It is as Lars de Wildt holds: by following the narrator's orders (and the game's official narrative), Stanley and the player move "along with the proposed behaviour of the system, by following the rules of the game."[102] Such a "naive player subject necessarily follows the system's proposed world view"[103] and becomes rewarded with a *simulation* of her dreams. This ending aligns the game with films such as THE MATRIX (The Wachowski Brothers, 1999) or DARK CITY which thematise life in a *simulacrum* and a world that "foregrounds governance."[104] Yet it also speaks to a more fundamental tendency in society that Michel Foucault describes in *Discipline and Punish: The Birth of the Prison* (1995) as the steady change from the practices of public *torture* as a "theatre of hell"[105] to a "system of punishment"[106] that works in disguise but nonetheless subjugates the individual.[107] Through *discipline*—by which Foucault means the "methods ... which made possible the meticulous control of the operations of the body, which assured the constant subjugation of its forces"[108]—the design of punishment extended not only to prison designs but also explained how society operates as a whole. These "general formulas of domination"[109] produce "'docile' bodies"[110] that are well equipped to execute the tasks of social demand.[111]

In light of these arguments, THE STANLEY PARABLE's official narrative virtualises a gameworld that can be considered a microcosm of the larger Panopticon

---

activities of the public. (Booker, *Impulse*, 79; Foucault, Michel. *Discipline and Punish: The Birth of the Prison. 2nd ed*, trans. Alan Sheridan [New York: Random House, 1995], 195-230).

102 de Wildt, "Precarious Play," 3.

103 Ibid.

104 Ibid., 9.

105 Foucault, *Discipline*, 46.

106 Ibid., 89.

107 Ibid., 7-8, 138, 264.

108 Ibid., 137.

109 Ibid.

110 Ibid., 138.

111 Ibid., 137-138.

of game design[112] and of empirical society. It refers to both the *meticulous rule system of a game* that subjugates the player by having her execute repetitive tasks and follow pre-established lines of orientation, and to the *confinements of a worker's experience* in a bureaucratic consumer society, which is geared to profit while neglecting the self-expression of the individual. The player and Stanley are thus *docile bodies* caught up in their respective Panopticons.

Figure 5: The futility of choice is illustrated in a mini-game that has the player repeatedly push buttons for four hours to save a baby and a puppy from a burning inferno. As such, the game plays with player expectations of a reward for conducting (menial) tasks.

THE STANLEY PARABLE (Galactic Café, 2013).

However, THE STANLEY PARABLE is not resigned to such a negative view of society and grants the player the possibility of a *counter-narrative* that depends on her actions: different choices result in different endings to the game. This counter-narrative begins once Stanley *chooses the door on his right* and acts contrary to the narrator's instructions. Infuriated by the player's decision to ruin his well-crafted story, the narrator attempts to bring Stanley back on track by closing off allies, rebooting parts of the game, or triggering *The Stanley Parable Adventure Line*, which guides the player to catharsis. It is another gaming trope the game plays with, and is familiar to players of FABLE 2 (Lionhead, 2008), in which *a*

112 de Wildt similarly views games as "power structures in a Foucaultian sense." (de Wildt, "Precarious Play," 5; cf. 5-9).

yellow line guides them through an open world to the next objective. It follows that breaking away from these pre-established pathways describes an *act of transgression*[113] and an attempt to escape the confines of the office building and the plot alike. Yet it is exactly this transgression that is about to fail in THE STANLEY PARABLE. None of the endings suggest a hopeful outcome for Stanley—one time he is stuck in a room with the sole option of committing suicide, while on another occasion he turns mad and winds up dead on the pavement outside the building.

Considering this helplessness, *utopian enclaves* are not discernible for Stanley. He is stuck within the procedures of bureaucracy and game design alike, within "a recurring, cyclical structure that endlessly repeats and occasionally introduces a random element"[114]—for with each failure the game begins anew. Hope thus lies solely with the player herself—and the game implies this in various instances. In one playthrough, a second narrator reminds the player that the only escape is to stop playing the game, while on another occasion—when the player does everything in her power to annoy the narrator—she ends up in a sequence where her ghost hovers over the paralysed Stanley. Consequently, it is only by refusing to play the games of larger power structures—and by breaking out of the *ludic contract* players have obliged themselves to follow in order to play a game[115]—that players may liberate themselves from them. In the vein of the classical dystopia, THE STANLEY PARABLE thus evokes the shocking insight that players might not be so far from Stanley—and that, if this is so, they should change their lifestyle by refusing to participate in bureaucratic madness.

2.3.2 The Critical Dystopia as a Liberating Route Towards a Better Future

The genre of the classical dystopia in video games can powerfully trigger an aesthetic response in the player, which is intensified (and personalised) by the medium's ergodic and spatial capabilities. Yet whereas games such as THE STANLEY PARABLE remain caught up in the pessimism they simulate, there is another variant of dystopia across the media that negotiates the poles of Utopia and Anti-Utopia

113 A hilarious, but powerful example to illustrate this point is when the player chooses to stay for a long period of time in a *broom closet.* The longer she stays there, the more infuriated the narrator becomes. Eventually, he reboots the level and boards up the broom closet, which suggests that even such an insignificant activity is not tolerated by the system.

114 Heron and Belford, "Co-Workers."

115 de Wildt, "Precarious Play," 10.

in a more neutral and ambiguous manner. Holding out the prospect of hope *within its diegetic storyworld* and exploring potential routes *through the nightmare* and towards a utopian horizon, *the critical dystopia* becomes of essential importance when discussing the VGD.[116] For in these "optimistic dystopias", the player becomes responsible for finding the "flaw" within a "seemingly stable" system, triggering a chain of events that will "lead towards its destruction."[117]

The critical dystopia is the dominant form of dystopia in video game fiction (402). It can already be observed in games from the mid-1980s and 1990s such as Capcom's STRIDER (1989), Midway's arcade shooter REVOLUTION X (1994), Micronet's ROBOTICA (1995), Revolution Software's BENEATH A STEEL SKY (1994), Origin Systems' CRUSADER: NO REMORSE (1995), and the optimistic sequels to their literary forebears NEUROMANCER: A CYBERPUNK ROLE-PLAYING ADVENTURE (Interplay Productions, 1988) and FAHRENHEIT 451 (Byron Preiss Video Productions, 1984). All of these games end on a hopeful or, at least, ambiguous notion, and their appearance thus works in conjunction with the critical dystopias of literature, which as "a new expression of the utopian imagination"[118] emerged in the political contexts of the 1980s and 1990s, in the wake of aggravated fundamentalist tendencies in modern society, environmental crises, and a renewed and solidified capitalism—in the form of a free market, regulated by "[m]ultinational corporations based in and supported by powerful nation-states."[119] With no clear opposition due to a weakened Left, utopian expression was on the decline, while dystopian fiction (mostly from the pen of SF writers) flourished. Especially the naissance of cyberpunk, which as a social (techno-punk) movement and literary genre "generated a usefully negative if nihilistic imaginary"[120] in its bewildering mixture of cybernetic prosthesis and imaginings about virtuality—to be found, for example, in William Gibson's *Neuromancer* (1984) and Ridley Scott's motion picture BLADE RUNNER (1982), which is based on Philip K. Dick's novel *Do Androids Dream of Electric Sheep* (1968).

However, the "authentically negative energy of cyberpunk"[121] in its first wave still yielded dystopias leaning towards the pole of Anti-Utopia, and it was not until a second wave of the genre that the critical dystopia began to appear. Starting with Pad Cadigan's *Synners* (1991)—*later than the critical dystopia in games*—novels

116 Baccolini and Moylan, "Possibilities," 239-240; Moylan, *Scraps*, 105ff., 147, 188.

117 Domsch, "Dystopian," 401; cf. 401-402.

118 Baccolini and Moylan, "Introduction," 7.

119 Moylan, *Scraps*, 184.

120 Baccolini and Moylan, "Introduction," 2.

121 Moylan, *Scraps*, 197.

written mostly by female writers[122] moved "beyond [the] nihilistic anxiety [of their forbears] into a new oppositional consciousness."[123] In what came to be known as the *dystopian turn*, the genre of dystopia experienced not only a revival but also a reformulation, opening "the door to a dystopian narrative that was, like its eutopian predecessors [referring to the *critical utopia*], *critical* in its poetic and political substance."[124]

As a form of narrative fiction, then, the critical dystopia allows for an *imaginative trial action* or *test run* in which its characters (directly) and the reader (imaginatively) participate.[125] Filled with horror on a journey through hell, they experience the dystopian society in all its particulars but, at the same time, are offered possibilities for social change and transformation.[126] In doing so, the critical dystopia offers *a way out* of the dystopian situation. It "opens a space of contestation and opposition" for its diegetic characters and maintains a "utopian impulse *within* the work."[127] This "radical openness,"[128] mostly to be found in its "ambiguous, open endings,"[129] combined with "an exploration of agency"[130] reveals "utopian trajectories against a seemingly overwhelming world system that is striving to achieve its historical goal of total external and internal exploitation of humanity and nature."[131] It is as Vieira maintains: "Dystopias that leave no room for hope do in fact fail in their mission,"[132] and the critical dystopia leaves fertile ground for utopian explorations in this respect.

---

122 Moylan classifies novels such as Kim Stanley Robinson's *Gold Coast* (1988), Marge Piercy's *He, She and It* (1991), and Octavia Butler's *Parable of the Sower* (1993) as critical dystopias. (Moylan, *Scraps*, 196). Peter Fitting includes films such as the Wachowski's THE MATRIX (1999) and Gary Ross' PLEASANTVILLE (1998). (Peter Fitting Peter. "Unmasking the Real? Critique and Utopia in Recent SF Films," in *Dark Horizons: Science Fiction and the Dystopian Imagination*, ed. Raffaella Baccolini and Tom Moylan [New York: Routledge, 2003], 160ff.).

123 Baccolini and Moylan, "Introduction," 3.

124 Ibid., 3; cf. 2-3; Moylan, *Scraps*, 103-104, 183-184, 186ff., 197-198.

125 Ibid., 106-107.

126 Baccolini and Moylan, "Introduction," 7-8.

127 Ibid., 7.

128 Ibid., 8.

129 Ibid., 7.

130 Ibid., 8.

131 Moylan, *Scraps*, 105.

132 Vieira, "Concept," 17.

As such, the critical dystopia is *not only* about pure survival in a dystopian society, as it also prompts a search for *alternatives* to the system.[133] This it does by locating *utopian enclaves* as spaces of collective resistance within its storyworlds—for to change the hegemonic order, individual resistance is not enough and a collective upheaval becomes imperative.[134] Correspondingly, Sargent defines the critical dystopia as

> a non-existent society described in considerable detail and normally located in time and space that the author intended a contemporaneous reader to view as worse than contemporary society but that normally includes at least one eutopian enclave or holds out the hope that the dystopia can be overcome and replaced with eutopia.[135]

Such an understanding of the critical dystopia becomes especially beneficial when discussing an ergodic medium. This is so because VGDs are often given "histories and futures that differ from their present conditions"[136] and encourage the player to influence the game's chain of events.[137] Consequently, they underline the element of "temporal development"[138] and most often lay the prospect of hope in the player's hands by confronting her with a gameworld that offers virtual possibilities to actualise—or at least gives her the opportunity to take the hopeful route. As such, the *search for utopian enclaves* within the gameworld (see Sargent) and the *ambiguous (hopeful) endings* of the critical dystopia that leave fertile ground for the betterment of society (see Baccolini and Moylan) become of particular interest for analysis.

Still, the differences between the classical and critical dystopia do not end here. An additional feature that distinguishes the forms is that the former often remains caged in a nostalgic look backwards,[139] while the critical dystopia strives

133 Moylan, *Scraps*, 189.

134 Baccolini and Moylan, "Introduction," 7-8; "Possibilities," 246; Fitting, "Unmasking," 161; Moylan, *Scraps*, 189-190, 193-194.

135 Lyman T. Sargent, "U.S. Eutopias in the 1980s and 1990s: Self-Fashioning in a World of Multiple Identities," in *Utopianism/Literary Utopias and National Culture Identities: A Comparative Perspective*, ed. Paola Spinozzi (COTEPRA/University of Bologna, 2001), 222.

136 Schulzke, "Virtual," 326.

137 Ibid.

138 Ibid.

139 Baccolini and Moylan, "Introduction," 5-6; Moylan, *Scraps*, 149.

"towards the unfulfilled, unachieved utopia."[140] In a rebellious look forward, the critical dystopia shows "the reader a road that must start in the present … from now-here."[141]

This orientation towards the future stands in sharp contrast to what Moylan classifies as *anti-critical dystopia*. Such texts

> remain in the camp of nihilistic or resigned expressions that *may appear to challenge* the current social situation but in fact *end up reproducing it* by ideologically inoculating viewers and readers against any form of anger or action, enclosing them within the very social realities they disparagingly expose.[142]

While some VGDs—such as Valve's HALF-LIFE 2 (2004), 2K Boston's BIOSHOCK (2007), and Deep Silver's METRO 2033 (2010)—are *critical* towards the society they attack in that they offer utopian trajectories for the player to find and follow, others—such as Sledgehammer Games' CALL OF DUTY: ADVANCED WARFARE (2014)—only pretend to scrutinise the contemporary present. As a matter of fact, the game can be classified as a prime example of the *anti-critical dystopia* (or *anti-utopia*) in video game fiction.

COD: AW involves the player in the dangers of private corporations assuming the role of monopolistic military powers that use warfare as a profitable business opportunity. In the year 2054, the U.S. government hands over most of its military interventions to the private corporation ATLAS. Led by the illusive father-figure Jonathan Irons (played by Kevin Spacey), ATLAS establishes an unprecedented military force that threatens to shatter the worldwide equilibrium of powers. Irons exploits the KVA attacks on nuclear power plants around the planet (a terrorist organisation under the command of Joseph Chkheidze, also known as Hades) and misleads the world population into believing the promises of his false Utopia—*hope*, *security*, and *progress*.

Correspondingly, COD: AW involves the player in an official narrative[143] and an agenda about profit and power that targets a variety of issues, including dicta-

140 Vita Fortunati, "Why Dystopia Matters," in *Dystopia(n) Matters: On the Page, on Screen, on Stage*, ed. Fátima Vieira (Newcastle-upon-Tyne: Cambridge Scholar Publishing, 2013), 29.

141 Ibid.; cf. 29.

142 Moylan, *Scraps*, 195-196; emphasis added.

143 In the first levels the player is part of ATLAS' strike forces and becomes a vital component of the game's official narrative. This is most prominent in the use and eventual

torships around the globe, U.S. foreign policy, allusions to Guantánamo Bay detention camp, the dangers of advanced war technologies, and the manipulative advertising campaigns of mass media. These promote stereotypical enemy images created by the U.S. military (in this case ATLAS) to make themselves seem glorious in victory—as a heroic figure the world desperately needs.

In the course of the game, however, a collective and worldwide resistance to Irons' agenda emerges. ATLAS declares war on the Western world and the player stands at the forefront of the counter-narrative. Notions of resistance notwithstanding, COD: AW remains trapped in the carnivorous plant of dominant ideology that it first seemed to criticise. When the player changes sides—from the private military corporation ATLAS to an international task force founded by the U.S. military—there is one thing that remains unchanged: her *following of orders* (for only then the game will proceed). Consequently, although game events reveal the truth about ATLAS and their false utopian promises, the player remains caught up in the lies of modern warfare and devolves into a faceless participant in the wars of greater powers. ATLAS, in fact, must be seen as an extrapolation of the U.S. military-industrial complex and the private corporations that financially back it—but this the player needs to decipher. Combined with a clichéd ending that glorifies the deeds of the U.S. military—reassuring players that *they* will be there if needed—the experience is robbed of its critical power.

On the grounds of these observations, COD: AW degenerates *into a propaganda machine for the U.S. military* and into an experience that solidifies current ideology. It can only be situated in the realms of dystopia if a deconstructivist counter-reading takes place—yet this is only possible if the player sees through what the game actually does with her. As a consequence, the game's true targets can be found in the player's pleasurable participation in the combat action and in her mindless following of orders—which robs her of the supposed agency she enjoyed during combat. Of course, such an experience may still trigger a militant response in the player by questioning her involvement in the devious plays of hegemonic orders. Whether this seeps through the clichéd story is another question, however.

---

confrontation with advanced warfare and its deadly efficiency. There is, for example, the exoskeleton that grants the player superior powers or other cunning technologies like the mute charge, a grenade that silences a small perimeter for the player to act in disguise, and the use of drones. Even more deadly is the bioweapon Manticore which is programmed to target specific ethnic groups, while sparing the lives of those enlisted in the ATLAS DNA databank.

This all leads to the conclusion that the *anti-utopia* in video games forecloses the potential for change and transformation and suggests that there is no alternative to the current ideological system by seducing the (naive) player into believing in its agenda. While this might be clear with COD: AW, there are other examples in which such a conclusion is more difficult to infer. The BIOSHOCK series especially has been the topic of heated debate in this respect, for the games' ambiguous endings push researchers to situate it on both the pessimistic and optimistic sides of the scale.[144] Aldred and Greenspan, for example, argue that BIOSHOCK only appears to be a critical dystopia (with its critique of the objectivist utopian proposal), while at its core the game tends towards the anti-utopian spectrum—by foreclosing "radically on any hope for another world" in all of its endings.[145] Thijs van den Berg expands on their claim by suggesting that while the VGD (in its entirety as a genre) achieves a viable critique of the capitalist system, it fails to imagine a world that goes beyond "dominant neoliberal ideology."[146] I am more hopeful in this regard, and a more nuanced interpretation of these games suggests a different conclusion—in which the player encounters utopian enclaves and ways out of the current system, as inconsequential as these may seem (see chapter V and Part III).

---

144 Packer, "Galt's Gulch," 215, 219-221; Maziarczyk, "Playable Dystopia," 247; Gibbons, "Wrap Your Troubles;" Domsch, *Storyplaying*, 156-158; Lars Schmeink, "Dystopia, Alternate History and the Posthuman in Bioshock," *Current Objectives of Postgraduate American Studies* 10 (2009), https://copas.uni-regensburg.de/article/view/113/137; Schulzke, Marcus, "The Bioethics of Digital Dystopias," *International Journal of Technoethics* 4, no. 2 (2013): 50, 56, https://www.igi-global.com/article/the-bioethics-of-digital-dystopias/90488; Martin T. Buinicki, "Nostalgia and the Dystopia of History in 2K's *Bioshock Infinite*," *The Journal of Popular Culture* 49, no. 4 (2016): 731, http://onlinelibrary.wiley.com/doi/10.1111/jpcu.12440/abstract

145 Jessica Aldred and Brian Greenspan. "A Man Chooses, A Slave Obeys: BioShock and the Dystopian Logic of Convergence," *Games and Culture* 6, no. 5 (March 2011):486; cf. 484, 486, 490, http://journals.sagepub.com/doi/abs/10.1177/1555412011402674. Similar to this interpretation, Buinicki views BIOSHOCK INFINITE (the third game in the series) as *anti-utopia* in Moylan's sense—and not as "dystopian narrative." (Buinicki, "Nostalgia," 734). This he justifies in that the game forecloses the potential for transformation due to the player's lack of agency to change game events or its pessimistic ending. (Ibid., 731-732, 734).

146 Thijs Van den Berg, "Playing at Resistance to Capitalism: *BioShock* as the Reification of Neoliberal Ideals," *Reconstruction* 12, no. 2 (2012): par. abstract, https://reconstruction.eserver.org/122/vandenBerg.shtml

Consequently, and to establish a clear grid to categorise the critical dystopia in video games, there is one last feature of the genre that remains to be addressed. This feature lies in the notion of the "adjective *critical*,"[147] which Peter Fitting understands as implying "an explanation of *how the dystopian situation came about* as much as *what should be done about it*."[148] This insight into the past enables the reader to understand the circumstances that led to the dystopian society and gives an "explanation of how it emerged from our present."[149]

Such an understanding of the critical dystopia is beneficial in a variety of ways. It first of all enables an *analysis of false utopian promises*, or "false utopian solutions to the dystopia of the present."[150] In this sense, COD: AW can be included in the category of critical dystopias (albeit with strong reservations), and more clearly the BIOSHOCK games. In these, the player experiences directly (in BIOSHOCK INFINITE) or with hindsight (in BIOSHOCK) how the former dream of Utopia may turn into a nightmare. The question of "how has it come to its present state"[151] thus permeates the player's imaginings, and both games uncover false (or misused) utopian promises[152] and could be classified as "*flawed utopia*[s]."[153] According to this understanding, then, the critical dystopia becomes "the alter ego of utopia … the Sancho Panza to utopia's Don Quixote, constantly challenging its impetus to transcend existing reality and pulling it back down to a more grounded sense of its own rootedness."[154]

Second, the critical dystopia seems particularly apt for application in video games because of the medium's ability to simulate storyworlds. The VGD uses this to its advantage and shows the potential of creating a *dynamic* and not static game- and storyworld by involving the player in both how the dystopian situation *emerged out of the empirical present* as well as *suggesting potential solutions* to it. In this regard, Schulzke claims that VGDs are more dynamic than their literary

147 Fitting, "Unmasking," 156.

148 Ibid., 156; emphasis added.

149 Ibid.; cf. 156.

150 Ibid., 164.

151 Maziarczyk, "Playable Dystopia," 240.

152 Packer claims that "Bioshock represents a counter-narrative to [Ayn] Rand's book Atlas Shrugged," (Packer, "Galt's Gulch," 215) from which the game draws inspiration, and to its underlying philosophy of Objectivism. (Ibid., 209).

153 Lyman T. Sargent, "The Problems of the 'Flawed Utopia'," in *Dark Horizons: Science Fiction and the Dystopian Imagination*, ed. Raffaella Baccolini and Tom Moylan (New York: Routledge, 2003), 225.

154 Davis, "Dystopia," 23.

forebears. Although he underestimates the dynamic storyworlds of the critical dystopia—which in comparison to the classical dystopia are exactly this—he is generally right about the player's increased involvement in the gameworld.[155]

This anticipates my discussion in Part II on the dynamic nature of the video game that enables a *space of possibility* which extends into the gameworld's *past* as well as into its *future*. Most VGDs show the player how their societies "originated and what specific processes gave shape to them. They call attention to the historical development of dystopia as a means of engaging players."[156] Schulzke mentions BIOSHOCK, which foregrounds "dystopia's temporal development" and "calls attention to Rapture's past by gradually revealing its long history through flashbacks and audio files narrated by the city's residents."[157] I will call such VGDs *critical dystopias*. Yet in order to qualify as such, both the society's past and, more importantly, the changeability of the future must be foregrounded. For Domsch, it is "[t]his sense of agency" and games' "general nature as systems" that "so strongly relates them to utopian and dystopian thinking: both are concerned with the extent to which human/player agency is able to influence and change the system."[158]

Now, before coming to a *typology of the video game dystopia* by dividing it into *four distinct subgenres*, there is one last point that needs clarification. It has been established that the critical dystopia lays the possibility for Utopia into the hands of its protagonists, and this role is now ascribed to the player. The question of hope is thus negotiated in a dialectic between the game's dynamic system and the player herself, who searches for utopian enclaves hidden within the gameworld and a strict set of rules. When found, these may be actualised into utopian horizons that show the potential for transforming the gameworld into a better place. However, this is not always for the player to decide, and there are a fair amount of critical dystopias that prestructure and outline such a venture. Such games include HALF-LIFE 2 (Valve, 2004), BIOSHOCK INFINITE (Irrational Games, 2013), and THE LAST OF US (Naughty Dog 2013, 2014), in which the route towards Utopia (or an ambiguous ending) is fixed and will be actualised by the player when completing the game. Such dystopias, then, shall be classified as *critical dystopias of variant I*.

On the other hand, there are dystopias that go even further. In games like BIOSHOCK (2K Boston 2007), METRO 2033 (4A Games, 2010), and FALLOUT 4

155 Schulzke, "Virtual," 315ff.

156 Ibid., 329.

157 Ibid., 326.

158 Domsch, "Dystopian," 401.

(Bethesda Game Studios, 2015), the player is not only ascribed the role of the *catalyst* but is given the *choice* to be one. Such games lay their outcome into the player's hands and hold the potential for a pessimistic, optimistic, or ambiguous ending. This is an important distinction when it comes to the VGD, for choice is one of the medium's defining characteristics. Consequently, the above-described games shall be classified as *critical dystopias of variant II*.

2.4 FOUR VARIANTS OF THE VIDEO GAME DYSTOPIA: A TYPOLOGY

I have been most careful in introducing Utopia to the realms of video game fiction. As in chapter I, the current chapter was dedicated to provide a solid ground for Part II in which I will lay the focus on the VGD and the player's involvement in it. In addition, I have addressed dystopia's traditional plot structure—the clash between official narrative and counter-narrative—and alluded to its importance for the VGD. Analysing this oppositional structure, one can determine the locus of hope within any dystopian narrative—and this becomes essential when classifying a certain game as either *classical dystopia* (where there is no hope for its diegetic characters) or *critical dystopia* (in which fertile ground for utopian deliberations remains within the bounds of the gameworld). These insights, then, point to the following important conclusions.

In the VGD, the prospect of hope depends on two interacting features: *the game's dynamic system* and *the player*. It is this dialectic, now, that results in primarily *four different types (or subgenres)* of the VGD.[159]

1. **The video game dystopia as anti-utopia:** This type seeks to deceive the player about its critical nature or involves her in pleasurable action to attenuate the system it represents. It can only be ascribed to the dystopian genre with great reluctance. Examples include CALL OF DUTY: ADVANCED WARFARE,

159 One needs to stress that any boundaries a typology establishes are, to a degree, fluid—which may aggravate the categorisation of a certain VGD. In addition, my typology serves a particular purpose and focuses on the VGD's plot framework and the player's involvement in the negotiation of hope. Consequently, other typologies that categorise the VGD according to ludic genres, sensorial involvement, or historical time period are imaginable.

CIVILIZATION (MicroProse, 1991),[160] and digital variants of MONOPOLY (Hasbro, 1935),[161] which are only classified as dystopia should the player see beyond their agenda of consolidating dominant ideology. Such dystopias are rare and similar to Moylan's pseudo-dystopias.

2. **The video game dystopia as classical dystopia:** In this type of VGD, the prospect of hope lies beyond the bounds of the gameworld. The counter-narrative results in failure and the gameworld's diegetic characters (and, figuratively, the player) are crushed by the dystopian regime. Examples include THE STANLEY PARABLE (Galactic Cafe, 2013), EVERY DAY THE SAME DREAM (Molleindustria, 2009), and I HAVE NO MOUTH, AND I MUST SCREAM (The Dreamers Guild, 1995). Contrary to the VGD as anti-utopia, this type aims to trigger a militant reaction in the player, who, although having failed in virtuality, is driven to action in the real world.
3. **The video game dystopia as critical dystopia variant 1:** In this variant of the critical dystopia, the prospect of hope lies within the bounds of the gameworld but is predetermined by the game's dynamic system. Here, the player follows a linear trajectory towards one or more utopian horizons (or ambiguous endings) and is assigned the role of a *catalyst* without enjoying the choice to be one. Moreover, many critical dystopias lay emphasis on how the dystopian situation came about by embedding information about the gameworld's past into it. This variant includes games such as WATCH_DOGS (Ubisoft Montreal, 2014), BIOSHOCK INFINITE (Irrational Games, 2013), and MIRROR'S EDGE (DICE, 2009).
4. **The video game dystopia as critical dystopia variant 2:** Finally, there is a most promising variant of the critical dystopia in video games which makes use of the medium's full possibilities and where the prospect of hope is negotiated between the game's dynamic system and the player. In these VGDs, the choice of becoming a catalyst depends on the player—and every playthrough

160 Mäyrä describes *Civilization* as "[i]deological simulation" (94) that involves the player in the inner mechanisms of colonisation that have her adopt the hegemonic point of view of Western cultures (specifically that of the U.S). (Frans Mäyrä, *An Introduction to Game Studies: Games in Culture* [London: SAGE, 2008], 94). The game therefore exerts "hidden influence on its player" and can be seen as a strategy of the status quo to justify their ideology. (Ibid., 98; cf. 95-101).

161 *Monopoly* can be described as pleasurably coaxing the player into the processes of a capitalist system and as "an enactment of the allures and disappointments of a zero-sum economy in which one gets rich by impoverishing one's neighbors." (Murray, *Hamlet*, 143).

may result in either an optimistic, ambiguous, or pessimistic ending (or actualised world). Examples of this most powerful variant include BLADE RUNNER (Westwood Studios, 1997), BIOSHOCK (2K Boston, 2007), DEUS EX: HUMAN REVOLUTION (Eidos Montreal, 2011), PAPERS, PLEASE (3909, 2013), METRO 2033 (4A Games, 2010), and FALLOUT 4 (Bethesda Game Studios, 2015).

Part II and III will extensively deal with types three and four of the VGD, and I will say no more about type one because of its borderline status. For the remainder of this chapter, then, I would like to extend on my *elaborations on the critical dystopia*—and specifically on type three, in which the PC's and player's route towards Utopia are fixed. For this purpose, and to further clarify how *dystopia's traditional plot structure* finds application in an ergodic medium, Valve's HALF-LIFE 2 serves as an ideal example. The game can be seen as one of the prime representatives of dystopia in video games and shall be looked at in detail. HALF-LIFE 2 sets the player on a journey through hell and involves her in the oppressive circumstances of its gameworld. Yet, for all the terror it portrays, hope glimmers at the edge of the horizon, setting the player at the forefront of the resistance and leaving her with an ambiguous ending.

2.5 THE RESISTANCE OF THE ONE FREE MAN: CITY 17 AS ORWELLIAN NIGHTMARE

HALF-LIFE 2 (Valve, 2004) lets the player enact the story of theoretical physicist Dr. Gordon Freeman in his struggle against human subjection to a merciless oppressive order. In what came to be known as the Seven Hour War, an alien race called the Combine invaded Earth and established a totalitarian regime on its surface. When the player takes control, the situation seems desperate, but a collective resistance instigated by Freeman (the PC and, by extension, the player) culminates in humankind's partial victory over its oppressors. This is a crucial aspect of the critical dystopia, which involves the player in a struggle for Utopia that may not be entirely won at the game's end, but has the player achieve a first glorious victory.

In the following, and as an extension to the theoretical part on dystopia, I will address the genre's typical *plot structure* and *framework*: the "typical dystopian conflict" that arises "between the established order and the potential dissident."[162] This allows first of all for a careful analysis of how dystopia's plot structure is

162 Moylan, *Scraps*, 112.

transferred to the video game medium, and a comparison between HALF-LIFE 2 and George Orwell's *Nineteen Eighty-Four* (1949) will further illuminate the issue. In HALF-LIFE 2 the player discovers a strange and unfamiliar world she has to make sense of and, therefore, assumes a role similar to Gordon Freeman (the protagonist of the story) to whom this world is also unknown. In the vein of classical dystopian fiction, both Freeman and the player undergo a *process of gradual realisation* that will lead them to a better understanding of the dystopian gameworld and to eventual rebellion against its system. To contribute to this agenda, HALF-LIFE 2 makes use of the genre-specific conventions of the first-person-shooter and grants the player agency through *ludic capabilities* such as navigating the gameworld, talking to NPCs (non-player characters), picking up certain objects (guns, ammunition, or supplies), engaging in combat, and expressing herself in creative ways by combining these abilities and toying with the environment, its object, and NPCs.

In addition, by discussing the clash between official narrative and counter-narrative, I will lay the focus on the *negotiation of hope* and address the player's involvement in it. HALF-LIFE 2, then, illustrates a critical dystopia of variant I in which hope is predetermined by the game's dynamic system—for although the player assumes the role of a catalyst, the choice to be one is not given. This first strategy of the implied player (whose additional strategies I will discuss in Part II) outlines a specific insight: by letting the player enact a terrifying vision of how her world could be and to a degree already is, HALF-LIFE 2 issues an unmistakable warning. It targets the ongoing threat of totalitarian regimes to the contemporary present and shows the ramifications of such regimes in an estranged gameworld.

2.5.1 The Player as Dissident and the Process of Coming to Awareness

Considered by many as *the* finest example of the first-person shooter genre (FPS), HALF-LIFE 2 has acquired a near messianic reputation for its excellent gameplay mechanics and a thrilling narrative experience. It first of all handed over the (visual/sensorial) perspective to the player and combined this strategy with a so-called silent-player-character (SPC). With the help of these discursive devices, the player's feeling of presence in the gameworld was intensified and the story appeared to be discoursed in an unmediated manner, setting the player at the centre of story creation (see chapter IV for a discussion of discourse devices). Of course, this had been done before in games such as DOOM (Id Software, 1993) and the first HALF-LIFE (Valve, 1998), but HALF-LIFE 2 managed to perfect this *feeling of presence* by how it discoursed its story. The game refrained from using non-interactive cutscenes, which would rip the player out of the immersion, and utilised modern discursive methods such as environmental storytelling techniques, scripted events, combat encounters, expressive characters and dialogue, and player action/navigation in a three-dimensional gameworld. It thus created an intimate experience of dystopia, considering the player's closeness to the action.

In addition, HALF-LIFE 2 is a forerunner of many dystopian video games in how it executes the genre's plot structure. As with many dystopian narratives in non-ergodic media, the game begins *in medias res*. The ominous G-Man has set Gordon Freeman and the player on their mission and into the nightmarish world where the events are about to occur. Once the player takes over Freeman's body, she finds herself in a train heading towards City 17. It passes a grey, industrial area reminiscent of Eastern Europe. People look terrified. They wear blue uniforms and hold black suitcases. From a citizen, the player learns that they are being relocated. But where to and to what purpose? Such questions are typical for the genre of SF and represent basic indeterminacies in the game- and storyworld the player ponders and that might be answered by subsequent events.[163] Right from the start, the player and Gordon alike thus experience the strangeness of this world, which hinders their understanding of it, as it creates "ideological distance" between the "elsewhere" of the gameworld and the player's "historical moment."[164]

This feeling of estrangement is underlined in that *Half Life 2* assigns the player a specific role the dystopian narrative normally reserves for its diegetic characters: the "dissident" or "misfit."[165] Most protagonists of dystopia are initially well-adjusted to their society and hardly see through its inner workings—but this state of mind is about to change. From their first sensations that something could be wrong with their world, dystopian protagonists incrementally come to see "the situation for what it really is"[166] and embark on an educational trip "from apparent contentment into an experience of alienation and resistance."[167] An illustrative example of such protagonists is Neo in THE MATRIX. Once the inception of the world's wrongness is placed, Neo inexorably seeks the truth, which develops from the initial realisation that the world as he knows it is somehow false to a revulsion and eventual resistance to it. Contemporary dystopian films like THE MATRIX or DARK CITY thematise this aspect closely and make viewers aware that "we are living in

163 Moylan, *Scraps*, 3.

164 Ibid., 112.

165 Ibid.

166 Ibid., xiii.

167 Baccolini and Moylan, "Introduction," 5.

an artificial or false reality as victims of sinister forces that control and manipulate us."[168]

Of course, there is a reason for the dissent's initial unawareness, which lies within the empirical present:

> Because we are socialized in a particular society and to an acceptance of its views, we are likely to be incapable of critical awareness of our situation, and we can define unfreedom as freedom, inequality as equality, injustice as justice.[169]

As Sargent illustrates, dominant ideology is well-equipped to rob people of critical distance and blind them to the truth of the situation. To elude such a trap, Utopia can offer an enlightening point of view to enable humankind "to break through the perspectives that tend toward the acceptance of the current situation, and this can be a shattering experience since it suggests that our current reality is simply wrong."[170]

It is this process of gradual realisation that, as Rudolf Inderst claims, the player will go through as well, taking on the role of the dissident and liberating herself from those ideologically inoculated perspectives that blind her. The result is a call to action in the gameworld—a form of "*gameplay dystopique*"—that may have the player affect change in the real world later on.[171] In HALF-LIFE 2, the player's understanding of the dystopian society continues when she arrives in City 17 and experiences a merciless order. Having disembarked the train, a Scanner (a flying surveillance robot) takes pictures of its passengers, and a voice coming from a large screen welcomes her to City 17.

> Welcome. Welcome to City 17. You have chosen, or been chosen, to relocate to one of our finest remaining urban centres. I thought so much of City 17 that I elected to establish my Administration here, in the Citadel so thoughtfully provided by our benefactors. I have been proud to call City 17 my home. And so, whether you are here to stay, or passing through on your way to parts unknown, welcome to City 17. It's safer here.[172]

---

168 Fitting, "Unmasking," 164.

169 Sargent, *Utopianism*, 113.

170 Ibid.; cf. 113.

171 Rudolf Inderst, "Stoppt den Endsieg! Das Videospiel Turning Point als Alternative History: eine digitale Fortsetzung Antinationalsozialistischer Dystopien," in *Contact, Conflict, Combat: Zur Tradition des Konflikts in digitalen Spielen*, ed. Rudolf T. Inderst and Peter Just (Boizenburg: Werner Hülsbusch, 2011), 180; cf. 180-182.

172 HALF-LIFE 2 (Valve, 2004), ch. Point Insertion.

It comes as no surprise that the key words Dr. Wallace Breen uses are *been chosen*, thus confronting the player with a society whose people lack self-determination. Already in the first minutes of the game, the dystopian mode of HALF-LIFE 2 is ubiquitous, and the player quickly composes a negative image of this society in her mind.

2.5.2 The Official Narrative and the Combine's False Utopia

HALF-LIFE 2 shines in letting the player explore an *official narrative* in a playful manner and in terrifying detail. On Entering City 17, a bleak and derelict picture awaits. The train station resembles a prison, with fences and barbed wire on both sides of the tracks. Police forces are searching incoming passengers and do not hesitate to use their truncheons. They call themselves Civil Protection (CP), but what they do is question, torture, and murder people. The player experiences these brutalities directly, as she is potentially hit and called in for questioning. Stepping into the interrogation room, terrifying expectations are aroused. Blood is spilled over the floor, and the CP officer asks for privacy. He switches off the surveillance cameras—yet, to the player's surprise, the man turns out to be Barney Calhoun, a member of the resistance, who will help Freeman escape.

By experiencing the events of HALF-LIFE 2, the player may discern fundamental parallels with George Orwell's *Nineteen Eighty-Four*, and the game clearly takes the novel as inspiration. It employs depressing spaces to raise awareness of the situation's severity and depicts City 17 as an inhospitable location, whose run-down houses are juxtaposed with the towering Citadel, the glorious and phallic headquarters of Dr. Wallace Breen's Earth administration. Indeed, City 17 explicates what Orwell's London imaginatively implies and illustrates a misanthropic environment at the centre of which the oppressive order pompously rules.

Even outside the cities, the dystopian mood is palpable. Barnacles (slug-like creatures that glue themselves to the ceiling and suck people in with their tongues) are infesting suburban slums and sewer tunnels, while beach areas are swarmed with ant-lions. Another threat is the radioactive contamination that has transformed the once-idyllic countryside into the undesirable place the player encounters. The dangerous scenery culminates in the mining town of Ravenholm, which before the Combine invasion thrived with industry and life. Now it has become a ghost town, marked by burning houses and HALF-LIFE 2's most iconic enemies: the headcrabs (a parasitic life form that attaches itself to human hosts. They burrow into their victim's brain and transform the human host into a zombie). People do not go to Ravenholm anymore, and only a lunatic by the name of Father Grigori awaits the player when she passes through.

Figure 6: The patriarchal order of the Combine, exemplified through the towering Citadel and Dr. Breen as the archetypal father figure.

HALF-LIFE 2 (Valve, 2004), ch. Point Insertion.

It is clear that the alien invaders of the Combine have plunged Earth into chaos and established a totalitarian regime on its surface. Like the Party in *Nineteen Eighty-Four,* they are organised according to some tight principles: an ideology. To promote their cause, the Combine employ Earth Administrator Dr. Wallace Breen as their representative—and, above all, it is the lies by which he justifies their ideology that are noteworthy: disguising the Combine's siege as a form of Utopia.

As such, Breen assumes a role common to dystopian fiction. He represents an archetypal *father figure* and the elusive leader of the dystopian regime. Similar to Zamyatin's *We,* Huxley's Brave *New World*, and Orwell's *Nineteen Eighty-Four*, then, HALF-LIFE 2 resembles a "patriarchal order,"[173] which is reinforced by the fact that the Combine are addressed with the euphemistic term *our Benefactors.* The analogy this creates leads back to Zamyatin who used the term in his novel to describe "the larger-than-life figure … who rules the One State."[174] As leviathans to the ant-like citizens of Earth, then, and symbolised through the phallic Citadel, the Combine's superiority over humankind is ubiquitous. Earth, in fact, was not

173 Ferns, *Narrating Utopia*, 115.

174 Ibid.

their first target and the hostile race had invaded a plethora of civilisations before—harvesting people (as their name suggests) and subjecting them to oppression.

To strengthen their rule on Earth, the Combine employ a technique common to totalitarian regimes. Through mind manipulation and propaganda, they create an *image of a Utopia* and lead people into accepting the false sense of safety and "womb-like security"[175] they promise. This shows itself in that the player is often fighting human NPCs, the Overwatch, who are suffering from memory replacement. They are physically superior soldiers enhanced through genetic and artificial augmentation—a technology offered by the generous Combine. As a result, it is only through this "suppressive use of technology and scientific advances" that the Combine are able to sustain their rule over Earth.[176]

In addition, Earth Administrator Dr. Breen propagates the false sense of safety through propaganda messages—so-called *Breencasts*—displayed via large screens and loudspeakers placed in key agglomeration areas, or through television and radio transmissions. Their function is clear: ensuring that people believe in their benefactors' caring nature. This brainwashing is further propelled by terms such as *our Benefactors* or *Civil Protection*, which recall *Nineteen Eighty-Four's* comprehensive falsifications of truths[177] and contribute to the fallacious sense of safety in HALF-LIFE 2. What further illustrates this point is that the falsification of history is enforced by substances the Combine have poured into the water supply. These make people forget the time before the invasion, and such a strategy serves the oppressive order a simple purpose: if you do not remember the past, how can you tell if it was better? So the present seems a little closer to Utopia.[178]

The image of womb-like security culminates in a suppression field the Combine have laid over Earth. It ensures that human procreation has dropped to zero—

---

175 Ibid.

176 Maximilian Heinrich, "The Role of Technology and Idea of Man in BioShock and Half-Life 2," in *Contact, Conflict, Combat: Zur Tradition des Konflikts in digitalen Spielen*, ed. Rudolf T. Inderst and Peter Just, (Boizenburg: Werner Hülsbusch, 2011), 185; cf 185.

177 In Oceania, the Party employs the techniques of *doublethink*—"the power of holding two contradictory beliefs in one's mind simultaneously, and accepting both of them"—and *newspeak*—the appropriation of language towards the needs of the hegemonic order. (Orwell, *Nineteen*, 223). This follows the purpose of reducing the diversity of expression and leads people into believing irrational equations such as 2+2=5. (Kumar, *Modern*, 292, 306, 322-325).

178 Ibid., 297, 324.

indeed, there are no children in HALF-LIFE 2. As such, the Combine have effectively obliterated the concept of the family from this world, and, with it, the last glimmers of hope. None of this is surprising, for the destruction of the private space is a common theme in dystopian fiction and links the game to novels such as P.D. James' *The Children of Men* (1992) in which humankind has become infertile.[179]

Thus an interesting image arises. In an inverted metaphor of the mother's womb, the suppression field functions as symbol of unfreedom and peril under the guise of safety and the promise of Utopia. Again, Dr. Breen justifies its necessity in one of his Breencasts:

> In order to be true to our nature, and our destiny, we must aspire to greater things. We have outgrown our cradle. It is futile to cry for mother's milk when our true sustenance awaits us among the stars. And only the universal union that small minds call 'The Combine' can carry us there. Therefore I say, yes, I am a collaborator. We must all collaborate, willingly, eagerly, if we expect to reap the benefits of unification. And reap we shall.[180]

To live up to the Combine's standards, the suppression field is of essential importance. Only when the day comes, Breen declares, when "we have mastered ourselves, the day we can prove we no longer need it", can it be shut down.[181] With the promise of a posthuman future where humankind will evolve to a higher state of being, the Combine have tricked them into oppression. Yet Earth's population is left no choice, and to survive at least for one generation, their sole option is adjusting to the reasonable and scientifically advanced Combine.

In the light of these findings, HALF-LIFE 2's gameworld resembles a *prison* the player wishes to escape from. Only few places are left untouched by the oppressive order, and this imagery links the game to dystopian fictions in which society is organised according to the principles of "conformity" and "uniformity,"[182] where every aspect is controlled by the ruling regime. For Ferns, it is especially "the extraordinary public character of life"[183] that the classical dystopia exposes.

179 The destruction of the family further links HALF-LIFE 2 to *Nineteen Eighty-Four*, where family spaces are characterised by a "pervasive atmosphere of fear and distrust." (Kumar, *Modern*, 315). And its members have "become in fact an extension of the Thought Police." (Ibid., 319).

180 HALF-LIFE 2 (Valve, 2004), ch. Anticitizen One.

181 Ibid., ch. Point Insertion.

182 Ferns, *Narrating Utopia*, 113.

183 Ibid., 112.

> Like the traditional utopia, dystopia portrays a society which is regimented and hierarchical—and also one where adherence to the societal ideal is ensured by an almost obsessive concern with surveillance, with the subjection of the individual to public scrutiny.[184]

Similarly, the Combine have elected *individuality* to be their prime enemy—for, as Dr. Breen recalls, humankind's "true enemy is instinct", and this he links to "superstition," "unreasonable impulses," and to "the basest of human urges: the urge to reproduce."[185]

With instinct being an inherent part of the human self, the Combine's targets of attack become clear: human individuality and the striving for liberty. Besides the suppression field, mercilessly stomping on human will, enforced conformity is noticeable in other aspects of the game. Like in *Nineteen Eighty-Four*, the Combine strive to "control not just the external but internal world of men,"[186] which is executed through the forces of the Civil Protection. In many ways, they are HALF-LIFE 2's *thought police*, wearing white gas masks and threatening black uniforms. People live in constant fear of the CP, as they intrude into their homes and torture their loved ones. What terrifies most, however, is the threat of abduction. Dissidents are transferred to the Citadel where they are murdered or held captive to undergo memory replacement. Given the Overwatch's omnipresence, every inch of life is spied upon, and the intrusion into the private sphere is aggravated through the usage of Scanners. These are flexible variants of Orwell's telescreens that take pictures of people's activities or track down enemies of the state. In short, in the world of HALF-LIFE 2, human spirit is broken, and the player encounters a highly regulated gameworld in which even cities and highways are numbered. But thankfully, it seems not everything is lost just yet.

2.5.3 The Counter-Narrative and the One Free Man as the Opener of the Way

> We now have direct confirmation of a disruptor in our midst, one who has acquired an almost messianic reputation in the minds of certain citizens. His figure is synonymous with the darkest urges of instinct, ignorance, and decay. Some of the worst excesses of the Black Mesa Incident have been laid directly at his feet. And yet unsophisticated minds continue

184 Ibid.

185 HALF-LIFE 2 (Valve, 2004), ch. Point Insertion.

186 Kumar, *Modern*, 306.

to imbue him with romantic power, giving him such dangerous poetic labels as the 'one free man', the 'opener of the way'.[187]

Figure 7: The counter-narrative begins with a minor but conceptually important event: with throwing a can at a CP officer.

HALF-LIFE 2 (Valve, 2004), ch. Point Insertion.

Walking through the alleys of City 17 and becoming involved in HALF-LIFE 2's official narrative, the player quickly becomes wary of the situation. On her trip through dystopia, she combines the gameworld's perspectives (including its spaces, conversations, combat encounters, and other plot events) and gradually comes to understand it on a deeper level. The first levels are thereby marked by the *absence of agency*. The player can only navigate this world, endure its injustice, and try to escape its reach by moving forward through the gamespace. This intimate experience of playing dystopia puts the player (similar to Gordon Freeman) in a state of revulsion and contributes to the urge to fight the Combine regime. The player's attempts to finish the game by taking the fight to the Combine are thus supplemented by her commitment to change dystopia for the better. In this respect, Dr. Breen confirms the existence of a *single disruptor*: Gordon Freeman and the player herself.

The counter-narrative in HALF-LIFE 2 begins with a minor but conceptually important event. When the player leaves the train at City 17's train station, she

187 HALF-LIFE 2 (Valve, 2004), ch. Water Hazard.

encounters a guard that commands her to throw a can into a nearby trash can. The player may comply with this request—which suggests a resigned and submissive stance—but she may also use her possibilities to act in creative ways: for example, if she throws the can at the Civil Protection officer. This action is optional but, when actualised, assumes substantial meaning. The player may be beaten in retaliation, without having the possibility to fight back, and so experiences the dystopia of HALF-LIFE 2 in a direct feedback loop. Such interactions point towards a militant/resistant attitude on the player's side and to a revulsion against the regime.[188]

Whatever the player chooses, in confronting her with the inner workings of a dystopian gameworld, she is led to the conviction that something can and must be done against it. The *counter-narrative* and *escape* from City 17 begins no later than when the player meets Barney Calhoun, a resistance member guised as CP officer who comes to the player's assistance. It is here that Freeman (a *telling name* to underline the opposition to the Combine regime) and the player acquire their first weapon. The *crowbar* has become notoriously popular amongst HALF-LIFE 2's appreciators and holds strong symbolism—a fact the Combine seem to be aware of. They are attentive to the single "disruptor", "the 'one free man'" they fear will become "the 'opener of the way'."[189]

The Combine's fear that even a "singular misfit" may tear "open the ideological veil of the system"[190] is an omnipresent theme in dystopian fiction. On his escape from City 17, the Overwatch relentlessly pursue Freeman by sending out troops, military helicopters, and dropships. Thankfully, all their efforts are in vain, and Freeman and the player manage to escape the city's confining spaces towards the countryside. This direction leads towards the picturesque spaces of nature and links Freeman's escape to that of dystopian dissidents such as Guy Montag's in *Fahrenheit 451* or the Savage's in *Brave New World.* Although nature spaces in HALF-LIFE 2 do not provide complete safety either (large parts of the gameworld have become contaminated by radiation or are infested by vicious creatures and Overwatch patrols), the countryside is where the resistance hideouts are located.

188 In this regard HALF-LIFE 2 shows its *satirical* nature. Besides the choice described above, the player may toy with the game system and go on a wild goose chase by provoking a CP officer and making him follow her on ridiculous routes through the environment. This sort of transgression naturally implies an opposition to the dystopian regime.

189 HALF-LIFE 2 (Valve, 2004), ch. Water Hazard.

190 Moylan, *Scraps*, 118.

Nature thus enables a space for utopian enclaves[191] and hideouts from the Combine regime—such as the research facility Black Mesa East, where Freeman meets key members of the resistance, including Dr. Eli Vance, Dr. Judith Mossman, and the charming Alyx Vance.

Figure 8: Alyx (to the right) is the player's companion and functions as the archetypal figure of the temptress.

HALF-LIFE 2 (Valve, 2004), ch. A Red Letter Day.

Alyx is a so-called companion character (the first of her kind in video games) and will accompany the player throughout most of the experience. In addition, she embodies an archetypal character of dystopian fiction, which is that of "the temptress."[192] In classical dystopian narratives, the *temptress* is most often a woman whom the main (male) protagonists meets and falls in love with—or establishes some sort of close relationship. The temptress functions as a *figure of guidance* as she charismatically helps the protagonist and reader gain insight into the dystopian

191 This is more apparent in HALF-LIFE 2: EPISODE 2 (Valve, 2007), whose events lead Freeman into a picturesque wilderness far away from the derelict city spaces of the main game. The game culminates in the battle at the White Forest base (a telling name indeed) and lead to a glorious victory for the resistance. Yet EPISODE 2 ends on an open, ambiguous note, for Freeman and Alyx are captured by Combine Advisors—thus fulfilling the necessities of the VGD as critical dystopia variant I.

192 Chad Walsh, *From Utopia to Nightmare* (London: Geoffrey Bles, 1962), 101.

situation of the fictional society.[193] Thereby, her methods of seduction are diverse and range from enthusiastic, curious inquiries about the world—such as Clarisse in *Fahrenheit 451* or Julia in *Nineteen Eighty-Four*, who both question the integrity of their society—to naive but nonetheless charming seductions in PLANET OF THE APES (Franklin Schaffer, 1968) or *The Time Machine* (H. G. Wells, 1895). In these, the female protagonists (Nova, Weena) foreground the society's paralysis through their juvenile behaviour and evoke in the protagonist a revulsion to the society at hand.

Alyx can be ascribed to the former methods of seduction. In HALF-LIFE 2 the player comes to know the woman when Alyx tutors her in the use of the Gravity Gun—the Zero Point Energy Field Manipulator. Here, the player also meets Alyx's robot pet Dog and is made familiar with the weapon in a game of fetch where Dog returns objects the player tosses into the air. Step by step, the player gets to know Alyx and (potentially) establishes a close relation to her and a *personal enclave* from the troubles of this world.

The reason for the player's emotional connection to Alyx has a couple of reasons. First of all, the game is displayed from a first-person point of view. Seeing the events through the eyes of Freeman, the player is addressed by Alyx, who in various conversations looks her directly in the eyes. Thereby, Alyx's facial animations enhance the feeling of connectedness. They range from serious, concerned, satirical, and quizzical expressions, to smiles and other charming emotions. Establishing such a connection to the temptress will propel the player to fight alongside her against the Combine—and although the player is not responsible for her safety (in combat Alyx will take care of herself), their mutual struggle strengthens the bond between them.

Thus far, the discussion of HALF-LIFE 2's counter-narrative implies a strong role on the player's part. Assuming the role of the *catalyst*, the player becomes responsible for laying the foundations of social transformation and is set at the forefront of the resistance—which began with the flight from City 17 and establishing a relation to Alyx, and continues in her intervention in the high security prison facility Nova Prospekt, where the resistance turns collective. Consequently, Freeman (and thus figuratively the player) have gone through a *crucial transformation*: from the single dissident who barely escaped the Combine's grasp, to the one free man and the glorious leader of a worldwide resistance that aims to overthrow the Combine regime. Their intervention in Nova Prospekt thereby serves humankind as an example: that something can be done.

193 Ibid.

As a result of these occurrences, City 17 has turned into a battlefield, for both the resistance and the Combine are gearing up for war. More than ever, disturbing noises mark the scenery. There are, for example, the resonating sounds of drop-ships and other noises that warn of Combine activity—such as the CP's intermittent radio chatter or horns and alarm bells. Most terrifying, though, are Striders and their haunting screams. These three-legged mechanical walkers are deadly enemies equipped with plasma guns and take their inspiration from H. G. Wells' tripods in *The War of the Worlds* (1898).

Despite all these adversary forces, the fight to overthrow the Combine looks promising, however brutal and costly it is. The resistance has advanced into City 17, and in a declaration of independence, they tear down the monitors used to display the Breencasts—a scene reminiscent of the Iraqi people tearing down the statues of Saddam Hussein. The time has come now, as Baccolini and Moylan would say, for the "climactic event that attempts to change society", for "the critical encounter that ensues when the citizen confronts, or is confronted by, the contradiction of the society."[194] Almost at the game's end, the player is thus confronted with dystopia's high priest, Dr. Breen, who tries to convince Freeman that the Combine's ways are those to be followed. Such an encounter is a vital part of dystopia's plot structure, and Kumar links it to Fyodor Dostoyevsky's legend of *The Grand Inquisitor* (1879–1880).[195] In this parable, Christ visits the city of Seville during the period of the Inquisition but is thrown in jail by the Grand Inquisitor, who explains to Christ why his arrival is superfluous and that he shall be sentenced to death.[196] The purpose of such an event is to illuminate the necessity of the hegemonic order and to lead the protagonist (or player) astray, in order to convince them of the regime's noble agenda and the necessity of its measures. Like Bernard Marx, then, who is enlightened by the World Controller Mustapha Mond about society's necessity, Winston Smith, who faces O'Brien in Room 101 and endures reconditioning, and Guy Montag, who eventually kills Captain Beatty, this encounter now lies in front of the player and occurs as follows.

After a brief conversation, the encounter transforms into a boss battle. Freeman pursues Breen to the Citadel's Dark Energy Reactor and prevents the opening of a portal to the Combine Overworld. Breen is supposedly killed in the blast, yet Freeman survives with the help of the G-Man, who freezes time and reclaims his

194 Baccolini and Moylan, "Introduction," 6.

195 Fyodor Dostoyevsky, *The Grand Inquisitor*, ed. Helena P. Blavatsky (Auckland: Floating Press, 2009).

196 Kumar, *Modern*, 120.

protégé before the Citadel blows up.[197] The phallic symbol of the Combine oppression now lies in ashes, and HALF-LIFE 2 ends on a hopeful though ambiguous note. A first glorious battle has been won, but the fight is not over yet.

2.6 DYSTOPIA'S PLOT FRAMEWORK AND THE VIDEO GAME DYSTOPIA

HALF-LIFE 2 lets the player enact a terrifying vision of how her world could be and in some places already is—with existing totalitarian regimes in China and North Korea and authoritarian ones in Russia, Turkey, and other parts of the world. To make the player aware of these problems, the game targets oppressive regimes and the lies by which they justify their ideology. Becoming involved in the events, the player experiences dystopia's nightmares in a playful trial action. As a stranger in an unknown place, she ventures into the estranged gameworld of HALF-LIFE 2 and eventually comes to see behind its wrongs.

Consequently, in the course of the game, Freeman and the player have gone through substantial transformations. From the unwary fugitives who use a wrench to crack open dystopia's ideological lock and come to see behind the lies of its oppressive order, to the powerful leaders of the resistance who free themselves from dystopia's confines and demolish the Combine regime with the use of the Gravity Gun as a technical augmentation of their powers (which also includes the Mark V HEV Hazardous Environment Suit). Although the player has rather quickly discerned the situation for what is (also because we are so accustomed to totalitarian dystopias), she is reminded of its severity and might draw unexpected connections to her empirical world.

In any case, the dystopia of HALF-LIFE 2 encourages the player to counteract totalitarian movements in the empirical world as she gains inspiration for her virtual endeavour. Especially the game's final sections incite her to do so, where the Gravity Gun displays a *powerful sense of agency* and functions as a symbol of

197 Indeed, one could argue that Freeman enjoys no agency because of his dependence on the G-Man, who sent him on the mission. Heinrich, "Role of Technology," 192. The G-Man could therefore be regarded as the puppeteer behind the curtain and the events of Half-Life 2 as a meta commentary on the player's lack of agency in FPSs or dystopia. In the light of the previous discussion, however, such a conclusion is inappropriate, specifically when having in mind the plot's internal logic and Freeman's efforts against the Combine regime. Claiming that everything is predetermined by a mysterious figure behind the curtains is in this case an easy excuse.

"empowerment" for the "creative use" [198] of science and technology as utopian enclave.[199] *Force and revolution are the suggested solution*,[200] and the player as dissident not only struggles for victory (which would be the result of an uncritical playthrough that focuses on the pleasures of combat) but also for human compassion and freedom from an oppressive regime. HALF-LIFE 2 thus offers the player several perspectives that justify combat against the Combine and endow it with an ethical quality by creating utopian enclaves on a personal level with the characters—first and foremost, the relation between Freeman and Alyx.

The analysis of HALF-LIFE 2 has primarily discussed a most integral part of dystopia's implied player, which has been described as the (critical) dystopia's *narrative framework* or *plot structure*. It has revealed that the oppositional structure of *official narrative* and *counter-narrative* is a perfect fit for the video game medium, because it sets the player within the dystopian gameworld and at the forefront of the resistance. In the vein of traditional dystopian fiction, HALF-LIFE 2 employs tropes and features that draw from Orwell's *Nineteen Eighty-Four*. These include: the beginning *in medias res*, the ideological distance created through estrangement, the claustrophobic mood of the gameworld and its misanthropic environment, the patriarchal order and its strict ideology that tricks people into believing the grand narratives of Utopia, and the archetypal father-figure who functions as the regime's mouthpiece. In combination, all these plot devices create a powerful official narrative and a dynamic framework for the player to participate in.

However, HALF-LIFE 2 would do no justice to the genre of the critical dystopia was it not to renounce its official narrative and give rise to a collective counter-narrative. Through Freeman's actions, a utopian horizon is actualised. Although this horizon is predetermined by the game system, it is also motivated by the process of realisation Freeman underwent—from initial unawareness and the mocking of the Combine regime to escape and revolution and to finding the enclave of

198 Heinrich, "Role of Technology," 186.

199 Science in HALF-LIFE 2 is displayed in a critical but not necessarily negative way and "as a double-edged sword." (Ibid). While the Combine's use of science can be regarded in a negative way as they use their technological superiority to suppress humankind, most human scientists are in fact saviours, endowing Freeman and the resistance with the means to fight back. (Ibid., 185-186).

200 Heinrich describes HALF-LIFE 2 as being built around "the logic of Charles Darwin's credo 'survival of the fittest', and that science and technology serve as extensions of the human species' capabilities and heighten adaptability in the Darwinian sense." (Ibid., 183).

human compassion and dialogue. This *coming-of-age* is moreover encouraged by the archetypal figure of the temptress, who helps both Freeman and the player comprehend the nature of the dystopian society and impels them to rebel against it. The player's journey thus led from the derelict spaces of City 17 towards the countryside, where utopian enclaves hide and prepare for revolution, and culminates in the encounter with dystopia's leader—closing the game on an ambiguous but hopeful note.

As a result of these observations, the VGD's plot structure creates a *malleable and dynamic framework for play* in which the player ergodically and imaginatively participates. It thus fulfils the requirements of what Grzegorz Maziarczyk has called "playable dystopias"[201] and involves the player in *a creative dialectic with the implied player* (which I will describe thoroughly in Part II). The result of this involvement is an *aesthetic experience* that translates into an *ethical response* by making the player reflect on the shortcomings of her empirical world and, potentially, inciting her to counteract dystopia. Such a reaction is certainly not unique to HALF-LIFE 2 and confirms Alexander Hall's hypothesis on dystopian Alternate Reality Games (ARGs). These possess a strong "utopian energy"[202] that comes to fruition in the act of play and evokes "the desire of gamers to seek out alternatives to the status quo."[203]

Hall is certainly confident about a game's potential to prepare the player for the task of confronting dystopia in the empirical world. Yet my descriptions of HALF-LIFE 2 came to a similar conclusion. By involving the player on both an

---

201 Maziarczyk, "Playable Dystopia," 237.

202 Hall, "Trying to Believe," 70.

203 Ibid., 80. The strategies of ARGs are close to the VGD's in that they ascribe the role of becoming "members of the resistance, which puts them in position to use the game as a means to a socially revolutionary end," to players. (Ibid). Thereby not only the player's physical interaction is of interest—although it "heightens the gamer's awareness of his or her ability to subvert the system in place" by encouraging "the kind of action necessary to move closer to the new social system through game-specific, diegetic action"—but also her imaginative interaction with it. (Ibid., 71). Dystopian ARGs are "cultural products" that force the player "to think historically," "to consider the present as the future's past" (Ibid.) and create cognitive "map[s] of the things which must change to avoid such a reality's being brought to fruition." (Ibid., 70). The desired real-world response is a "call to action against the aspects of culture that are dystopian themselves" (Ibid., 80) and a facilitation of "political action through gaming." (Ibid., 69).

ergodic and an imaginative level, the game created a private experience but also gave her the possibility to imaginatively connect the virtual events to her empirical surroundings. In other words, the player became involved in what Steffen Walz calls a "play-ground" that "feature[s] inherent conflicts and … goals for player-heroes to achieve in that they exhibit word-flaws [sic] [world-flaws] or imbalances that the player must overcome in order to turn dystopia into a regular, if not heterotopian world."[204] Through these *lessons on agency* in virtuality, the player is made susceptible to utopian alternatives in her empirical reality and is driven towards their realisation.

HALF-LIFE 2's warning is thereby only one of many communicated by the VGD—and the forthcoming chapter will clarify additional ones. Before moving on to this issue, however, let me close with the following assertion: besides the many ways dystopian gameworlds entice their players, the most important form of involvement can be found in the player's *aesthetic involvement* in a game. This emancipated involvement brings forth an intimate comparison between the gameworld and the player's empirical surroundings and gives rise to the newness art promises: an *aesthetic effect* that can only be experienced in and through the act of play.

Consequently, and as result of these finding, let me formulate four hypotheses on the VGD that are both general (applying to all variants except for the anti-utopia) and specific (describing the variant at hand more closely).

1. *The video game dystopia represents a virtual society outlined in minute detail which the game designer(s) intended a contemporary player to view as considerably worse than her or his empirical present. It either entraps the player ergodically in the gameworld—suggesting no solutions to dystopia (classical dystopia)—or holds out hope for the betterment of society in one or more utopian enclaves situated in the gameworld (critical dystopia). This is often (though not always) accompanied by granting the player an insight into how the dystopia came into being (critical dystopia).*
2. *The primary locale of the dialectic between hope and despair is the genre's plot framework: the clash of the official narrative and counter-narrative. Depending on the type of dystopia, the counter-narrative may end in despair and in imaginative resistance on the player's side (classical dystopia) or on a hopeful though ambiguous notion—which is either predetermined by the*

204 Steffen P. Walz, *Toward a Ludic Architecture: The Space of Play and Games* (ETC Press, 2010), 145.

game's system (critical dystopia variant I) or negotiated by the player (critical dystopia variant II).

3. *To retain the notion of the society's hypothetical possibility, the VGD works according to the logic of extrapolation and estrangement. In doing so, it emits a persuasive intent that is made tempting through the sensualising power of fiction and the participatory (ludic) affordances of the video game medium, which give rise to the player's ergodic and imaginative involvement.*
4. *As a consequence of such diverse qualities, the VGD shows the potential of issuing a powerful warning to the player and of triggering an aesthetic response in her, which will have her see the empirical world with different eyes and drive her to action in the most pessimistic times.*

3 Warning, Effectiveness, and Targets of the Video Game Dystopia

The pleasures of play are many, and there can be no strict delineation of what players may find aesthetically alluring during each and every moment of their gaming experience. Such a statement necessarily demands explanation, and it might be beneficial to start with the player's *supposed pleasures* of the previously described HALF-LIFE 2. In the game, the player is sent on a mission against a merciless order and comes to experience City 17 and its surrounding areas. These he explores in detail on a linear journey and plot to counteract the Combine regime through force and combat. By doing so, he meets several members of the resistance with whom he potentially establishes close relations (first and foremost, Alyx) and becomes emotionally attached to the narrative he helps to create.

It is easy to discern that the pleasures of HALF-LIFE 2, as I have described them, fit well with Gordon Calleja's six distinct but interwoven forms of player involvement: kinesthetic, spatial, narrative, ludic, shared, and affective.[1] Notwithstanding these possibilities, the previous chapter also hinted at a form of subversive involvement that resulted from the player's comparison of the game events to his empirical surroundings. This *aesthetic (or emancipated) involvement* is universal to the appreciator's engagement with representational art and due to an artwork's perspectival arrangements and the resulting interaction processes with the participant. In HALF-LIFE 2 these perspectives included the gameworld, which is ruled by a brutal regime, the combat actions against it, the countryside as a locus of utopian enclaves, and the relations/dialogues between the main characters. Closing the blanks between them, the player comes to see the ethical justifications for the combat and the necessity of such a resistance. Still, I also hinted at the possibility of an *uncritical playthrough* that focuses on the pleasures of combat

1 Calleja, *In-Game*, 37-38.

and a world in which the use of guns is generally accepted, while downplaying the plot occurrences and characters.

Such a playthrough is a viable option and ingrained in the implied player (see the discussion on player types in chapter IV). Yet in order for HALF-LIFE 2 to exercise its full (subversive) effect as fiction, a more contemplative interaction is needed from the player. This begs a more general question, namely, *what are the prerequisites for a video game dystopia's effectiveness*; that is to say, *how does it outline an aesthetic effect for the player to experience*? In this regard, the subversive potential of VGDs as a vehicle of social criticism and transformation has been widely acknowledged by video game scholars in different ways.[2] Lars Schmeink, for instance, talks about a "utopian moment within the agency of the player" which "strikingly offers more actual utopian impulse than a linear narrative ever could."[3] Aldred and Greenspan argue that dystopian games "provide a meaningful site" on which to explore societal issues against a "dialectic of utopian and dystopian alternatives."[4] However, not all VGDs (or supposed VGDs) adhere to these claims, as was described with the example of CALL OF DUTY: ADVANCED WARFARE, which remained caught up in pleasurable combat and a clichéd ending that confirmed rather than disapproved of such drastic measures without giving them further context.

Consequently, in order to fully elaborate on this aspect a more thorough investigation is needed in which I will suggest that there are primarily two constituents necessary for a VGD's effectiveness: 1) a game's *aesthetic complexity*, by which I mean its degree of openness or multifacetedness that allows for a diverse richness of playthroughs, imaginings, and interpretations—in the VGD this *complexity of*

---

2 Tulloch, "Ludic Dystopias;" Nyman and Teten, "Lost and Found," 3; Hall, "Trying to Believe," 78; Schulzke, "Virtual," 326, 331; "Bioethics," 49,56; Domsch, "Dystopian" 408; Maziarczyk, "Playable Dystopia," 253; Schmeink, "Dystopia;" Walz, *Toward a Ludic Architecture*, 139; Gibbons, "Wrap Your Troubles;" Packer, "Galt's Gulch," 209-210, 219; Aldred and Greenspan, "A Man Chooses," 479-480; Shannon OB. Wilder, "A Narrative of Synthetic Fear: Virtualizing Dystopia in a Gaming World" (PhD diss., University of Georgia, 2006), 26, 130, https://oatd.org/oatd/record?record=oai%5C%3Augakr.libs.uga.edu%5C%3A10724%5C%2F23760; Ryan Lizardi, "*BioShock*: Complex and Alternate Histories," *Game Studies: The International Journal of Computer Game Research* 14, no. 1 (August 2014), http://gamestudies.org/1401/articles/lizardi; Grant Tavinor, "*BioShock* and the Art of Rapture," *Philosophy and Literature* 33, vol. 1 (2009): 95, https://philpapers.org/rec/TAVBAT

3 Schmeink, "Dystopia."

4 Aldred and Greenspan, "A Man Chooses," 479.

perspectives is anchored in the plot structure of *official narrative and counter-narrative*; 2) a certain player type that, given the diversity of player preferences in games, shows the necessary willingness to engage with a work of art on a complex level and in a self-aware manner. This *emancipated player* may savour all of the above-mentioned forms of involvement but, at the same time, subjects the enacted events to scrutiny and takes pleasures in the experience of meaning.

To prove these points will not be entirely possible within the confines of this chapter, and its results are rather to be seen as a premise to Part II and III, where I will go into detail about the implied player's dialectic with the emancipated player. Nonetheless, three important steps will be taken. The chapter will begin with a closer look at Utopia's universal *function as warning* to then expand the discussion to the VGD and the genre's *effectiveness in issuing a warning*—for not all dystopian games fall into the category of the VGD, because of their lack of aesthetic complexity. Lastly, a categorisation of the VGD's *real-world targets* will be made, together with the application of dystopia's *traditional plot structure* (official narrative and counter-narrative). This will give a good overview of the genre's diversity. Focusing on an enemy not uncommon to dystopian fiction, all targets discussed here—the threat of oppressive regimes, the capitalist world system, and the dangers of science and technology—centre on the universal theme of *human nature's dark side*, which shall be overcome by the player in a process of emancipation. Of concern here will be the perspectival arrangements of the VGD at hand and how they contribute to its aesthetic complexity.

3.1 THE VIDEO GAME DYSTOPIA'S PRIMARY FUNCTION AS WARNING

Considering the results of the previous chapters, it has become clear that there can be no deliberations on dystopia without regarding utopia, and the entire philosophy essentially shares a common denominator. To recapitulate, Utopia was described as setting the appreciator in a precarious situation and playing with humankind's innate fear of losing something of value. Jameson has called this phenomenon *a disruption of the present*, which radically breaks with the ideological chains of contemporary society to imagine a considerably altered now.[5]

Utopia, therefore, primarily functions as a *warning* to humankind. It attempts to raise awareness of societal issues and to arouse the individual from stupefaction and the paralysed state of the well-adjusted citizen—which, in the best case, may

5 Jameson, *Archaeologies*, 228, 233.

transform him into an active agent who gradually changes the world.[6] This persuasive attempt was described as being inherent to all of utopian and dystopian fiction, and it comes as no surprise that many scholars (as described in chapter II) regard both genres as being "part of the same project"[7] though differing in strategy. While utopian fiction closely investigates contemporary ills by evoking imaginings of a better future in the appreciator, dystopian fiction prefers a therapy of the worst kind to arouse his attention.[8] Both genres, however, clearly aim at the construction of an improved system. Otherwise, as Booker reminds us, dystopian fiction would be futile.[9] What follows is that in focusing on any one of these genres, its specific strategies of issuing warnings become of interest. These will shed light on the player's involvement in the respective genre and illuminate nuances of the types of dystopia I have described before. In the following, I will thus lay emphasis on the genre of dystopia—and observing it more closely reveals primarily *two kinds of warnings* that are of interest for the discussion of the VGD.

The first type of warning excites considerable anger in the player. It is characteristic of the *VGD as classical dystopia* and triggered by a gameworld that confronts the player with radical hopelessness and a confining system of rules from which there is no escape. Such dystopias are similar to what Moylan calls "dystopias of resignation."[10] They "embrace an anti-utopian pessimism" which, in the worst case, reinforces the player's "preference for the status quo or [helps] to … produce their capitulation to it as all hope for change is shattered."[11] Effectively, such a variant tends towards Anti-Utopia but still shows the potential of triggering in the "resistant reader" (or player) a subversive response, "by choosing to get and perhaps stay angry, and even to fight back rather than lapse into abject nihilism or trendy irony."[12] In fact, are works such as Orwell's *Nineteen Eighty-Four* (1949) or Galactic Cafe's THE STANLEY PARABLE (2013) not two of the most effective dystopias,[13] exciting considerable anger in the reader/player, who is confronted with a malignant story- or gameworld?

---

6 Vieira, "Concept," 6, 17.

7 Booker, "Impulse," 15.

8 Seeber, *Selbstkritik der Utopie*, 228.

9 Booker, "Impulse," 15.

10 Moylan, *Scraps*, 181.

11 Ibid.

12 Ibid.

13 The opinion is shared by Schulzke, who claims "that some of the most effective … dystopias in this sense are those that do not preserve any possibility of hope." (Schulzke, "Virtual," 10).

The second type of warning also excites anger but does so in a more intricate manner. Similar to the resigned variant (the classical dystopia), the critical dystopia triggers a dismissive attitude towards the gameworld but, in doing so, retains a horizon of hope within its bounds—thus giving the player an incentive and, more importantly, the possibility (or agency) to change the situation. This type of warning can be compared to Moylan's "militant dystopias,"[14] whose effect on the reader he describes as follows:

> Although fear and outrage may inform the writings of such texts, the creative encounter with the realities of the social systems leads not to doubt and despair but to a renewed and focused anger that can be tempered with radical hope and visions.[15]

In the militant VGD (*the critical dystopia*), the creative encounter with the social system is intensified in comparison to non-ergodic forms of dystopia and leads to the arousal of fictional anger[16] towards the dystopian regime. Certainly, the classical dystopia may achieve this effect as well, but the critical dystopia goes further by giving the player ample opportunity to express his emotions through ergodic effort and to fight dystopia in virtuality (through ludic capabilities, as described with the example of HALF-LIFE 2). In other words, whereas the player's endeavours in the classical dystopia are exhausted, the critical dystopia allows him to focus his anger and effectively work towards Utopia. This is not to say that a reader's or viewer's involvement in dystopia is devaluated, or that such anger cannot emerge in these non-ergodic forms. However, it is the *possibility to react to emotions on an intradiegetic level* that the VGD allows, because of the medium's diverse layers of involvement that extend beyond that of a novel or film (as described at the beginning of this chapter). Consequently, the fictional anger accumulated in playing dystopia finds an *outlet* in the player's actions towards change in the virtual society. Often, this journey towards Utopia is, however, predetermined by the game's system (the critical dystopia variant I), yet this nonetheless gives the player the sensation of *actively doing something* against the dystopian regime.

There are, however, other instances in which the utopian horizon lies virtualised within the game system, awaiting the player's actualisation. In such cases,

---

14 Moylan, *Scraps*, 181.

15 Ibid.

16 I agree with Walton in that emotions triggered by representational art are fictional and not real, for they lead to different reactions in the appreciator compared to those in real situations. (Walton, *Mimesis*, 195ff).

the VGD makes use of its full potential, as both failure and success become an option. The *critical dystopia of variant II* encloses the player within the confines of a hegemonic order but also invites him to escape it. By metaphorically lock-picking his way out of "the iron cage of Anti-Utopia,"[17] he transforms into an "agent … capable of effecting a real transformation of the global social and economic system."[18] Still, the player's route towards Utopia can be difficult, and the failure to attain it remains a constant companion. This is apparent in games such as BIOSHOCK (2K Boston, 2007) or METRO 2033 (4A Games, 2010), where the player is confronted with the precarious situation of potentially losing something of value. In these, the tiniest mistake may lead to the loss of Utopia, and any playthrough can result in a pessimistic, optimistic, or ambiguous outcome—that is to say, in a classical *or* critical dystopia. It would be a mistake, however, to determine a game's narrative genre via the outcome of a playthrough; rather the *potential of certain options* is the decisive factor.

As a result, while the VGD of resignation (classical dystopia) ergodically entraps the player in the gameworld (see THE STANLEY PARABLE) and the VGD as anti-utopia continues to foreclose imaginative possibilities (see COD: AW), the militant version of the VGD (the critical dystopia) does not halt at "a rattling of the bars"[19] and involves the player in a precarious situation. Being confronted with the choice of becoming a catalyst for social change and transformation, the player holds the possibility of acting upon solutions proposed by the game and of actualising the utopian horizon through imaginative and ergodic effort. Such a task is by no means easy and entails *responsibility* on the player's part. It requires considerable effort and the choices of an *ethical player* who searches for utopian enclaves within the system and actualises them out of the conviction that something can and must be done to counteract dystopia. As Sicart claims:

> [E]thical agents … create and practice their virtues and thereby become a better human being. … The model ethical player sees an opening in the game system, which might be a tease to play morally, and follows it. This player explores the gray zones of decision in *Fallout 3* ... sits down and waits for the child to finish dinner in *Heavy Rain* (Quantic Dream 2010), plays *Unmanned* (Pedercini 2011 [sic] to understand the mundane ethic of drone warfare … .[20]

---

17 Moylan, *Scraps*, 154.

18 Levitas and Sargisson, "Utopia in Dark Times," 16.

19 Jameson, *Archaeologies*, 232.

20 Sicart, *Beyond*, 78.

Like an emancipated player, thereby, ethical players need "to be empowered as ethical agents"[21] by the game design—this is to say, inscribed in the game system.[22] They cannot express themselves creatively if the game focuses on intense combat action without giving the ethical player additional perspectives on the enacted events, or the possibility of acting in a different, pacifist manner (which games such as METRO 2033 or FALLOUT 4 allow). As such, Sicart's observations underline my initial hypothesis that for a game to exercise its full subversive effect, both the aesthetic complexity of the game at hand and the willingness of an ethical or emancipated player become necessary prerequisites.

I will go into further detail in this regard in chapter V, yet for now these claims lead to an unsurprising question: is a VGD only effective when the player chooses to play in an ethically correct manner? What if he embarks on an unethical route with games such as INFAMOUS: SECOND SON (Sucker Punch, 2014) or the FALLOUT series?[23] I object to such a conclusion and suggest that even if one fails to actualise Utopia in one playthrough—or plays in a destructive manner—the experience this creates may still evoke a positive reaction in the player, if he ponders his actions within the virtual diegesis. Dystopia, then, seems to fit into the requirements of an *emancipated player*: a critical player who seeks pleasures in subversive experiences and tries to obtain a transcendental vantage point on the events by examining the encountered and co-created perspectives.

Consequently, if both prerequisites are met—the aesthetic complexity of a game and the open-mindedness of a critical player—the lessons of virtuality can be valuable (as are those of non-ergodic fiction) and beneficial for the player's conduct in the empirical world. Thereby, the new dystopias in video game fiction foreground problematic trends in the designer's empirical reality, and by letting the player tinker with alternative scenarios, involve him in creative solutions for these issues. As *playful trial actions* in estranged gameworlds, they create a precarious situation in which there is something to *lose*—and by this I do not only mean the prospect of a virtual future. By involving the player in the inner workings and "underlying logic"[24] of a dystopian system and having him work towards the utopian horizon (successfully or not), the player may lose certain ideologically infused perspectives on the real world. The inception begins with a terrifying but

21 Ibid., 78.

22 Ibid., 77-79.

23 FALLOUT (Interplay Entertainment, 1997); FALLOUT 2 (Black Isle Studios, 1998); FALLOUT 3 (Bethesda Game Studios, 2008); FALLOUT: NEW VEGAS (Obsidian Entertainment, 2010); and FALLOUT 4 (Bethesda Game Studios, 2015).

24 Schulzke, "Virtual," 10.

enlightening experience in virtuality (leading to a partial restructuring of the player's habitual disposition) and continues to exercise its effect in real life. Having learned from the lessons of the VGD, the player comes to see empirical reality for what it is and may be inclined to work towards Utopia in real life. Such a *warning and incentive to action* shall be seen as the VGD's *primary function*.

Given the magnitude of such a claim, the question of the VGD's effectiveness needs to be explored further—and the remainder of this study will continue to assess it. For now, I wish to extend the discussion to the realms of representational art and to the transmediality of the dystopian genre. Then I will come to detailed prerequisites concerning dystopia's hypothetical possibility and shocking proximity to the empirical world. Finally, I will apply these deliberations in the analysis of various VGDs in terms of their respective real-world targets and diversity of perspectival arrangements (or lack thereof).

3.2 The Video Game Dystopia's Effectiveness in Issuing Warnings

It has been established that the VGD virtualises a negative society that foregrounds the problems of the designers' and players' empirical reality. These negative trends have been extrapolated into the future, or to some other alternative scenario, to show them in a magnified, distorted form. The VGD sets the player within this defamiliarising scenario and into an estranged gameworld he has to make sense of and involves him in a struggle against dystopia's ruling order. The genre thus holds the potential of issuing a powerful warning—but not every dystopian game (or supposed VGD) manages to do so.

The hypothesis is that the effectiveness of a VGD depends on both the game's aesthetic complexity and the player's willingness to experience meaning. Such deliberations are not new to the medium, and the aesthetic quality of games has been discussed in academia and the general public since their beginnings. In this respect, Markus Schulzke brings the discussion to the VGD. He argues that while the video game utopia has "generally failed … [to] perform a critical function because of their tendency of mirroring the real world,"[25] the VGD tends "to be very effective in presenting critical themes."[26] Although there is much truth to

25 In fact, whenever a game mirrors empirical reality, it does not count as any type of utopia/dystopia, because of the lack of extrapolation and estrangement necessary for the genre.

26 Schulzke, "Virtual," 315.

Schulzke's assertion (and he justifies it with various examples), one has to scrutinise such a claim further—which leads the investigation back to the questions of *effectiveness* and *aesthetic complexity*. Such questions are regularly asked when it comes to comparatively new media,[27] but they are important to pose for any kind of representational art—be it a new or an established form.

To approach the issue, it is beneficial to describe why the genre of dystopia is so popular in games but also which games stand out through their effectiveness in issuing warnings and which fail to do so and why. Being a transmedial genre, dystopia is essentially open to related genres—and the VGD makes no exception. Whether it is games belonging to the greater category of SF or its subgenres—post-apocalypse, cyberpunk, steampunk, posthuman—these settings invariably offer thriving grounds for the VGD. All of these worlds have a powerful allure for both the game medium and the players, and this is primarily due to three reasons.

To begin with, it is undeniable that the dystopian narrative is an ideal setting for action, since it justifies the player's opposition to masses of enemies with a dangerous regime. It thus caters to the tastes of first- or third-person shooter aficionados (or related action genres) and works in tandem with the quantitative explosion of FPSs in the last two decades—which promise high sales figures and revenue.[28] On this level, it equals a *power fantasy* that evokes humankind's drive to alter things through violent behaviour. Second, the dystopian setting has an alluring effect on the player, as do SF worlds. They speak to humankind's urge to explore faraway worlds and different cultures, and represent terrifying yet enticing escapes from the contemporary present. However, not every game that virtualises a negative scenario can be included into the genre of the VGD, for as Booker notes:

> It should be emphasized, however, that the mere fact that a novel or film features a grim future does not make the work dystopian. To be dystopian, a work needs to foreground the oppressive society in which it is set, using that setting as an opportunity to comment in a critical way on some other society, typically that of the author or the audience.[29]

27 Booker poses the same question for *young-adult dystopias*, where it is uncertain whether these new forms of dystopia are critical or that their "more positive conclusions" and "the fact that they often focus more on plot and character than on exploring the characteristics of their dystopian societies" robs them of their "effectiveness." (Booker, "On Dystopia," 14).

28 Maziarczyk, "Playable Dystopia," 235.

29 Booker, "On Dystopia," 5.

This is an important aspect and functions as premise for the third reason, which postulates the VGD as a reaction to the precarious times of the 21st century. It thereby magnifies these beyond recognition in order to scrutinise empirical reality.

Given the above described trichotomy, it follows that dystopian games can be diverse in their focus and not all will show the amount of aesthetic complexity necessary to fulfil the requirements and necessity of points two and three. Hence, when analysing a dystopian game and ascribing the status of a VGD to it, three requirements are of importance: 1) The society at hand stands in direct or analogic *extrapolation* from the designers' empirical reality—thereby excluding fantasy games. This is a necessary condition, because a gameworld based on cognitive estrangement remains in the realm of the believable—and, thus, the threat of dystopia seems palpable in both virtuality and the empirical world. 2) The plot foregrounds the "social workings"[30] of the society at hand and makes use of dystopia's narrative framework: the clash between official narrative and counter-narrative, which involves the player in a creative encounter with the dystopian regime and invites him to struggle against it. Only then does the VGD go beyond the mere depiction of a negative world and refrains from (mis)using the dystopian setting as "a decoration out of the toolkit of science fiction stereotypes"[31] to justify its spectacle.[32] 3) To achieve these points, the VGD intertwines its fictional setting with gameplay mechanics and creates participatory experiences that hold subversive potential.[33]

Many dystopian games fulfil the above-mentioned requirements and can be endowed with the status of a VGD. Gameworlds such as City 17 (HALF-LIFE 2), Columbia and Rapture (BIOSHOCK INFINITE, BIOSHOCK), Neo Paris (REMEMBER ME), the post-apocalyptic Russian metro tunnels (METRO 2033, METRO LAST LIGHT), or the floors of a random office building (THE STANLEY PARABLE) illustrate critical dystopian gameworlds the player will traverse and whose ludic experience shows the potential of triggering a subversive response in him. This is not to say that these games lose their entertaining allure: they are virtual dystopias that oscillate between *pleasure* and *terror*[34]—and successfully combine the layers of

30 Domsch, "Dystopian," 397.

31 Ibid.

32 Frelik, "Video Games," 236.

33 Domsch, "Dystopian," 396-397.

34 Wilder, "Synthetic Fear," 16.

entertainment and aesthetic education.[35] This multifacetedness elevates them beyond the mass of popular culture mediocrity into the status of what Umberto Eco has called *multi-layered artefacts*.

Ascribing the status of a multi-layered artefact to a VGD is an important step, for such a status designates a vital premise to the player's aesthetic response to it. This is so because multi-layered works of fiction (such as postmodern texts) show aesthetic complexity in that they allow for a variety of readings. In this respect, Eco argues that an adventure story pertaining to this category can either be read for entertainment purposes—as Iser would say: the reader involvement centres on "the level of plot"[36] and affective emotions—or, and on an additional plane, can be understood on the level of concept—here, reader involvement extends to "the level of significance."[37] Of course, this may only be a simplification of the diverse pleasures fiction evokes. But it foregrounds a game of *distance* and *proximity* that is fundamental to the appreciator's involvement, which oscillates between the poles of "distanced investigation and vital participation,"[38] between the desires for "knowledge"[39] and "the gaze for illusion"[40] and the frenzy of "spectacle."[41]

This dialectic between the player's reflection of the enacted and his exploration of and interaction with the gameworld (in both ergodic and imaginative terms) will be discussed thoroughly in Part II. Still, it is important to hint at it now, for the implications are considerable. If the player's involvement in the dystopian

35 Similarly, Parker argues that that many VGDs are so-called "prestige games" (2) that attempt "to reconcile art and commerce in order to generate both economic and cultural capital." (Felan Parker, "Canonizing *Bioshock*: Cultural Value and the Prestige Game," *Games and Culture* [2015]: 3, http://gac.sagepub.com/content/early/2015/08/28/1555412015598669.abstract). Frelik confirms this observation for the SF genre in that he states that "[a]lthough most SF video games are, of course, far more invested in providing entertainment than in raising players' consciousness regarding global problems, it is the genre's future-oriented mode of perceiving the world that elevated the game's problem solving, an integral part of so many video games, to an actual engagement with the world outside the game." (Frelik, "Video Games," 236).

36 Iser, *Act*, 123.

37 Ibid.; cf. Umberto Eco, *Nachschrift zum >Namen der Rose<*, trans. Burkhart Kroeber (München: Deutscher Taschenbuch Verlag, 1986), 76-82.

38 Rancière, *Emancipated*, 5.

39 Ibid., 2.

40 Ibid., 3.

41 Ibid., 4.

gameworld comes to full fruition in the life-giving tension between critical reception and illusive immersion, the initial hypothesis that both a VGD's *aesthetic complexity* and the player's willingness to engage with a game on an *emancipated level* are the determining factors for a VGD to fulfil its primary function as warning. This function is anchored in the VGD's underlying structure (the implied player) and enables a form of play that involves the player in *pleasurable, affective* yet also *subversive* and *regenerative* experiences. Revealing this structure's intricacies as a *perspectival network that composes the game* remains a vital task of this study—but before pursuing this road, the VGD's areas of investigation remain to be clarified.

For this purpose, the following and last section of this chapter will deal with the genre's range of *real-world targets* and how the plot framework of the *official narrative* and *counter-narrative* contributes to its warning (or fails to do so). Of particular interest here will be the *perspectival diversity* of the game at hand, which alongside its plot structure may include the destructiveness of the gameworld, the emotional bonds with the characters, and the ludic action—whether it revolves around combat or other forms of interaction. The aim is to show a general overview of the dystopian genre in video games and to establish a *categorisation* of its most recurrent points of attack (without being able to cover every single VGD in existence, for such research would go beyond the scope of this study).

3.3 TARGETS OF THE VIDEO GAME DYSTOPIA

It is undoubtedly difficult, if not impossible, to paint a complete picture of the historical moment and the potential directions in which it moves.[42] Still, utopian and dystopian fiction have invariably attempted to do so, with varying results. According to Jameson, Utopia works like a "*foreign body* within the social ... a space in which new wish images of the social can be elaborated and experimented on."[43] The process of attaining such an insight and finding society's "fundamental ill" is a cumbersome one, and there are primarily two parties involved: "the material" and "the vocation."[44]

---

42 Jameson, *Archaeologies*, 14.

43 Ibid., 16; emphasis added.

44 Ibid., 14.

The *material* refers to the historical moment and to “the specific situations and circumstances under which [the] … composition”[45] of utopias (and dystopias) becomes possible. Out of these “building blocks,”[46] the Utopian creates his vision of a better (or worse) future. Philosophers, filmmakers, or game designers all have to bring a specific toolset: the ability to identify a certain problem and “the inventive ingenuity with which a series of solutions are proposed and tested.”[47] This ability is what Jameson calls *vocation*: the talent of “the outsider … for seeing overfamiliar realities in a fresh and unaccustomed way” and of reducing reality to a comprehensible model.[48]

The Utopian, then, takes “delight in construction” and is motivated by the prospect of offering “a simple, single-shot solution to all our ills”[49]—a solution that “the social situation … must admit of … or at least of its possibility.”[50] However, it is this “obsessive search”[51] for Utopia that can be dangerous, because one may be tempted to force the proposed solution onto humankind (a fact that Sargent explicitly warned about, see chapter I). Indeed, Utopians these days seem to be aware of the issue. For there has been change in the conception of the better (or worse) future, as “utopia no longer aspires to change the world at a macro-level … and is focused now on operating at a micro-level.”[52] The question remains whether such ‘micro’ solutions are sufficient to overcome an “endangered today” and the dystopian situation we are currently facing.[53]

With these stumbling blocks in mind, the Utopian embarks on a venture “to extirpate this specific root of all evil from which all the others spring.”[54] Having found it, or so he believes, the Utopian shares his vision in “dramatic or aesthetic forms”[55]—if the Utopian is an artist and chooses fiction as a means of communication. To do so, he makes use of his ability to condense “social totality … [to] a

45 Ibid., 11.

46 Ibid., 14.

47 Ibid., 11.

48 Ibid.; cf. 11.

49 Ibid.

50 Ibid., 12.

51 Ibid., 11.

52 Vieira, “Concept,” 22.

53 Darko Suvin, “Theses on Dystopia 2001,” in *Dark Horizons: Science Fiction and the Dystopian Imagination*, ed. Raffaella Baccolini and Tom Moylan (New York: Routledge, 2003), 187; cf. 187.

54 Jameson, *Archaeologies*, 12.

55 Ibid., 13.

small-scale model … on which the fundamental tendencies and the lines of flight can more clearly be read."[56] In a simplification of reality, and with eagle eyes, the Utopian reveals troublesome tendencies that plague his present, and, consequently, Jameson comes to the important conclusion: "that Utopian space is an imaginary enclave within real social space, in other words, that the very possibility of Utopian space is itself a result of spatial and social differentiation."[57]

Given this construction process, utopian and dystopian fiction is the artist's way to cope with the neurosis caused by empirical reality. Even though Jameson's observations focus on the genre of utopia, it is feasible that his insights hold true for dystopia as well—which brings the investigation back to the question of the vision's totality. In this respect, Kumar pointed out that, similar to utopia, dystopia has fragmented and narrowed its focus. Unlike Orwell or Huxley, who managed to compose an alternative society that centred on "a dominant trend or central principle in contemporary developments," recent dystopias "have had highly specific targets" and have generally failed to shock their readers "with a compelling sense of a march into a nightmarish future."[58] Kumar underlines this claim in a recent article where he maintains that "it was never the function of dystopia to give a complete picture of the world" but "to pick out the most distinctive and novel features of the time and to present them in the form of an imaginatively realised society."[59] Sometimes, however, specific trends have far-reaching effects—as with Huxley, who chose to attack consumerism in *Brave New World*, thereby touching a nerve that continues to resonate in the 21st century.[60]

These observations notwithstanding, the question remains whether dystopia's picture is partial—focusing on specific issues—or whether it manages to capture a dominant trend that plagues social totality. The VGD, for the most part, continues the trend of specific targeting, as it issues focused warnings that communicate

56 Ibid., 14. This facet connects the genres of utopia and dystopia to the procedural representations of *simulation*, which is generally understood as a simplification of empirical reality (see chapter V). In the words of Salen and Zimmerman: "simulation cannot depict every aspect of something; it has to choose a very small subset of characteristics around which to build its representation." (Salen Katie and Eric Zimmerman, *Rules of Play: Game Design Fundamentals* [Cambridge, Mass.: MIT Press, 2004], 423).

57 Jameson, *Archaeologies*, 15.

58 Kumar, *Modern*, 422; cf. 422.

59 Krishan Kumar, "Utopia's Shadow," in *Dystopia(n) Matters: On the Page, on Screen, on Stage*, ed. Fátima Vieira (Newcastle-upon-Tyne: Cambridge Scholar Publishing, 2013), 22.

60 Ibid.

greater contemporary problems to the players: the dangers of oppressive regimes, life in bureaucratic consumer capitalism and the consequent unfreedom of the individual, manmade environmental problems, or the dangers of science and technology. What is not so obvious, however, is that within these targets, a certain enemy not uncommon to dystopian fiction emerges: the hideousness of *(human) nature* and the inability to evolve past vital shortcomings. The following categorisation of VGDs will therefore illuminate the genre's points of attack and describe how these centre on the above-mentioned general theme. By doing so, I will illustrate how the VGD entraps humankind in a circular loop of interconnected events based on basic needs from which the player tries to free himself in a process of emancipation.

"The potentiality for constructive human effort exists, but it contends against the deeper inheritance of our animal nature."[61] In *On the Origin of Species* (1859, 1999), Charles Darwin describes the evolution of species in a process of natural selection in which the survival of a race is dependent on how well it adjusts to its surroundings.[62] This *survival of the fittest* was often regarded by proponents as "a triumphant scientific vindication of the law of progress, natural and social"[63] and was held responsible for humankind's potential to evolve past vital shortcomings and to reach Utopia. Besides such optimistic takes on Darwin's theory of evolution, there were more pessimistic, if not outright destructive, interpretations of it.[64] One of the most famous can be found in the works of T. H. Huxley[65] which, for Kumar, essentially boils down to the insight

> that the evolutionary process was blind, arbitrary and frequently hideous. There was no discernible purpose in evolution, and nothing to justify a belief in progress. The 'survival of the fittest' not only was the same thing as the 'survival of the best' – as humans understood that term – but was frequently its antithesis. If aggression and selfishness suited the conditions of the time, they would be selected by nature as against tolerance and altruism.[66]

---

61 Kumar, *Modern*, 177.

62 Charles Darwin, *On The Origin of Species* (New York: Bantam Dell, 1999).

63 Kumar, *Modern*, 175.

64 Ibid., 175ff.

65 Thomas H. Huxley, "Evolution and Ethics (the Romanes Lecture, 1893)," in *Evolution and Ethics 1893-1943*, ed. T. H. Huxley and J. S. Huxley (London: Pilot Press, 1947).

66 Kumar, *Modern*, 176.

Seen in the context of Social Darwinism, Huxley's negative take on the evolution of species can be transferred to human society (as one can observe today), suggesting "an unending process of struggle and strife, out of which the 'fittest' and 'best' would emerge in successive waves."[67] In this line of sight, man is unable to escape his animal nature and resembles a hideous beast: "he was selfish, acquisitive, aggressive, even murderous and predatory, like the wolf."[68]

Now, it is humankind's inability to free themselves from basic urges and needs—such as aggression, greed, selfishness, mistrust and fear of the Other, and the longing for dominance/power—that can be held responsible (at least partially) for a variety of interrelated issues that have plagued the empirical world for centuries. This major theme and its resultant issues are targeted by the VGD. Most often, thereby, the genre not only warns of potential futures dominated by such problems but, in addition, sends the player on a quest for agency and invites him to overcome these issues in a playful trial action.

3.3.1 The Ongoing Threat of Oppressive Regimes: Anti-Totalitarianism

To begin the discussion, I wish to direct attention to the offshoot of *oppressive regimes*, two of which I have addressed earlier—COD: AW as an unsuccessful example and HALF-LIFE 2 as a successful one. The issue still is highly topical in the contemporary world, and with instances of theocratic regimes in the Middle East, authoritarian ones in Turkey and Russia, and totalitarian regimes in China and North Korea, the VGD assumes an important role in addressing these problems. Moreover, supposed democracies in the U.S. and Europe (for example, Italy), continue to debilitate the individual's freedom, and the recent political and cultural swing to the right in these regions further aggravates the situation. It is as Moylan puts it: "We live in a world shaped by capitalism in its global stage, generally subject to authoritarian power (be it soft or hard), be it wrapped in an aura of democracy or served straight in varying degrees of overt control."[69]

67 Ibid., 383.

68 Ibid.; cf. 383.

69 Tom Moylan, "Step into the Story," in *Dystopia(n) Matters: On the Page, on Screen, on Stage*, ed. Fátima Vieira (Newcastle-upon-Tyne: Cambridge Scholar Publishing, 2013), 42.

Dystopian fiction has warned of the dangers of oppressive regimes (soft or hard) ever since its inception and foregrounds the "opposition between social control and individual desire."[70] Novels such as H. G. Wells' *The War of the Worlds* (1898), Jack London's *Iron Heel* (1908), Robert Harris' *Fatherland* (1992), and films like Kurt Wimmer's EQUILIBRIUM (2002), Karyn Kusama's AEON FLUX (2005), and Gary Ross' THE HUNGER GAMES (2012) focus on the consequences of ideological extremes. Most effective is Orwell's *Nineteen Eighty-Four* (1949) as the ultimate vision of totalitarianism and a potential future that, as O'Brien puts it, pictures "a boot stamping on a human face—for ever."[71] The VGD figuratively follows in these footsteps, with varying results. Still, there are a variety of exceptions that go beyond the pleasures of combat and focus on ethical problems the player has to solve.

One example is Lucas Pope's PAPERS, PLEASE (3909, 2013), which involves the player in the mechanisms of a totalitarian regime and sets him in a precarious situation between duty and death. The player assumes the job of an immigration inspector and is responsible for safeguarding the country of Arstotzka from terrorists, spies, or smugglers, but often encounters people begging for help. This task confronts him with ethical dilemmas, since granting entrance to immigrants without proper documents puts the PC's family at risk.

Although PAPERS, PLEASE evokes the nostalgia[72] for a classical dystopia—in its 16-bit mood, which recalls Orwell's *Nineteen Eighty-Four*—it can be included in the realms of the critical dystopia of variant II, for a utopian horizon may pave

70 Booker and Thomas, *Handbook*, 65.

71 Orwell, *Nineteen*, 280.

72 In this respect, Mateusz Liwinski argues that PP combines the lectures of dystopia with a nostalgic look backwards at the 16-bit area of gaming (the 1980s), which he describes as "the utopian impulse of returning 'home' to the childhood gaming experience." (Mateusz Liwinski, "Nostalgia for Dystopia: Critical Nostalgia in Lucas Pope's *Papers, Please*," in *(Im)perfection Subverted, Reloaded and Networked: Utopian Discourse across Media*, Peter Lang Edition, ed. Barbara Klonowska, Zofia Kolbuszewska, and Grzegorz Maziarczyk [Frankfurt a.M.: Internationaler Verlag der Wissenschaften, 2015], 227). This nostalgia "is a craving for a place or state which is currently idealized yet unattainable" and "also entails the recognition that the remembered state was only reachable in the past." (Ibid.; cf. 224-227). Such an insight is interesting, because it links the game to the nostalgia of the classical dystopia, which, for the most part, situated the utopian enclave in the prosperity of ancient times. PP expresses this desire of a "space of familiarity and security." (Ibid., 227). It thus harbours a particular wish: not only for the 16-bit era but for a future Utopia.

the way to a new order. The official narrative involves the player in the bureaucratic routine of an unnamed immigration officer whose job revolves around the task of inspecting documents of immigrants or interrogating them. For every processed individual, the player receives 5 credits, the sum of which he can use to provide for his family: for rent, health care, food, and so on. If he makes a mistake, however, penalties are enforced, which deprive the player of the much needed money.

At this point, the regime's penetration into the individual's freedom makes itself apparent. In Arstotzka, every niche is controlled by the government through various Ministries such as the Ministry of Information, Arstotzka's secret police, or the Ministry of Justice, which pursues supposed terrorists. Consequently, the player recognises that the job is more demanding than initially believed. In the course of the 31 days, he will encounter immigrants whose fates confront him with ethical dilemmas. There is, for instance, the story of Elisa Katsenja and Sergui Volda—a guard stationed at the player's border station. Eventually, Sergui will tell the player that he fell in love with a woman in Kolechia and that for them to be together, she needs access to Arstotzka, although she has no proper documents. However, showing compassion puts the PC and his family's life at risk—but this is also where the counter-narrative begins. Starting on an individual level of resistance, the counter-narrative soon becomes collective once the player chooses to work with a resistance group called The Order of the EZIC Star (but he may ignore them as well).

The resulting constellations give rise to *twenty different endings*, most of which result in the PC's punishment—he is either condemned to hard labour for disobeying the rules or killed for helping EZIC. On day 31, however (if the player manages to reach it), EZIC attempts to overthrow the government, and the player may contribute to this cause when he accomplishes at least four out of five missions for EZIC and refrains from shooting intruders. Having actualised this ending, he is offered a home and contributes to building a new Arstotzka. Although seemingly a 'good' ending, it is ambiguous—for EZIC's revolution was brutal, and it remains uncertain whether they succeed in building a better Arstotzka. In addition, there are endings where the player flees to Obristan (with or without his family) and those in which he fights EZIC and contributes to the consolidation of the status quo.[73] In the latter case, he sides with the official narrative and is rewarded with an endless code that enables him to play the game in an infinite loop.

73 Liwinski, "Nostalgia," 232.

Figure 9: Inside the office of the Immigration Officer in PAPERS, PLEASE.

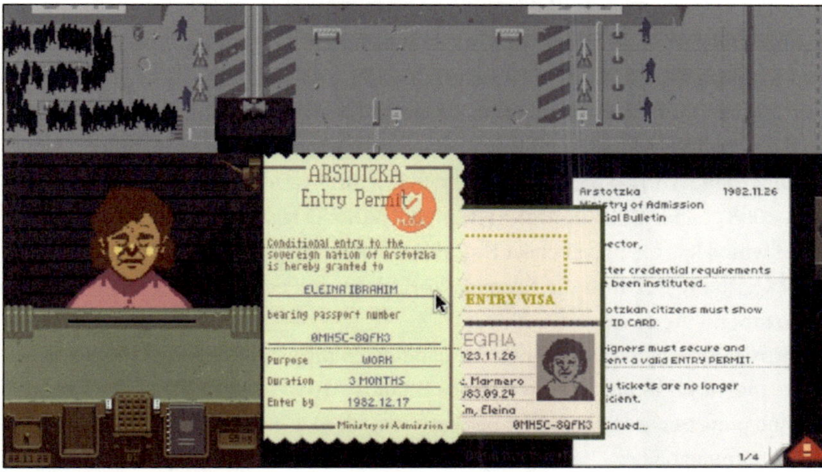

PAPERS, PLEASE (3909, 2013).

PAPERS, PLEASE is a magnificent example of the VGD as critical dystopia variant II, because it not only warns of the dangers of totalitarian regimes and the mind-numbing tasks of bureaucracy but also offers a possibility space in which utopian horizons lay hidden. To actualise these, the player has to work hard, for it is not easy to fulfil the inspector's tasks. Even the tiniest mistakes are punished and, in combination with the repetitive tasks of a bureaucratic labourer, the player might ignore the immigrants' dilemmas[74]—thus processing human beings as if they were files and showing no sympathy for personal catastrophes (which links bureaucracy to the mechanisms of oppressive regimes).

Besides PP, there is a variety of other games that have more or less successfully targeted totalitarianism, for example RÉPUBLIQUE (Camouflaj, 2012), whose regime entraps so-called Pre-cals in the mysterious facility of Metamorphosis where they are educated. In addition, there is the METRO series,[75] which in an unsettling way targets the hideousness of human nature and the inability to evolve past cravings for violence and power, illusive father-figures, and resultant totalitarian regimes. Although involving the player in brutal ludic action—and, at the same time, because of it—the games succeed in conveying the atrocities of human conflict to the player and centre on humankind's innate fear and suspicion of the

74 Ibid., 230, 232.

75 METRO 2033 (4A Games, 2010) and METRO: LAST LIGHT (4A Games, 2013).

Other. Their effectiveness, thereby, is due to the employment of several perspectives that give the player additional points of view on events (for example, ethical deliberations of characters). This sheds a different light on the ludic encounters and foregrounds the player's hideous nature in combat. He is thus reminded that through such a behaviour, the nuclear genocide that led to the post-apocalyptic world is about to repeat itself. But he may also embark on a different, ethical route. Similar to the METRO series are TURNING POINT: FALL OF LIBERTY (Spark Unlimited, 2008)[76] and WOLFENSTEIN: THE NEW ORDER (MachineGames, 2014), the latter of which lets the player enact B. J. Balzkowicz's nightmarish delusion in which Nazi Germany has won the Second World War and established a totalitarian regime around the globe.

It is noticeable that many of the candidates discussed here are FPSs that involve the player in the act of war. Kaos Studios' HOMEFRONT (2011), for instance, lets the player enact the conflict between two ideologies: the liberal United States and communist Korea. Set in the year 2027, HOMEFRONT revolves around the expansionist policy of a reunited Korea that declares war on the U.S.—which is plagued by unrest due to economic problems and a deadly avian influenza. The Korean invasion therefore proceeds without interference, and only a small resistance group continues to fight the invaders. Part of this group is the player, who in the final battle for San Francisco sacrifices his PC's life to direct an airstrike, resulting in America's partial victory over their oppressors.

HOMEFRONT's unofficial sequel, HOMEFRONT: THE REVOLUTION (Dambuster Studios, 2016) proceeds similarly, as do other games such as the RESISTANCE series.[77] In these, an alien race called the Chimera have invaded Earth by planting a virus in central Russia, which transforms human beings into fellow aliens. Similar to H. G. Wells' Martians in *The War of the Worlds* (1898), the invaders metaphorically represent the threat of communism and its oppressive ideology. With the help of their superior technology, the planet has fallen into their hands, and it is up to the player (and the resistance) to win a war against a superior enemy.

One may also enlist the GEARS OF WAR series here (Epic Games, 2006, 2008, 2011, 2013; The Coalition 2016), the KILLZONE trilogy (Guerrilla Games, 2004, 2009, 2011), and its sequel KILLZONE: SHADOW FALL (Guerrilla Games, 2013)—though it is debatable whether they are all effective examples of the VGD. At this point, it is useful to remember that for a VGD to be effective, involving the player

76 Inderst discerns a subgenre of VGDs that targets oppressive regimes: the anti-national socialist dystopia. (Inderst, "Endsieg," 182).

77 RESISTANCE: FALL OF MAN (Insomniac Games, 2006); RESISTANCE 2 (Insomniac Games, 2008); and RESISTANCE 3 (Insomniac Games, 2011).

in intense, pleasurable combat is not enough. Although such games are able to trigger fictional anger in the player towards the aggressor at hand, a careful juxtaposition of ideologies is lacking—which risks blatant criticism. Rather, what is needed is a diversity of perspectives that offer various points of orientation for the player and make him question the combat encounters and ideologies—as is done successfully in METRO 2033, BIOSHOCK INFINITE, and FALLOUT 4.

There are, however, other examples which do not fall in the ludic genre of the FPS such as the action-adventure BEYOND GOOD AND EVIL (Ubisoft Montpellier, 2003). Tending towards the genre of the fantasy but retaining a SF core of analogic extrapolation, BEYOND GOOD AND EVIL proceeds in the manner of a critical dystopia variant I. The year is 2435, and an alien force called the DomZ has invaded the mining plant of Hillys. To make things worse, the oppressive regime of the Alpha Section is established on the planet. Reassuring the people of Hillys to defend them against the DomZ, it comes out that the Alpha Section is colluding with the alien invaders. Meanwhile, a resistance movement called the IRIS Network attempts to retake the planet, and the player takes control of a single dissident. Jade is a photographer and becomes a member of IRIS. Together they confront the DomZ's High Priest and end the game on a hopeful, but ambiguous note—for the future of Hillys remains unknown.

In sum, BEYOND GOOD AND EVIL is effective in warning of the threat of totalitarian regimes because it offers the player various perspectives on the gameworld. Some of these are conveyed through likeable characters, such as Pey'j or Jade, who make the player care about this society and give him the ethical incentive to do something about it. Such a perspectival diversity does not focus solely on combat action (which in many effective VGDs is still present) but foregrounds ethical dilemmas and imaginings/actions about how to solve them—using force or other creative methods accessible to the player.

Table 1: Video game dystopias that primarily target oppressive regimes (authoritarian, totalitarian, theocratic).

HALF-LIFE 2 (Valve, 2004)
CALL OF DUTY: ADVANCED WARFARE (Sledgehammer Games, 2014)
PAPERS, PLEASE (3909, 2013)
RÉPUBLIQUE (Camouflaj, 2012)
METRO 2033 (4A Games, 2010); METRO: LAST LIGHT (4A Games, 2013)
TURNING POINT: FALL OF LIBERTY (Spark Unlimited, 2008)
WOLFENSTEIN: THE NEW ORDER (MachineGames, 2014)
HOMEFRONT (Kaos Studios 2011);

HOMEFRONT: THE REVOLUTION (Dambuster Studios, 2016)
RESISTANCE series: RESISTANCE: FALL OF MAN (Insomniac Games, 2006); RESISTANCE 2 (Insomniac Games, 2008); RESISTANCE 3 (Insomniac Games, 2011)
GEARS OF WAR series: (Epic Games, 2006, 2008, 2011, 2013; The Coalition 2016) KILLZONE trilogy (Guerrilla Games, 2004, 2009, 2011); KILLZONE: SHADOW FALL (Guerrilla Games, 2013)
BEYOND GOOD AND EVIL (Ubisoft Montpellier, 2003)
INFAMOUS: SECOND SON (Sucker Punch Productions, 2014)
ODDWORLD: ABE'S ODYSSEY (Oddworld Inhabitants, 1997)
THE SABOTEUR (Pandemic Studios, 2009)
DISHONORED and DISHONORED 2 (Arkane Studios, 2012, 2016)
CRACKDOWN (Realtime Worlds, 2007)
BIOSHOCK INFINITE (Irrational Games, 2013)

3.3.2 Capitalism and Its Momentous Consequences

VGDs often target more than one issue, and it can be difficult to separate different points of attack. This is also the case with the multifarious phenomenon of *capitalism*, which stands in close relation to the threat of oppressive regimes, for it continues to deprive the individual of self-determination and establishes a world system that is based on the unequal distribution of wealth. Alongside classics such as Frederik Pohl and C. M. Kornbluth's *The Space Merchants* (1953), Ray Bradbury's *Fahrenheit 451* (1953), and the film BRAZIL by Terry Gilliam (1983), the greatest representative to target the issue is Aldous Huxley in *Brave New World* (1932). In contrast to Orwell, who firmly believed the world to be heading towards a totalitarian future, Huxley discerned in capitalism the major culprit of social unrest in his times. In *Brave New World*, there is "no atmosphere of fear and paranoia at all,"[78] as was the case in Orwell's London. To attain their goals of "COMMUNITY, IDENTITY, STABILITY,"[79] the Controllers refuse to employ brute force but rely on clever mind control. As the Director of Hatcheries and Conditioning states, it is mental conditioning that "is the secret of happiness and virtue – [the] liking what you've *got* to do."[80] With its mind-numbing forces, then,

78 Kumar, *Modern*, 260.

79 Aldous Huxley, *Brave New World* (London: Chatto & Windus Vintage, 2004), 1.

80 Ibid., 12; cf. Kumar, *Modern*, 258ff.

capitalism has created a form of oppression that, unlike classic dictatorships, works in disguise. In *Brave New World Revisited*, Huxley formulates this as follows and predicts a future that has now become a reality: "By means of ever more effective methods of mind-manipulation, the democracies will change their nature … [to] a new kind of non-violent totalitarianism," a sort of "democracy and freedom in a strictly Pickwickian sense."[81]

It is a now familiar story that the world system of capitalism can be held responsible for many interrelated issues, and the VGD targets it on a grand scale. Like in *Brave New World, The Space Merchants,* or *Fahrenheit 451*, one of the sub-targets is *consumerism*, which is a strategy of capitalism "to stupefy the populace by saturating their minds with useless information"[82] and coaxing them into buying goods they do not need. Given these deliberations, it is quite ironic that a medium that generated approximately 100 billion U.S. dollars in revenue in 2016 (and is predicted to top this in 2017)[83] functions as a viable resistance to capitalism, in the face of such unbounded consumerism. It is a central ethical problem that needs to be addressed here—that in buying these games, the player contributes to the capitalist production machinery that regulates the game industry. Nonetheless (and the same is true for Hollywood blockbusters and commercially successful literature), it is only through reaching a broad audience and making them aware of the central problems of their times through playful interactions that criticism may evoke counteraction in the empirical world.

Consequently, the VGD addresses various issues that arise out of the capitalist world system through both big budget AAA games and smaller indie productions. It thereby proceeds in a subtle manner and addresses these issues in an estranged way through gameplay. Such a mode of involvement has the player confronted with a gameworld that infuses its perspectives (and the potential ones created by the player) with traits of social norms, conventions, and ideologies known from the empirical world. It magnifies these, however, and distorts their interrelations by redistributing them among the game perspectives (see chapter V). The player thus steadily realises that the gameworld is not so far from the capitalist world system of his empirical surroundings.

---

81 Aldous Huxley, *Brave New World Revisited* (London: Chatto & Windus Vintage, 2004), 145-146.

82 Booker and Thomas, *Handbook*, 69.

83 "The Global Games Market Reaches $99.6 Billion in 2016, Mobile Generating 37%," *Newzoo*, accessed September 25, 2017, https://newzoo.com/insights/articles/global games-market-reaches-99-6-billion-2016-mobile-generating-37/

As such, games like the BORDERLANDS series,[84] THE STANLEY PARABLE (Galactic Cafe, 2013), FALLOUT 4 (Bethesda Game Studios, 2015), BIOSHOCK (2K Boston 2007), and BIOSHOCK INFINITE (Irrational Games, 2013) foreground a specific type of player that may succumb to the enticing pleasures of capitalist processes. This ludically oriented player aims to win the game and finds pleasure in collecting points or other tools—weapons, ammunition, upgrades, treasures—he does not necessarily need (see chapters IV, V, VI). The above-mentioned games are effective because they offer perspectives, such as dialogues or character fates, which scrutinise the player's behaviour and make him aware of the mind-numbing, destructive processes he participates as he neglects the needs of others.

A second point of attack are *multinational corporations*. These are the puppeteers behind the curtain of the capitalist production machinery and can be held responsible for the ever-increasing gap between the rich and the poor. In the VGD, corporations are often the *antagonists* against which the player competes. Powerful, illusive men (such as Andrew Ryan, Handsome Jack, Pirandello Kruger, and Hugh Darrow) have displaced the rulers of oppressive regimes—and, consequently, the VGD follows in the steps of dystopian fictions since the 1990s that have laid their focus on the corporate machinery that dehumanises its victims.[85] In the BORDERLANDS series, for example, corporate oppression can be found with Hyperion and ATLAS. These corporations are the prime weapons manufacturers on Pandora, a desert planet where human greed has become the norm. Various Vault Hunters embark on explorations to excavate the riches of Vaults, and the player participates in this survival of the fittest and in the estranged mechanisms of consumer capitalism which drives a wedge between social groups.

Besides the BORDERLANDS corporations, there are a variety of other examples like the infamous Tyrell Corporation in BLADE RUNNER (Westwood Studios, 1997), the Camerata in TRANSISTOR (Supergiant Games, 2014), ATLAS in COD: AW, Fontaine Futuristics and Ryan Industries in BIOSHOCK, and the corporations to be found in the DEUS EX series[86] such as Versalife, Sarif Industries, Tai Young Medical, and the Belltower Associates.

This list could go on (see the table at the end of the section), but it can already be discerned that the ramifications of a capitalist world system are comprehensive.

---

84 BORDERLANDS 1, 2, and THE PRE-SEQUEL (Gearbox Software, 2009, 2012, 2014); and TALES FROM THE BORDERLANDS (Telltale Games, 2014-2015).

85 Booker and Thomas, *Handbook*, 72.

86 DEUS EX (Ion Storm Austin, 2000), DEUS EX: INVISIBLE WAR (Ion Storm Austin, 2004), DEUS EX: HUMAN REVOLUTION (Eidos Montreal, 2011), and DEUS EX: MANKIND DIVIDED (Eidos Montreal, 2016).

So far, I have described the impact of a consumer culture orchestrated by multinational corporations that reaches "out to every human being on the planet by means of images, sounds, and smells and by intensively reaching into every psyche."[87] "The protocol of this hegemonic order, of course, has its official narrative sequence: construct new consumers, capture new markets, increase profit, merge, increase power in the state apparatus and the society at large, take over more, make more profit."[88] One of the consequences of such a system is the aforementioned suppression of the individual, and I now wish to describe the related issue of *bureaucratic consumer capitalism* in selected games.

One example in this respect is EVERY DAY THE SAME DREAM (2011), a classical dystopia by the indie developer Molleindustria. Here, the player takes on the role of unnamed labourer and plays out his nightmarish delusions of bureaucratic consumer capitalism—a dream hiding at its core the wish for Utopia. The game begins with the ringing of an alarm clock as the player engages in his daily routine. He switches it off, puts on a black suit and a white tie, takes his briefcase, and passes his wife—who is preparing a breakfast he never gets to eat. In the meantime, the flashing television underlines the dreamlike character of the events and the mind-numbing forces of mass advertising. Not all of these actions are mandatory, but for now the player is playing the official narrative. It continues with him rushing to work, being stuck in traffic, and entering an office building where he sits down in his work cubicle. The place evokes imaginings of an army of faceless workers, for the game switches the size of the visible gamespace as the player moves forward: from initially displaying five cubicles to widen the shot to seven and ten.

EDTSD thus involves the player in the menial tasks of a bureaucrat and resembles dystopian films such as Terry Gilliam's BRAZIL (1985) or the game THE STANLEY PARABLE (see chapter II). When the working day ends, the player awakens the next morning in his apartment and resumes the servile routine for as long as he follows the described steps. This looping plot structure can, however, be interrupted, and a counter-narrative emerges within this world's absurdity.

Right after leaving the apartment, the player meets an old lady in the elevator with a message that makes him ponder: "5 more steps and you will be a new man."[89] The counter-narrative thus begins when the player refuses to put on his suit and goes to work in his underwear. He will be fired once he arrives and the day restarts. Yet it is one of the five steps that has the player transgress the official

87 Moylan, *Scraps*, 170.

88 Ibid.

89 EVERY DAY THE SAME DREAM (Molleindustria, 2009).

narrative and invite him to explore the gameworld in creative ways. There is, for example, a homeless person who leads the player to a cemetery, or a cow he may pet if he decides to stop his vehicle and continue by foot. Finally, the last step involves walking through the office space towards a green exit sign. The player reaches a rooftop where he jumps to his death. Having completed the five steps—with all their ambiguity and the imaginings they evoke—the gameworld is now empty and the player's path leads back to rooftop where he witnesses an NPC jumping to his death (who is similar in appearance to the PC).

Figure 10: The counter-narrative in EVERY DAY THE SAME DREAM *revolves around the renunciation of the capitalist system through creative interactions.*

EVERY DAY THE SAME DREAM (Molleindustria, 2009).

EDTSD confronts the player with a pessimistic vision of life where the individual's worth lies in contributing to the income of big companies and where the only exit from the system is death. With the enacted events resembling the images of a dream, however, the game can be classified as a classical dystopia—and *not* an anti-utopia. It reminds the player through bright colours marking certain objects that it is not too late to escape the system and to seek refuge beyond its confines. Some of these objects are utopian enclaves such as the red stop sign, the cow's pink belly, the orange leaf plummeting to the ground, or the green exit sign. They encourage the player "to step outside the game," "to stop playing mechanically" and to search for interesting ways of interaction.[90] Indeed, most of the five steps distract the PC (and player) from his mind-numbing work and illustrate a journey to nature and to paying attention to the world's particulars. Consequently, even

90 Sicart, *Beyond*, 73; cf. 73.

though there is no hope for its nameless protagonist, EDTSD may trigger in the resistant player a focused anger against the capitalist world system. Similar to THE STANLEY PARABLE, the game thus functions as "a reflection on the repetitive tasks of modern labour and the wish to avoid routine."[91]

I have only sampled the complexity of the problem, but one can discern capitalism's ability to stomp on human will and convert the populace into obedient servants. Besides limiting human agency in the mode of oppressive regimes, the world system of capitalism can be held responsible for several *ecological issues* that plague the planet—which is the last point I wish to stress. According to Lucy Sargisson, the causes of climate change in ecological dystopian fiction can often be traced back to "hierarchical and exploitative societies characterised by such traits as mass selfishness, egoism, greed, consumerism, civic irresponsibility and a collective stupidity."[92] In addition, there is the problem of overpopulation, a theme that is addressed in the 1973 film SOYLENT GREEN by Richard Fleischer and the novel *The Space Merchants*.[93]

Because such problems often get out of hand, ecological issues are frequently discussed in post-apocalyptic futures where nature has made a forceful return and where the capitalist system was mainly responsible for the downfall of the old order. This can be discerned in an overwhelming number of games such as RAGE (id Software, 2011), S.T.A.L.K.E.R: SHADOW OF CHERNOBYL (GSC Game World, 2007), TOM CLANCY'S THE DIVISION (Massive Entertainment, 2016), and MAD MAX (Avalanche Studios, 2015)—not all of which qualify as typical VGDs. As Booker warns: although post-apocalyptic fiction is related to the genre of dystopia, it most often focuses on the struggle for survival in the aftermath of a widespread catastrophe such as nuclear war, an infectious plague, or environmental disasters. Out of this catastrophe, however, different societies and forms of dystopian regimes might emerge. Yet, unfortunately, post-apocalyptic tales often fail to foreground the particulars of the imaginary societies they describe or depict.[94]

Such is the case with Telltale Games' THE WALKING DEAD series,[95] which does *not* qualify as a VGD. In contrast to this game series, there are VGDs in which societies have emerged in the aftermath of the apocalypse, a famous example of which is the FALLOUT series. Here, the nuclear holocaust has paved the way

---

91 Ibid.

92 Sargisson, "Matter," 40.

93 Booker and Thomas, *Handbook*, 70; Moylan, *Scraps*, 170.

94 Booker, "On Dystopia," 5.

95 THE WALKING DEAD (Telltale Games, 2012); THE WALKING DEAD: SEASON 2 (Telltale Games, 2013-2014); and THE WALKING DEAD: SEASON 3 (Telltale Games 2017).

for different factions that are described in detail and are built around a specific ideology. In addition, the series deals with the environmental consequences of the nuclear catastrophe[96]—and FALLOUT 4 specifically goes into detail on this matter, with nature being shown as a sublime phenomenon and in an unsettling but utterly beautiful manner.

A game that places even more value on nature's return after the cataclysm is Ninja Theory's ENSLAVED: ODYSSEY TO THE WEST (2010), a critical dystopia of variant I set in a world devastated by war and where nature has reclaimed the planet. The tranquil scenery is misleading, however, as so-called mechs—remnants of a bygone age—roam the lands and enslave the human population. The game's events revolve around the journey of Monkey and his temptress Trip, who forces him to help her return to her village by placing a slave headband on him. The headband ensures that Monkey follows Trip's orders, and if she dies he shares the same fate. The counter-narrative in ENSLAVED is thus at first a forced one—at least for Monkey (a telling name for humankind's servitude to the machines) and the player who takes control of him. But soon the temptress will work her magic. During the game, Monkey and Trip establish a near romantic connection—and with Monkey now following Trip out of 'free' will, they confront dystopia's high priest, Pyramid. It is revealed that it was he who was controlling the mechs and slaves to protect humankind from the new environment. Pyramid did so by feeding them visions of the ancient world and entices Monkey into believing him. The utopian enclave is thus neither actualised by the player nor the PC, but Trip undertakes this action. Through a shutdown of the system, she kills Pyramid and frees the slaves by refusing to be caught up in memories of an old, paralysed order, and chooses nature as mankind's new destiny.

As so often in post-apocalyptic narratives, hope is thus located after the cataclysm "in the future renewal of Earth and its plant and animal species,"[97] which is established after the capitalist order. Consequently, the games described above—and also Naughty Dog's THE LAST OF US (2013, 2014) (see chapter VII)—fall into the category of eco-fiction and scratch the surface of the ecotopia: a genre that was established in the early 1980s and focuses on a society based on ecological principles and technology.[98]

All together, these conclusions underscore that the VGD does not capitulate to capitalism (contrary to Packer's observations mentioned in chapter II). It works against the anti-utopian claim "that this is the way things *are*, and that change is

96 Domsch, "Dystopian," 406–407.

97 Ferreira, "Biodystopias," 50.

98 Kumar, *Modern*, 405-407.

neither possible nor desirable"[99] and attempts to break away from it or, at least, attenuate its extent. The VGD thus fulfils Jameson's demand of Utopia's political function

> to concentrate on the break itself: a mediation of the impossible, on the unrealizable in its own right. This is very far from a liberal capitulation to the necessity of capitalism, however; it is quite the opposite, a rattling of the bars and an intense spiritual concentration and preparation for another stage which has not yet arrived.[100]

It is effective through a diversity of perspectives that constitute a particular game and that question the player's involvement in the estranged, capitalist processes he participates in. These may include dialogue lines or character fates that illuminate the true nature of the gameworld (ENSLAVED: ODYSSEY TO THE WEST) or juxtapositions that contrast natural worlds to old orders dominated by capitalism (ENSLAVED, THE LAST OF US). In addition, there are satirical elements that question the integrity of the gameworld events (BORDERLANDS series) or nuanced processes that align the act of play with the servile routines of a capitalist world system (EVERY DAY THE SAME DREAM). As such, the VGD targets capitalism on a grand scale, and even AAA productions may fulfil the requirements of social criticism and transformation.

Table 2: Video game dystopias that primarily target the capitalist system (corporations, bureaucracy, and environmental pollution).

THE STANLEY PARABLE (Galactic Cafe, 2013)
THE LAST OF US (Naughty Dog, 2013, 2014)
BIOSHOCK series: BIOSHOCK (2K Boston 2007); BIOSHOCK 2 (2K Marin, 2010); BIOSHOCK INFINITE (Irrational Games, 2013); BURIAL AT SEA: EPISODE 1 and 2 (Irrational Games, 2013, 2014)
BORDERLANDS series: BORDERLANDS 1; 2; and THE PRE-SEQUEL (Gearbox Software, 2009, 2012, 2014); TALES FROM THE BORDERLANDS (Telltale Games, 2014-2015)
BLADE RUNNER (Westwood Studios, 1997)
TRANSISTOR (Supergiant Games, 2014)

99 Ferns, *Narrating Utopia*, 232.

100 Jameson, *Archaeologies*, 232-233.

Call of Duty: Advanced Warfare (Sledgehammer Games, 2014)
Deus Ex series: Deus Ex (Ion Storm Austin, 2000); Deus Ex: Invisible War (Ion Storm Austin, 2004); Deus Ex: Human Revolution (Eidos Montreal, 2011); Deus Ex: Mankind Divided (Eidos Montreal, 2016)
Perfect Dark (Rare, 2000)
Oddworld: Abe's Odyssey (Oddworld Inhabitants, 1997)
Shadowrun (Beam Software, 1993)
Syndicate series: Syndicate (Bullfrog Productions, 1993); Syndicate American Revolt (Bullfrog Productions, 1993); Syndicate Wars (Bullfrog Productions, 1996); Syndicate (Starbreeze Studios, 2012)
Red Faction series: Red Faction (Volition, 2001); Red Faction II (Volition, 2002); Red Faction: Guerrilla (Volition, 2009); Red Faction: Armageddon (Volition, 2011)
Fallout series: Fallout (Interplay Entertainment, 1997); Fallout 2 (Black Isle Studios, 1998); Fallout 3 (Bethesda Game Studios, 2008); Fallout: New Vegas (Obsidian Entertainment, 2010); Fallout 4 (Bethesda Game Studios, 2015)
Every Day the Same Dream (Molleindustria 2011)
Rage (id Software, 2011)
S.T.A.L.K.E.R: Shadow of Chernobyl (GSC Game World, 2007)
Tom Clancy's The Division (Massive Entertainment, 2016)
Mad Max (Avalanche Studios, 2015)
Enslaved: Odyssey to the West (Ninja Theory, 2010)

3.3.3 On the Dangers of Science and Technology

With Enslaved I touched on an integral theme of utopian and dystopian fiction that leads back to Bacon's 1627 scientific utopia *New Atlantis*.[101] With Bacon as the "inspirer of … [Utopia's] fundamentally expansive and dynamic character," *science* became an integral part of attaining a better world.[102] However, the initial

101 Kumar, *Modern*, 29ff.

102 Ibid., 30; cf. 30.

Renaissance and Enlightenment dream of Utopia was damaged (if only temporarily) by the atrocities of the two world wars, in which technological marvels were used for mass murder.[103] As a consequence, even the SF of the 20th century—which for the most part retained an optimistic core—could not turn a blind eye to the horrible events.[104] What Kingsley Amis called the *New Maps of Hell* quickly became a dominant trend in the SF of the 1950s, depicting humankind's fears of a future marked by scientific and technological excess.[105] Yet one should recall that it is never science as such that can be blamed but rather its potential "applications."[106]

The range of VGDs that address the (mis)use of science/technology is too vast to address here, and only an excerpt will be taken into consideration. The history begins with the idea of the "artificial man"[107] and goes back to the roots of the SF genre that, arguably, began with Mary Shelley's *Frankenstein; or, The Modern Prometheus* (1818).[108] Out of Frankenstein's "hubris"[109] and the thirst for "dangerous or forbidden knowledge,"[110] the monster was created, which drew from and inspired a range of artificial beings. Kumar names the golem, the android, the robot, and the cyborg, but one could easily extend the list to modern versions of posthuman composites or Artificial Intelligences (AIs). What is common to these artificial men and women is that they inspire terror and recall the dangers of scientific abuse—of refusing to take on responsibility for the created creatures.[111]

It would be a mistake, however, to incorporate any game that includes zombies (or other posthuman composite) into the realms of the VGD, because many of them fail to address the societal implications of the monsters. Nonetheless, there are plenty of games that qualify, such as the CRYSIS series (Crytek, 2007, 2008, 2011, 2013), Westwood Studios' BLADE RUNNER (1997), or the "Biodystopia"[112] BIOSHOCK (2K Boston 2007).[113] Also, the DEUS EX series, which belongs to the

103 Ibid., 42ff., 380ff.; Vieira, "Concept," 18.

104 Kumar, *Modern*, 385.

105 Amis Kingsley, *New Maps of Hell* (London: Penguin Books, 2002); Kumar, *Modern*, 403.

106 Ibid., 254.

107 Ibid., 114.

108 Jameson, *Archaeologies*, 1.

109 Kumar, *Modern*, 113.

110 Ibid., 112.

111 Ibid., 113ff.

112 Ferreira, "Biodystopias, 49.

113 Schulzke, "Bioethics."

cyberpunk/posthuman genre, deals with the benefits/dangers of augmentation technology and the related threat of AIs (see chapter V). Moreover, there are games such as NEUROMANCER (Interplay Productions, 1988), PORTAL 1 and 2 (Valve, 2007, 2011), I HAVE NO MOUTH, AND I MUST SCREAM (The Dreamers Guild, 1995), or REMEMBER ME (Dontnod Entertainment, 2013). The latter game addresses the related phenomenon of mingling with the human brain and is similar in theme to Christopher Nolan's film INCEPTION (2010) or Philip K. Dick's "We Remember for you Wholesale" (1966)—better known in its film version TOTAL RECALL (Paul Verhoeven, 1990). REMEMBER ME is a critical dystopia of variant I that takes place in the futuristic Neo Paris in 2084 (one hundred years after Orwell's London), where the corporation Memorize offers the possibility of erasing unwanted memories. The protagonist Nilin is a so-called Errorist who discovers her ability to extract/remix memories. Consequently, the player is able to mingle with the minds of characters, and this created perspective and the reaction of the subjects will be a shocking revelation to him in that it illustrates the technology's potential.

From Frankenstein's monster, then, the leap "to the atom bomb is not a big one,"[114] to the "fearful monster that had burst loose from the control of its masters."[115] In the 1950s, the fear of nuclear annihilation was omnipresent, and dystopian fiction thematised it widely. Similarly, there are a fair number of games that address the ramifications of a world ravaged by radiation and the related issue of automation—which in the 1950s "was beginning to make human labor obsolete, while at the same time turning people into machine-like automatons, living thoroughly scripted, regulated lives."[116] These include, for example, S.T.A.L.K.E.R: SHADOW OF CHERNOBYL (GSC Game World, 2007), METRO 2033, THE LONGEST JOURNEY series,[117] and the aforementioned FALLOUT games (the two latter also address the fear of automation). Similar in sarcastic tone to the film DR. STRANGELOVE: HOW I STOPPED WORRYING AND LOVED THE BOMB (Stanley Kubrick, 1964) is FALLOUT 3 (Bethesda Game Studios, 2008), where the city of Megaton was established around an undetonated nuclear warhead—a telling name, indeed. The bomb creates a beautiful metaphor of this world's dilemma, in which a faction called the Children of the Atom established a religious cult around

114 Kumar, *Modern*, 113.

115 Ibid., 389.

116 Booker and Thomas, *Handbook*, 69.

117 THE LONGEST JOURNEY (Funcom, 1999); DREAMFALL: THE LONGEST JOURNEY (Funcom, 2006); and DREAMFALL CHAPTERS: THE LONGEST JOURNEY (Red Threat Games, 2014-2016).

it. The player becomes involved in this problem and can decide between disarming the bomb (thus reassuring the safety of Megaton's citizens) or blowing it up (and receive an apartment in the nearby Tenpenny Tower).

Finally, I wish to end this part on the dangers of science/technology with the issue of *surveillance*, a thematic often considered synonymous with Orwell's *Nineteen Eighty-Four*. The intrusion into the private sphere is still highly topical in the modern world, not only since Edward Snowden's revelations about the machinations of the NSA, which have temporarily startled the populace. Again, there are a vast number of VGDs that tackle this issue, most often combined with some form of oppressive regime or corporation: from Camouflaj's RÉPUBLIQUE (2012) and REMEMBER ME, to THE STANLEY PARABLE and both MIRROR'S EDGE games—MIRROR'S EDGE (DICE, 2008) and its resurgence in 2016 with MIRROR'S EDGE: CATALYST (DICE).

In the original MIRROR'S EDGE, the player becomes involved in a society in which the dream for a comfortable life has led to a police state and where the power over the hegemonic order is negotiated between the state and private corporations. The game takes place in a vast urban area called The City, which awaits the player in majestic white, punctuated with an orange and blue colour palette. This 'clean' Utopia comes at a price, however. Creative lifestyles, bodily pleasures, and all forms of digital communication are regulated, and only a resistant group of parkour artists do not tolerate this state of affairs. They call themselves Runners and transport packages of non-digital data across the city's rooftops. MIRROR'S EDGE is thus built around a counter-narrative in which the player (in the role of protagonist Faith) relies on "spatial agency" and "the ability to make the avatar move through the gameworld in incredible ways."[118] The game warns about a hegemonic order's ability to intrude into the private sphere and involves the player in a resistance where the City's rooftops become the new undergrounds—ensuring a "free flow of information"[119]—and where the player's artistic mobility is juxtaposed with a world that is strictly organized. This renunciation of the hegemonic order and the city's sterile white is underlined by the PC's ability to discern strategic routes through the labyrinths of rooftops. The so-called Runner Vision clearly marks potential ways in red and offers the player utopian enclaves and possibilities of action he may grasp within the confines of the system.

Similar in theme to the MIRROR'S EDGE games is WATCH_DOGS (Ubisoft Montreal, 2014), which belongs to the genre of the cyberpunk and confronts the

118 Domsch, "Dystopian," 403.

119 Ibid.

player with a chilling vision of the near future. In WATCH_DOGS, fictional Chicago has become one of the most surveilled cities in the world (just like the actual Chicago), because of a newly installed operating system. Developed by the Blume Corporation, the Central Operating System (ctOS) is used to create a smart city based on technological innovation—and thus a future Utopia. Thereby, the system does more than handle the city's infrastructure. In addition to controlling and optimising traffic flow, improving the health care system, and promoting environmental awareness, the ctOS facial recognition software reduces crime by calculating the severity of potential offences—similar to Dick's Precogs in "The Minority Report" (1956). With the promise of security, health, comfort, and limitless connection, the citizens of Chicago are coaxed into believing in this technological Utopia—and one can see them wandering the streets with their heads down, entrapped in the cyberspace of their phones.

To operate on such a level of efficiency, the ctOS is gathering vast amounts of data—about income, age, and history of disease, to citizens' precise locations. In WATCH_DOGS, information means power, and this is where the supposed Utopia fails. In a story of corporate intrigue, corrupt individuals, and data heists, various parties use the ctOS to their benefit. One of these is Dermot Quinn, who is involved in cybercrime and human trafficking and is close to the Blume Corporation and Chicago's mayor Donavan Rushmore.

In the midst of these intrigues, the player takes control of the hacker Aiden Pierce, whose niece was accidently killed by Quinn's men. His story is thus one of personal revenge and his development remains shallow. Although Pierce meets and is aided by Clara Lille (the supposed temptress and member of the hacker group DedSec) and Raymond Kenney (a former ctOS software engineer), he remains caught up in his thirst for vengeance. In the meantime, there is no sign from hacker group DedSec, which could have led to a collective counter-narrative and perspective on the events.

Consequently, through the absence of critical characters and plot developments, reflective distance is left to the player in his ludic ability to tinker with the ctOS by using his in-game smartphone. Chicago, as such, is at the player's mercy—for the hacking possibilities are vast and include control of Chicago's electric grid, accessing people's personal information and bank accounts, or intruding into their homes through cameras in their Xbox Kinects. As a consequence, the player becomes a voyeur who witnesses the citizens of Chicago playing games or engaging in sexual fantasies. Becoming involved in such a trial action is pleasurable without a doubt, but it is also thought-provoking by having the player ponder these possibilities in real life, and what it means to become a victim.

Figure 11: Quinn's human trafficking auction in WATCH_DOGS.

WATCH_DOGS (Ubisoft Montreal, 2014).

All in all, WATCH_DOGS is a critical dystopia of variant I that fails to unfold its subversive potential because the counter-narrative and critical character/plot perspectives remain in the background. The protagonist Pierce shows little change and is caught up in personal revenge; it is only at the game's conclusion that his attitude changes. Nonetheless, having experienced Chicago in the act of play and closing the blanks between the perspectives of his co-creation (the hacking and tinkering with the ctOS), the player may ponder the ramifications of such technology and what it means if corrupt individuals get their hands on it.

Table 3: Video game dystopias that primarily target the dangers of science/technology (environmental hazards, posthumanism, cyberspace, surveillance, and post-apocalypse).

ENSLAVED: ODYSSEY TO THE WEST (Ninja Theory, 2010)
CRYSIS SERIES (Crytek, 2007, 2008, 2011, 2013)
BLADE RUNNER (Westwood Studios, 1997)
BIOSHOCK series: BIOSHOCK (2K Boston 2007); BIOSHOCK 2 (2K Marin, 2010); BIOSHOCK INFINITE (Irrational Games, 2013); BURIAL AT SEA: EPISODE 1 and 2 (Irrational Games, 2013, 2014)
NEUROMANCER: A CYBERPUNK ROLE-PLAYING ADVENTURE (Interplay Productions, 1988)
PORTAL 1 and 2 (Valve, 2007, 2011)

Deus Ex series: Deus Ex (Ion Storm Austin, 2000); Deus Ex: Invisible War (Ion Storm Austin, 2004); Deus Ex: Human Revolution (Eidos Montreal, 2011); Deus Ex: Mankind Divided (Eidos Montreal, 2016)
I Have no Mouth, and I Must Scream (The Dreamers Guild, 1995)
Remember Me (Dontnod Entertainment, 2013)
S.T.A.L.K.E.R: Shadow of Chernobyl (GSC Game World, 2007)
Metro 2033 (4A Games, 2010); Metro: Last Light (4A Games, 2013)
Fallout series: Fallout (Interplay Entertainment, 1997); Fallout 2 (Black Isle Studios, 1998); Fallout 3 (Bethesda Game Studios, 2008); Fallout: New Vegas (Obsidian Entertainment, 2010); Fallout 4 (Bethesda Game Studios, 2015)
The Longest Journey series: The Longest Journey (Funcom, 1999); Dreamfall: The Longest Journey (Funcom, 2006); Dreamfall Chapters: The Longest Journey (Red Threat Games, 2014-2016)
République (2012 Camouflaj)
The Stanley Parable (Galactic Cafe, 2013)
Mirror's Edge (DICE, 2008); Mirror's Edge: Catalyst (DICE, 2016)
Watch_Dogs (Ubisoft Montreal, 2014); Watch_Dogs 2 (Ubisoft Montreal, 2016)
NieR: Automata (Platinum Games, 2017)
Horizon Zero Dawn (Guerrilla Games, 2016)
Enter the Matrix (Shiny Entertainment, 2003); The Matrix: Path of Neo (Shiny Entertainment, 2005)

3.3.4 (Human) Nature as the Main Culprit

It seems that the main culprit for humankind's inability to attain a better society can be found in the hideous elements of human nature. Ever since More's *Utopia*, whose "citizens have to be controlled—educated to do right,"[120] the dark parts of

120 Ferns, *Narrating Utopia*, 42.

the human self have thwarted the dream of Utopia—the desire for power, "to consume, eliminate, and destroy the Other."[121] Whether it is oppressive regimes in the form of authoritarian states, or under the veil of democracy, inglorious leaders or corporate oppressors who use scientific advancements to their favour, they all are united by the VGD, which calls for emancipation from them.

The larger theme of the hideousness of human nature can be discerned in most of the VGDs enlisted in this chapter—and, specifically, with games such as the FALLOUT or METRO series, which deal with the difficulties of attaining Utopia because of who we are. In the line of these ideas, I wish to lay focus on two final examples. The first is the classical dystopia I HAVE NO MOUTH, AND I MUST SCREAM, which is similar to that in Dante's Inferno in the epic poem *La Divina Commedia* (ca. 1304-1321). Here, the supercomputer AM exterminated humankind except for five individuals and created a world in which agency is absent, as a punishment for their sins. The second example is Platinum Games' MAD WORLD (2009), which is similar in thematic to the films THE PURGE (James DeMonaco, 2013) and GAMER (Mark Neveldine and Brian Taylor, 2009). In MAD WORLD, the fictional Varrigan City has turned into the setting of the ultra-violent television game show 'Death Watch' in which contestants engage in a brutal survival of the fittest. Originally created to satisfy humankind's lust for violence, it is revealed that the games benefit pharmaceutical companies. MAD WORLD thus stands as prime example of dystopia's major theme and involves the player in brutal carnage, letting him enact his unconscious desires.

Table 4: Video game dystopias that primarily target human nature.

BIOSHOCK series: BIOSHOCK (2K Boston 2007); BIOSHOCK 2 (2K Marin, 2010); BIOSHOCK INFINITE (Irrational Games, 2013); BURIAL AT SEA: EPISODE 1 and 2 (Irrational Games, 2013, 2014)
METRO 2033 (4A Games, 2010); METRO: LAST LIGHT (4A Games, 2013)
FALLOUT series: FALLOUT (Interplay Entertainment, 1997); FALLOUT 2 (Black Isle Studios, 1998); FALLOUT 3 (Bethesda Game Studios, 2008); FALLOUT: NEW VEGAS (Obsidian Entertainment, 2010); FALLOUT 4 (Bethesda Game Studios, 2015)
I HAVE NO MOUTH, AND I MUST SCREAM (The Dreamers Guild, 1995)
MAD WORLD (Platinum Games, 2009)

121 Levitas and Sargisson, "Utopia in Dark Times," 25.

3.4 THE DIVERSITY OF THE PERSPECTIVAL NETWORK AS PRECONDITION FOR DYSTOPIA'S EFFECTIVENESS

Playing dystopia can be depressing at times, especially if you are playing these games for years consecutively. Nonetheless, the VGD achieves artistic merit in engaging the player with a cathartic experience and in the struggle for Utopia. This chapter has revealed interlocking themes and targets apparent in the VGD and illustrated the genre's diversity. Its effectiveness in issuing warnings has thereby been linked to a game's aesthetic complexity and to the amount of differing perspectives that involve the player in a reflective process.

These perspectives are outlined by dystopia's narrative framework of official narrative and counter-narrative, which seems to withstand transfer from different mediums. Perspectives such as the gameworld events, processes, and vicious characters introduce a game's hegemonic order as a palpable threat and serve as a justification for the (combat) action against it. This is done either in an explicit fashion—by setting the player in direct resistance to waves of enemies (RESISTANCE, HOMEFRONT, GEARS OF WAR)—or more implicitly—by involving the player in the underlying processes of the society at hand (BORDERLANDS, BIOSHOCK). As a result, the way the player acts against the official narrative can be as bland as shooting his way through waves of enemies or converge in creative counter-narratives and ludic expressions. This is the case in the player's artistic use of space in MIRROR'S EDGE and the way this gives rise to utopian enclaves, the hacking in WATCH_DOGS and tinkering with the ctOS, the solving of ethical dilemmas in PAPERS, PLEASE by exposing the life of the PC's family to risk when helping others, or the creative ways to interact with a bland society in EVERY DAY THE SAME DREAM, which gives rise to nature enclaves. In addition, likable temptresses and round characters like in BEYOND GOOD AND EVIL, ENSLAVED: ODYSSEY TO THE WEST, and HALF-LIFE 2 give the player private reasons to counteract dystopia, for he potentially establishes close relations to them.

All in all, it can be argued that the perspectival network of an effective dystopian game fuels the player's imaginings and ergodic actions and results in creative, ethical solutions to dystopia. Depending on which of the perspectives are foregrounded (either by the game or the player's actions), different constellations enable the player to close blanks in a variety of ways. These insights, then, shall stand as a premise for Part II, where the player's aesthetic response to dystopia shall be treated in detail and further prove the hypotheses and results of Part I.

Part II: Playful Trial Actions in Estranged Gameworlds

Preface to Part II

Part I revealed significant aspects concerning a potential genre of the video game dystopia and has paved the way for further observations on it.[1] In establishing a solid classification of dystopia's subgenres, I have laid the groundwork for deliberations on and categorisations of the VGD. These include: the pseudo-dystopia or anti-utopia (which can be seen as a deceptive strategy of the status quo), the classical dystopia (that entraps the PC in a pessimistic world from which there is no escape), and the two variants of the critical dystopia (that most often disclose dystopia's past and offer the player one or several utopian enclaves to actualise or follow).

Hope, consequently, lingers at the VGD's core, pushing it at times into a considerable utopian direction (if one can look past the misfit of the anti-utopia). This is also reflected in dystopia's traditional plot structure of official narrative and counter-narrative, which finds ample application in the video game medium. The player, thereby, assumes the role of the dissident, and later, the catalyst. While at times it is clear that the world she plays in is a negative one, a process of gradual realisation can still be discerned—at least for the PC. It leads the player through knowing that something is wrong, to seeing it clearly, then to discerning connections between the fictional and empirical world, which are shockingly minimal.

1 Part II illustrates a *substantial expansion* of my conference paper *The Emancipated Player*, which I presented at the 2016 1st International Joint Conference of DiGRA and FGD. The focus has turned to *dystopia*, but the deliberations on representational art and the dialectic between implied and emancipated player remain conceptually similar. (Gerald Farca, "The Emancipated Player," *Proceedings of the First International Joint Conference of DiGRA and FGD* 13, no. 1 (2016), http://www.digra.org/digital-library/publications/the-emancipated-player/).

With such an incentive, the VGD is primarily about taking action and finding potential solutions to dystopia, which in the critical dystopia of variant II is allotted to the player.

Critique, it seems, is not enough for the VGD, although it certainly addresses a plethora of themes: the threat of oppressive regimes, capitalism, the misuse of science, and the dark side of human nature. Yet what the VGD suggests more pressingly is that something can be done—at least in virtuality—and that in learning from these lessons, the player's empirical world may not yet be lost.

To unravel the complex mechanisms behind such an aesthetic response and its effect on the player will be the main focus of Part II, which will lay the emphasis on the phenomenology of playing dystopia. Such deliberations can never be isolated from discussions of narrative and fiction, and I will therefore move into the realms of the video game narrative as a form of representation, which I use synonymously with fiction. Consequently, uncovering the deep structure of the implied player becomes of prime importance: the affordance and appeal structure of a game that offers all those predispositions for the game to exercise its effect—an aesthetic effect experienced in and through the act of play. As a structural construct, the implied player outlines the empirical player's participation in the gameworld by involving her in a creative dialectic with a comprehensive perspectival system (including the player's sensorial perspective, the gameworld, the plot framework, and player actions). As such, the implied player not only guides the player's involvement but also triggers structured acts within her, thereby altering her habitual dispositions by allowing for new experiences to pervade her mind.

Playing dystopia—and, in general, any aesthetically complex VGN—can thus be precarious, for it creates a space for play in which the fictive has outlined a gameworld that is permeated by distortions and doubling, while the imaginary takes hold of the player and finds an outlet in her psyche, imaginings, and ergodic actions. This is especially so for the genre of dystopia, since its gameworld appears strange to the player, though it is surprisingly familiar in some parts. As such, it is only through the player's act of ideation, of closing the blanks between the perspectives she encounters and co-creates, that the new reality of the gameworld can be understood. These perspectives have been taken out of their familiar context—from norms, conventions, and processes of the empirical world—but are magnified through extrapolation and rearranged in an unfamiliar manner in the dystopian gameworld, which is often taken to an extreme. They consequently confront the player with various games of estrangement that are similar to, though also differ in crucial respects from, those the appreciator has come to know in non-ergodic dystopian SF.

In other words, dystopia's implied player invites the emancipated player into a creative dialectic and dance with a distorting and nightmarish dream world. It offers a space for play and trial action that involves her in a constant feedback oscillation between two worlds (the fictional and the empirical world): an experience which may trigger a cathartic and emancipatory response in the player.

4 Towards the Implied Player

"What are games? What do they consist of? What are they in relation to similar phenomena?"[1] Video game studies have often focused on describing what it means to *be* a game and what aspects make games *different* from non-ergodic forms of representational art. This perspective is beneficial, as it sheds light on the ontological dimension of the medium, which tries to clarify basic questions such as those of Espen Aarseth, above, and discusses whether video games should be seen as forms (or show elements) of non-digital games, hypertexts, simulations, narrative fictions, films, performances, dramas, virtual artefacts, and so on.[2] Partly, this was necessary (and still is) to describe the peculiarities of the video game medium and designated a first inevitable step towards establishing a new field of study dealing specifically with digital games. In turn, a stubborn and uncritical demarcation of video games as one of these forms neglects the multifarious nature of the phenomenon. The initial narratology vs. ludology[3] debate stands as

1 Aarseth, "Ontology," 484.

2 For example: See Wardrip-Fruin and Harrigan for different approaches to video games such as drama and performance, ludology, simulation, hypertext, space and narrative, and so on.

3 The ludology vs. narratology argument was a fictitious straw man debate that hit video game studies at its beginnings in the early 2000s. While the former approach relied on the analysis of games as simulations or dynamic processes that are organised according to a set of rules (without acknowledging that games would be able to outline interesting narratives), the latter approach tried to describe the phenomenon as a form of narrative. In focusing on specific aspects, both perspectives neglected the complete picture of the phenomenon. A first attempt to reconcile the seemingly opposite poles was attempted by (Gonzalo Frasca, "Ludologists love Stories too: Notes from a Debate that Never took Place," *Proceedings of the 2003 DiGRA International Conference: Level up* 2, [2003]: 91-99, http://www.digra.org/wp-content/uploads/digital-library/05163.01125.pdf).

a reminder of the one-dimensionality that views of video games may show.[4] Hence, Aarseth's recent attempt to clarify

> the 'ludologist' position [which] was not, as has been claimed, 'to see the focus shift onto the mechanics of gameplay' (Jenkins 2001) [Jenkins, *Architecture*, 118] but to emphasize the crucial importance of *combining* the mechanical and the semiotic aspects and to caution against and criticize the uncritical and unqualified application of terms such as 'narrative' and 'story' to games.[5]

In fact, there has often, if not invariably, been a distinction made between "two elementary senses or 'layers' in the concept of game: (1) *core*, or game as gameplay, and (2) *shell*, or game as representation and sign system."[6] While it is true that such a separation between core and shell may be useful for analysis and for laying the focus on specific parts of a game, their mutual complementation must not be neglected. Especially if the researcher narrows her focus to one of these interlocking aspects, it can have detrimental effects on the understanding of the video game phenomenon as a whole and on the player's experience of a game. What further complicates the matter is Mäyrä's use of the term *shell*, which runs the risk of having a depreciative connotation—this is just one example in which scholars (implicitly or explicitly) try to lay the focus on the inner mechanics of a game, its rules and algorithms, while neglecting semiotic aspects of the gameworld. Faced with such a variety of ingredients in a game, it would be fatal to reduce a player's experience to either one of these aspects. It is therefore only through the combined analysis of *mechanics* (the game as system) and *semiotics* (the sign system of the gameworld) that the act of play can be properly described. As Aarseth formulates: "Mechanics and semiotics together make up the *game object*, which is a type of information object, and when a player engages this object the third component, gameplay, is realized."[7] Naturally, for the purpose of analysis, mechanics and semiotics can be regarded separately, but it is indisputable that they only signify as a whole.

---

4 For example: Jesper Juul's argumentation in *Half-Real*, that games cannot be seen as stories (Jesper Juul, *Half-Real: Video Games between Real rules and Fictional Worlds* [Cambridge, Mass.: MIT Press, 2005]), or Ian Bogost's in *Persuasive Games*, where he claims that meaning-making in games primarily depends on the medium's procedural abilities. (Bogost, *Persuasive*).

5 Aarseth, "Narrative," 130; emphasis added.

6 Mäyrä, *Introduction*, 17; emphasis added.

7 Aarseth, "Ontology," 488.

The discussion of ontology in video game studies has been beneficial, however difficult it has been and still is, and offers a gateway to regarding games as the diverse forms of entertainment and aesthetic artefacts they are. To reach such a conclusion, one has to first see the ills of one-dimensional approaches to video games and, second, to start seeing them as a multifaceted phenomenon. This view has carved its way into the minds of many scholars—including Aarseth, Tosca, Calleja, Domsch, and Ryan—who address a certain but most common subtype of the video game. To them, many video games are *hybrid* forms between non-digital games and narratives.[8] Aarseth, for instance, underlines "the composite makeup" of "the story-game amalgams"[9] and Domsch states that "[s]ome things are played as games, and some things are read as narrative, and sometimes, a thing is both. The latter is what is called *storyplaying*."[10] Similarly, for Susana Tosca the game-narrative hybrids only signify if their two intertwining parts—the "story" and the "action / procedures"—are regarded as inseparable parts similar to the concept of yin and yang.[11] While these three scholars foreground the aspects of game and narrative in their descriptions, Calleja throws virtual environments into the equation: "games nowadays are in fact extended virtual environments which contain a game or multiple games within them."[12] Finally, Ryan invites the approach of

> ludo-narrativism that studies how the fictional world, realm of make-believe, relates to the playfield, space of agency. By connecting strategic dimensions of gameplay to the imaginative experience of a fictional world, this approach should do justice to the dual nature of video games.[13]

In these views, then, video games are hybrid forms of many things, the most prominent variant of which offers the player a virtual environment and fictional space for *narrative play*. This variant I want to refer to as the *video game narrative*, a

---

8 The current trend of many video games to incorporate literary aesthetics into their formula points to a departure from traditional games. This was indirectly (that is, without the scholar sharing this opinion) alluded to in Juul's *classic game model*, where storytelling was given the status of a non-game. (Juul, *Half-Real*, 44; Dominic Arsenault, "Narratology," in *The Routledge Companion to Video Game Studies*, ed. Mark J. P. Wolf and Bernard Perron [New York: Routledge, 2014], 481).

9 Aarseth, "Narrative," 130.

10 Domsch, *Storyplaying*, 3.

11 Tosca, "*Amnesia*," 120; cf. 119-120.

12 Calleja, *In-Game*, 14-15.

13 Ryan, *Avatars*, 203.

large and diverse genre to which I will also ascribe the *video game dystopia*. Consequently, when talking about video games (or games), and if not specifically described otherwise, I will invariably be referring to these hybrid forms.

Though video game studies have made huge leaps forward in respect to game ontology, much remains to be said. Instead of embarking on this path, however, I choose to adjust the question of what it means *to be a game* to *what it is like to play a game*, and what actually *happens to the player during the act of play*. I thus enter the realms of *phenomenology*, which many scholars have entered before, with different concerns that can roughly be divided into three different branches (it should be noted that most of them are interested in a 'textual' analysis of response rather than an empirical investigation of player reactions).[14] The first branch describes the related phenomena of *immersion*, *presence*, and *involvement* that are fundamental to a player's experience of a game. Janet Murray, for example, described the concept of immersion in 1997 as "the physical experience of being submerged in water. ... the sensation of being surrounded by a completely other reality ... that takes over all our attention,"[15] while Alison McMahan illustrates the feeling of presence in 2003 as one "of being there."[16] Alongside other takes on this issue, such as Marie-Laure Ryan's in 2001,[17] there are recent attempts to elaborate on what it means to become involved in a game. These include, for example, Carl Therrien's discussions of immersion in 2014[18] or Gordon Calleja's 2011 theory on player involvement—to which I have alluded before and that I will primarily follow.[19]

In addition to these, a "*corporeal turn*"[20] occurred in game phenomenology, with studies focusing on the *avatar's (or PC's) relation to both the gameworld and the empirical player*. Important works in this respect were conducted by Rune

---

14 Fahlenbrach and Schröter offer a similar and detailed subdivision of player response theories in game studies. (Kathrin Fahlenbrach and Felix Schröter, "Game Studies und Rezeptionsästhetik," in *Game Studies: Aktuelle Ansätze der Computerspielforschung* [Köln: Herbert von Harlem Verlag, 2015], 166-174).

15 Murray, *Hamlet*, 98.

16 Alison McMahan, "Immersion, Engagement, and Presence: A Method for Analysing 3-D Video Games," in *The Video Game Theory Reader*, ed. Mark J.P. Wolf and Bernard Perron (New York: Routledge, 2003), 68.

17 Ryan, *Narrative*.

18 Therrien, "Immersion."

19 Calleja, *In-Game*.

20 Fahlenbrach and Schröter, "Rezeptionsästhetik," 170.

Klevjer, who considers the player's involvement as a form of "fictional and vicarious embodiment."[21] As such, "[t]he relationship between the player and the avatar is a prosthetic relationship; through a process of learning and habituation, the avatar becomes the extension of the player's own body."[22] Daniel Vella builds on these conclusions, but focuses on the mutual influence of avatar and player. He describes a "playable figure" that "encapsulates both the fact that the entity is taken on and 'played out' by the player … but also the fact that it remains a figure in its own right."[23] Lastly, Brendon Keogh is interested in the "physical"[24] but "unconscious, embodied engagement between player and videogame, where the videogame is touched, seen, heard, and ultimately understood through a perceiving located (and augmented) body."[25] This claim he explains in "how the player embodies the videogame" but also "how that embodiment is always already constituted by the videogame."[26]

The phenomenological theories on embodiment can, according to Fahlenbrach and Schröter, be integrated into the vast branch of "Cognitive Game Studies"[27] such as Bernard Perron's works on emotions created in playing horror games[28]—but these are of minor importance for my deliberations here. Contrary to these theories, I focus on the player's *aesthetic experience* in the act of play and how a game's mechanisms outline the player's involvement—that is to say, the *structure* that affords play in the first place and the player's experience of meaning. This structure I refer to as the *implied player*: a dynamic framework of play that offers the empirical player a specific role (or roles) to be performed and functions as a trajectory towards catharsis. My approach in this chapter is thus both a *phenomenological* and *structuralist* one—inspired by theories of fiction/aesthetic response and narratology—which understands a game and its world as a system of perspectives that borrow elements from the empirical world but rearrange them in an unfamiliar manner as representations in order to have the player make connections between the two realities. By analysing these preconditions of play's aesthetic effect, the chapter opens up several threads that will be answered in chapter V, which

21 Klevjer, "Avatar," 9.

22 Ibid., 10.

23 Vella, "Ludic Subject,"10.

24 Keogh, "Play of Bodies," 15.

25 Ibid., 17.

26 Ibid., 19.

27 Fahlenbrach and Schröter, "Rezeptionsästhetik," 170.

28 Bernard Perron, *Silent Hill: The Terror Engine* (Ann Arbor: University of Michigan Press, 2012).

will lay the focus on the empirical player's dialectical communication with the intersubjective framework of dystopia's implied player. Through such an analysis of play's underlying mechanisms (with play seen in ergodic and imaginative ways), it will not only be possible to comprehend how the VGD exerts a certain effect on the player but also to draw conclusions about the ontology of the game-narrative hybrids alluded to above.

Given the video game medium's multifacetedness, then, it makes sense to begin with a theory of representational art. By means of this approach, I avoid the mistakes one-dimensional approaches to video games have made in neglecting to address the vital similarities between games and other forms of art. What this does *not* mean is that I consider the player's experience equal to that of reader, spectator, viewer, or appreciator—and becoming involved in a virtualised storyworld (or gameworld) and playing its contents shows an aesthetic of its own. Yet there are also crucial similarities that must not be overlooked, ones which most fundamentally revolve around the appreciator's and player's dialectic with a work of art and a potential storyworld (if that is the case). For these reasons, Kendall Walton's theory on representational art or fiction, and the appreciator's communication with the latter in terms of make-believe becomes strikingly beneficial as a starting point.

These general deliberations will then be supplemented by related theories from game studies (for example, Klevjer, Tavinor, Aarseth, Ryan, Domsch) and refined by integrating Jacques Rancière's thoughts on an emancipated spectator's involvement in plays, Lubomír Doležel's recent conception of fiction as a semantic communication between the work of art and the appreciator, and Wolfgang Iser's groundbreaking theory of aesthetic response. In the latter, the reader engages in an imaginative dialectic with the structural concept of the implied reader, which is described as a system of perspectives which creates various indeterminacies (gaps/blanks) for the reader to fill in or close. These outline the empirical reader's imaginative involvement in the literary work by evoking acts of ideation in her—that is to say, the creation and continuous revision of images by resorting to her real-world knowledge to close the blanks.

From these theoretical manoeuvres, I wish to arrive at a *unified theory of aesthetic response* and a *phenomenology of art experience* in VGNs in general and the VGD in particular. For only if one regards fiction in terms of *a semantic communication* and as *a functional concept* can the relation between the gameworld and the player's empirical surroundings be properly understood, and the effect this experience has on the player. As Doležel puts it:

> Fictional texts are composed by actual authors (storytellers, writers) using the resources of an actual human language and destined for actual readers. They are called fictional on *functional grounds*, as media for making, preserving, and communicating fictional worlds. They are stores of fictionality within the world of actuality, where the products of the writers' imaginations are permanently available to receptive readers.[29]

4.1 VIDEO GAMES AS FORMS OF REPRESENTATIONAL ART AND FICTION

There has been much debate about whether video games count as forms of representational art[30]—and even Marie-Laure Ryan, who normally tends towards the narrative pole of the spectrum, poses this question. She thereby distinguishes between "[*r*]*epresentational*" and "[s]*imulative*"[31] forms of narrative and argues that "[w]hile the simulation machine [to which she allots the video game] cannot by itself be called a narrative, each of its individual runs produces images of a world that undergoes change as the result of events."[32] Consequently, video games "may not be stories, but they can be machines for generating stories."[33] Indeed, although video games differ in their form of discourse, they still, and most magnificently, create worlds—and thus, virtual representations with which the player can interact. As such, I see no reason why they should not fall under the umbrella category of representational art and agree with Grant Tavinor who comes to the conclusion that games that create virtual worlds such as "*The Elder Scrolls V: Skyrim* (Bethesda Game Studies, 2011) can be considered a ... representational artifact."[34] However, then TETRIS (Nintendo, 1989) would also count as representation, as the game simulates some form of visual, if only rudimentary, world on the screen.[35]

29 Doležel, *Heterocosmica*, 28.

30 Tavinor, *Art of*, 44ff.

31 Ryan, *Avatars*, 13.

32 Ibid., 188.

33 Ibid., 188-189.

34 Tavinor, "Art," 59.

35 Murray views the game as an allegory on the "perfect enactment of the overtasked lives of Americans in the 1990s—of the constant bombardment of tasks that demand our attention." (Murray, *Hamlet*, 144). Also, Sebastian Möring offers an insightful discussions of Murray's interpretation. (Sebastian Möring, "Games and Metaphor – A Critical Analysis of the Metaphor Discourse in Game Studies," [PhD diss., IT University of Copenhagen, 2013], 229-230, 233-234, https://en.itu.dk/~/media/en/re

The question of representation may not be so easy to answer, and one has to pursue such a claim further—an enterprise that leads to the question of *fictionality*. It comes as no surprise that the consensus on this topic in video game studies is no more unanimous than that of narrative. There are primarily two schools of thought that radically differ in opinion: "ludo-fictionalism and ludo-realism."[36]

> The ludo-fictionalist school, inspired by Kendell Walton's radical and influential *Mimesis as Make-Believe* (1990), on the one hand, sees games, game objects, and game worlds as fictional, as 'props in a game of make-believe.' For them, the rules may be real, but the discursive elements and actions are fictional (Juul 2005; Bateman, 2011). The ludo-realist school, on the other hand, sees game objects and game events as real, or at least closer to reality.[37]

Building on this distinction, Aarseth allocates himself to the school of ludo-realism and explains his ontologically interested position in terms of the status of in-game objects and the player's usage of them. Thereby, it must be noted that his focus rests on multiplayer games and online worlds in which "players typically treat important in-game objects much the same ways as they treat their extra-ludic property."[38] It is especially the real-world value of these "ludic objects" (often in-game objects can be traded and sold for large amounts of real-world money) that makes them different from "fictional objects" and which leaves them "on an entirely different ontological level, in the same category as digital word processing documents … and money in our digital bank accounts."[39]

In addition to this, Aarseth goes on to downplay the importance of make-believe for competitive multiplayer games such as COUNTER-STRIKE (Valve, 1999). Here, in-game objects fail to function as props in the Waltonian sense and could rather be compared to "sports equipment" in that what is of importance are the object's "capabilities"[40] and not the imaginings they evoke.

---

search/phd-programme/phd-defences/2013/20130929-full-thesis-sebastian-moering-itupdf.pdf?la=en).

36 Aarseth, "Ontology," 491.

37 Ibid.

38 Ibid.

39 Ibid.

40 Ibid.

> But there is no need for make-believing when players shoot each other in *Counter-Strike* … ; they are manipulating nonphysical, informational guns that shoot non-physical, informational projectiles and when their avatars are hit, they do not have to make-believe that they are eliminated. This happens, factually, in the game machine, entirely independent of the players' imagination, just like a pinball when it drops below the reach of flippers.[41]

Admittedly, there is a certain degree of truth to this claim, especially when discussing specific types of games such as TETRIS, pinball, or competitive multi-player games like the above-mentioned COUNTER-STRIKE. One could agree that the primary function of such games is to engage the player in sports-like competitions where the fictional quality of the gameworld becomes of secondary importance (at least, its visual aspects). Nonetheless, such thinking robs video games—and specifically the genre of the VGN—of one of their essential qualities: in focusing on the internal mechanisms of a game, one runs the risk of neglecting its *imaginative-evocative* qualities. This would be a serious mistake, and one that fundamentally underestimates or misunderstands the power of representation, or fiction (two terms that I am using interchangeably for the sake of reducing complexity).

Indeed, it seems that Aarseth's trouble lies with the concept of fiction in general, or a certain understanding of it.[42] In another essay, he suggests a three-part segmentation of games into the ontological layers of "the real, the virtual and the fictional" and argues that gameworlds are "composites"[43] made up of these layers. His argument thereby runs as follows: to begin with, game events are *real* and not fictional, in that "we really win or lose" when playing a game, and labyrinths (out of which the gameworld is composed) are real at least "in a conceptual sense."[44] Labyrinths are also "virtual in the physical sense"[45] just as other game objects—such as characters or items—with which the player can interact. Continuing his discussion of purely *virtual* existents, Aarseth mentions doors that can be opened and closed and dragons as instances of animated characters. These are "neither physically nor conceptually real, but merely simulated" and "can typically be acted upon in ways that fictional content is *not* acted upon."[46] The *fictional* elements of a game he then reduces to objects such as doors with which the player

41 Ibid., 491-492.

42 Klevjer, "Avatar," 83-84.

43 Aarseth, "Fiction."

44 Ibid.

45 Ibid.

46 Ibid.

cannot interact and that "are merely textures on the walls that look like doors, but whose function is purely decorative."[47] Combined with the notion of fiction as "illusory, fabricated,"[48] Aarseth's argument shows a dismissive, "radical"[49] stance towards the concept of fiction he employs—and he underestimates the importance of make-believe to many games (going as far as negating these games' function as props).

Before hastily dismissing Aarseth's conclusions, however, one must concede that there is nonetheless huge benefit to be found here. First of all, Aarseth directs attention to the need to critically investigate "the status of fiction in games,"[50] and suggests that not all video games fall under the category of fiction. Secondly, he offers a vital starting point for further deliberations on the issue. This he does by coming to the conclusion that "games are not fictions, but a different type of world, between fiction and our world: the virtual."[51] Rune Klevjer is aware of these issues and offers an insight that will become vitally important to my deliberations. Instead of excluding the concept of fiction from virtuality, as Aarseth does, Klevjer claims that virtuality—following Ryan and Walton—is in fact what connects non-ergodic fictions to games. This he explains by referring to Aarseth's example of doors the player can open and those she cannot open. Instead of distinguishing between virtual and fictional artefacts here (and thus between different ontological levels), Klevjer argues that both function as *props* in the Waltonian sense. They differ, however, in that one of these props is a "model," "a dynamically reflexive prop"[52] and "a functional representation (the expression of a process in terms of a material or logical structure) ... that prescribes as fictional the *changes* that we effect in it," whereas the other prop is "perceptually reflexive."[53] This conception implements a "form of agency"[54] as a discursive act into the realms of fiction and

47 Ibid.

48 Ibid.

49 Klevjer, "Avatar," 84.

50 Aarseth, "Fiction."

51 Ibid.

52 Klevjer, "Avatar," 78. For Klevjer, especially the avatar is "a dynamically reflexive prop in relation to its environment." (Ibid., 87). This he explains in that "[e]mbodied make-believe is premised upon an *environment* within which the participant can become an acting body. Mediated by the avatar, the environment becomes our tangible world, our habitat." (Ibid., 88).

53 Ibid., 77; cf. 77-78, 82-85, 115.

54 Ibid., 78.

serves well to describe the player's oscillation in the act of play between ergodic and imaginative actions.[55]

As such, fiction is seen in a "*generative*"[56] way that constructs realities and becomes useful for considering games as virtual environments that show fictional qualities—for many outline diverse narratives and offer possibilities for player interaction that go beyond those of traditional fictions. Video games, so it seems, are thus similar to what Michel Foucault has called *Other spaces* or *heterotopian spaces*.[57] They are *playgrounds* which are, "formally speaking, demarcated from everyday space" but that nonetheless stand in intimate connection with it: "a heterotopian other feeding from and mirroring the everyday."[58] What this means is that the concept of fiction does not lose its validity—at least for the genre of the VGN. As microcosms within the real world and other spaces of estranged, artificial nature, video games occupy a limbo state between the fictional and the empirical world, drawing from both, but, at the same time, showing the player their results in a refracted mirror (see chapter V). To explain this specific quality, Aarseth's notion of fiction remains problematic, yet this offers the possibility for clarification and a finer granularity of the topic.

I therefore want to explore fiction in games in a different sense, moving away from its ontological dimension and towards the concept's *function as a phenomenological experience*. This quality manifests itself in a specific, aggravated relation to the empirical world, and in order to establish the connection between the gameworld and the real world, the player has to exert efforts of ideation—an enterprise that will eventually give rise to something new and the aesthetic effect of play. Fiction, then, shall not be seen as something illusory, fabricated or fake, but in a Waltonian sense as a powerful means of involving the player in games of make-believe that fuel her imagination in diverse manners and which hold the possibility of influencing her actions in both the virtual and the real world. The concept of virtuality I employ does thereby not exclude fiction but embraces it.[59] For the gameworld the player encounters and interacts with, exhibits virtual prop-

55 Ibid., 44.

56 Ibid., 87.

57 Michel Foucault, "Of Other Spaces: Utopias and Heterotopias," in *Rethinking Architecture: A Reader in Cultural Theory*, ed. Neil Leach (London: Routledge, 1997), 350-356.

58 Walz, *Toward a Ludic Architecture*, 143; cf. 135, 136, 143.

59 For Tavinor, the "virtual and the fictional" are "somewhat overlapping categories" that are conceptually related. (Tavinor, *Art of*, 44; cf. 46).

erties (the gamespace) as well as showing inherent fictional qualities (the storyworld), but in a different sense than Aarseth has proposed. These sensualise the abstract space of the virtual and its underlying rules and evoke in the player diverse imaginings.[60] To prove my claims, Walton's concept of fiction offers a promising starting point (especially if augmented by Doležel and Iser's observations on the phenomenon, which go into more detail). Here, fiction becomes a matter of *attitude* and holds the *function* of immersing the player into the occurring events, involving her in a playful manner in a virtualised storyworld (or gameworld). As Ryan remarks, "fictionality is not a property inherent to a certain media but a specific use of the media for which the concept is valid."[61] Walton's theory, she continues, thereby offers "a basis for a transmedial theory of fiction"[62] and seems extremely valuable for a discussion of the phenomenon in games.[63]

4.2 THE DIFFERENT GAMES WE PLAY WITH FICTIONS

In his seminal work *Mimesis as Make-Believe: On the Foundations of the Representational Arts* (1990), Kendall Walton meticulously describes the appreciator's involvement in representational works of art which for him are synonymous with works of fiction in a specific way.[64] The concept of fiction is thereby used in a very broad and inclusive manner.[65] Certainly, one could argue that Walton's scope

60 Domsch similarly argues that a storyworld "is the fictional world in which the structure of the game and its rules as well as the actions of the player within it are given meaning" (Domsch, *Storyplaying*, 27-28)—and that to better understand a game's events, players semanticise "its abstract properties (rules)." (Ibid., 19).

61 Ryan, *Avatars*, 37.

62 Ibid.

63 Klevjer, "Avatar," 25, 29; Tavinor, *Art of*, 40.

64 Walton justifies his *synonymous use* of *fiction* and *representation* in ascribing "an extension both broader and narrower than it is usually understood to possess" to the latter term. (Walton, *Mimesis*, 2). Similarly, works of fiction shall not be limited "to human artifacts" and a use of the term "[f]ictional representation" could lead to the implication that this "category is a species of a larger class of 'representations,' understood to include 'nonfictional' as well as 'fictional' ones." (Ibid., 3). Consequently, and because Walton "know[s] of no better" term, he assigns *representation* a specific, fictional use in his work. (Ibid).

65 Ibid., 3, 72.

is too wide, but he nonetheless manages to describe the phenomenon in a persuasive and useful manner.

To engage with representations (or fictions), the appreciator willingly participates "in a game of make-believe in which the appreciated work is a prop."[66] This is Walton's main thesis and premise for the appreciator's engagement with fiction. As stated above, his scope is broad. For the American philosopher, make-believe assumes a "nearly universal"[67] role common to all cultures and is intimately linked to children's games of make-believe and a *specific* form of imagining.[68] He therefore exemplifies:

> In order to understand paintings, plays, films, and novels, we must look first at dolls, hobbyhorses, toy trucks, and teddy bears. The activities in which representational works of art are embedded and which give them their point are best seen as continuous with children's games of make-believe themselves, and I shall argue that representational works function as props in such games, as dolls and teddy bears serve as props in children's games.[69]

If fiction designates such an all-inclusive category, why then not include video games as well? The question is valid and on the surface seems easy to answer.[70] To be certain, however, one has to take a detour, which begins with a brief but telling example.

"Let's call that stump a bear."[71] This is Walton's famous example in which he describes children's experience of embarking on a fictional adventure. For this purpose, they make-believe that a real-world stump they encounter confronts them with a bear in their game.[72] The stump thus functions as a *prop*, and this prop "generate[s] fictional truths independently of what anyone does or does not imagine."[73] This logic, Walton continues, is based on so-called *principles of generation*

---

66 Ibid., 190.

67 Ibid., 11.

68 Ibid., 12.

69 Ibid., 11.

70 In fact, many scholars in game studies have tried to answer it, with differing results. For example: Klevjer, "Avatar;" Tavinor, "Art"; *Art of*; Aarseth, "Fiction;" "Ontology;" Ryan, *Avatars*; *Narrative*; Domsch, *Storyplaying*; or Aaron Meskin and Jon Robeson, "Fiction and Fictional Worlds in Videogames," *Philosophy of Computer Games* (2009), https://www.academia.edu/244532/Fiction_and_Fictional_Worlds_in_Videogames

71 Walton, *Mimesis*, 23.

72 Ibid., 21ff.

73 Ibid., 38.

and on "a certain convention, understanding, agreement on the game of make-believe."[74] In other words, if players *agree* that all stumps *are* bears, then in the fictional world they *imagine* all stumps *are* bears. If this is so, it follows that to engage with fiction the appreciator willingly accepts the conventions of a work world (and often storyworld) and follows certain "*rules* about what is to be imagined in what circumstances."[75]

However, such deliberations run the risk of conflating the notions of *what is imagined* by the appreciator and *what is fictional*. Walton is aware of this pitfall in that he argues that although

> [b]eing fictional and being imagined are characteristics that many propositions share ... it would be a serious mistake simply to identify the fictional with what is imagined. What is fictional need not be imagined, and perhaps what is imagined need not be fictional."[76]

In a footnote, he explains the difference: "For any imagining, we might recognize a fantasy in which what is imagined is fictional. But it need not be fictional in the 'world' the imaginer is mainly concerned with – e.g., that of a game of make-believe."[77] With this claim, Walton wants to stress that what is fictional is not determined by the imagination but by the work of art itself. Imagine a game of make-believe in which a stump is covered by branches and moss. Even though the players do not recognise it (they do not imagine a bear to hide in the forest), the rules of the game prescribe it nonetheless, if they were agreed upon beforehand—that is to say, "[f]ictionally a bear is lurking in a thicket."[78]

Fiction, as such, can be seen as *a specific mode of the imagination* that creates realities and in which people most often, if not invariably, show an open-minded attitude towards the work they are confronted with by respecting its rules and integrity. Samuel. T. Coleridge has most famously called this attitude "the willing suspension of disbelief"[79] and which Murray—giving it a more positive connotation—turns around into the "active creation of belief."[80] So it can be said that the

74 Ibid.

75 Ibid., 40; emphasis added.

76 Ibid., 37

77 Ibid.

78 Ibid.; cf. 37.

79 Samuel. T. Coleridge, "Biographia Literaria," *Project Gutenberg* (2004), ch. XIV, http://www.gutenberg.org/files/6081/6081-h/6081-h.htm

80 Murray, *Hamlet*, 111.

appreciator's open-minded attitude is crucial to creating the fictional space in the first place, but equally important is the work of art itself.

According to Walton, fiction fundamentally differs from nonfiction in that it occupies the *function* of "*prescribing imaginings*"[81] about the work appreciators are confronted with, while the latter, to put it simply, does not.[82] Walton takes as an example of nonfiction Charles Darwin's *On the Origin of Species*, about which he claims that the "book itself does not prescribe believing"[83] in the same way Swift's *Gulliver's Travels* does, which counts as representation in the specific Waltonian sense. While in the former, believing is up to the reader (she can choose to approve of Darwin's observations or not—that is, the work does not mandate believing, although Darwin certainly intended it), fictional works *are* to be believed (it is *mandatory* for the appreciator to believe), and she generally does so without questioning the contents.[84] Consider William Gibson's famous opening line of *Neuromancer* (1984): "The sky above the port was the color of a television, tuned to a dead channel."[85] Nobody engaging with the novel would doubt the truth of this utterance,[86] and a simple dismissal of it would lead to a breakdown of the reader's immersion. The unlikeliness of such disbelief is easily explained. Fiction, so the general consensus, is not about deceiving or lying to the appreciator[87]—instead, the reader takes the novel's storyworld to include a plethora of *fictional truths*. Consequently, one can say it is *Neuromancer*-fictional that the sky above the port resembles a certain colour and that Case partially lives in the matrix. These truths are specific to *Neuromancer*'s fictional storyworld and are not to be confused with any real-world truths, though they show a certain connection, or relation, to them.[88]

---

81 Walton, *Mimesis*, 91; emphasis added.

82 Ibid., 58, 70-71.

83 Ibid., 71.

84 Ibid., 70-71.

85 William Gibson, *Neuromancer* (New York: ACE Books, 1984), 1.

86 An exception to the norm might be when people know that they are dealing with a form of *unreliable narration*. Still, even then they would refrain from questioning the integrity of a storyworld, but only question the narrator's explication of it (although a distinction is certainly difficult to make).

87 Richard M. Sainsbury, *Fiction and Fictionalism: New Problems of Philosophy* (London: Routledge, 2010), 11.

88 Walton, *Mimesis*, 41, 60, 62.

Hence it follows that for a work of art to be included in the category of fiction, not only the appreciator's attitude towards it is important but even more so the work's *function*.

> Works of fiction are simply representations … *whose function is to serve as props in games of make-believe* … however minor or peripheral or instrumental this function might be … ; only what lacks this function entirely will be called nonfiction.[89]

Walton's insight thus goes hand in hand with other observations on fiction, specifically with those that determine the concept on the basis of so-called *fictive intentions*. Sainsbury, for instance, defines fiction as follows: "a fiction is either the product of fictive intentions, or, though it starts as serious narrative, it rightly comes to be treated as a work to which make-believe, not belief, is the appropriate response."[90] This view correlates well with Walton's in that Sainsbury lays emphasis on the work's function while not excluding the appreciator's attitude and response to it. He, however, narrows his claim by stating that when seeking to understand fiction, the work of art becomes more important than the appreciator's perspective, as "consumers are fallible."[91] "Whether something is fiction is determined by how it came into existence and in particular by the aims and intentions of the producer"[92]—and, in this respect, mainly the fictive intentions are of importance. With these, "the utterer intends a potential audience to make-believe something."[93]

How, then, can one distinguish between fiction and nonfiction, if the aims of the producer are unknown to the appreciator (or whether visual representation in games is intended as decoration or goes beyond that in holding specific fictional quality)? I reject Sainsbury's notion, at least partially, and instead opt for the consideration of *both function and attitude*[94] in determining whether a certain work can be called fiction or not. I will thus speak about fiction in terms of a *functionalist approach*, following Walton, Doležel, and Iser. Here, it is not so much the

89 Ibid., 72; emphasis added.

90 Sainsbury, *Fictionalism*, 21.

91 Ibid., 5.

92 Ibid., 5-6.

93 Ibid., 8.

94 Admittedly, Sainsbury is aware of this fact: "In seeking to understand what fiction is, we can look either to the producer or the consumer or to some combination." (Ibid., 5). As mentioned before, he however delimits his claims by stating that "consumers are fallible." (Ibid).

question of truth that is the decisive factor in determining whether a work of art shall be called fiction or not—for the appreciator generally accepts these worlds for what they are—but rather the *effect on her* becomes of prime importance, and how this effect is outlined or triggered by the work. Fiction can therefore be described as a *communication of a special sort between the work of art and the participant* and "is primarily a semantic phenomenon located on the axis 'representation (sign)—world.'"[95] As such, "fictional worlds are accessed through semiotic channels and by means of information processing. Because of the semiotic mediation, accessibility is a bidirectional, multifaceted, and historically changing commerce between the actual and the fictional"[96]—and for this communication to occur, it is crucial that the participant knows with what she is engaging, for this knowledge will change her perception.

A brief example will illustrate my claims. Consider the player of so-called *newsgames*[97] or *documentary* games[98] and how she relates to these games' contents. Now, imagine the player of a virtualised storyworld and do the same thing. Both players, it is clear, "will bring different attitudes and expectations"[99] to the games. While the former player assumes a direct connection between virtual and empirical world (the response to these games is believe or disbelieve), the latter player first has to make sense of what she encounters (the response to these games is make-believe). This is so because video games that project a fictional world involve the player in vivid games of estrangement and postulate an indirect, hindered connection between the virtual and the empirical reality—forcing the player to exert effort in connecting the dots. Things fall neatly into place if one regards fiction to require a specific kind of effort, which is that of *ideation*. In this sense, fiction does not work against reality (nor can it be seen as its opposite) but rather designates "a reformulation of an already formulated reality, which brings into the

95 Doležel, *Heterocosmica*, 2; cf. Iser, *Act*, 53-54.

96 Doležel, *Heterocosmica*, 20.

97 Bogost et al. claim that newsgames refer to "a broad body of work produced at the intersection of videogames and journalism." (Ian Bogost, Simon Ferrari, and Bobby Schweizer. *Newsgames: Journalism at Play* [Cambridge, Mass.: MIT Press, 2010], 6).

98 For Aarseth documentary games "refer to events and existents in our world (e.g. in our history), they do not fictionalize but document." (Aarseth, "Ontology," 491). In addition, Domsch mentions so-called "realist' games" that, similar to realist fiction, are never able to achieve complete realism, although that is the intent. (Domsch, *Storyplaying*, 16; cf. 16-17).

99 Ryan, *Avatars*, 51.

world *something that did not exist before*"[100] and, therefore, becomes "a means of telling us something about reality."[101]

Fortunately, players usually know perfectly well what they are dealing with, and video games that virtualise a storyworld are able to engage them in a fruitful communication with the worlds they create (especially the genres of SF and utopia). Before going into further detail here and discussing Iser and Suvin's take on fiction in chapter V, the lesson to be drawn is the following: not all video games can be considered fictions, so I explicitly reject Meskin and Robson's claim that "all videogames fall into the category of walt-fictions"[102] as an oversimplification. However, the vast majority of video games do qualify—including the VGN and the VGD—and these are of interest here.[103]

For such games, the attitude of make-believe is of essential importance, but this attitude "does not admit degrees."[104] The appreciator either make-believes or she does not. Make-believe therefore stands in stark contrast to "analog" theories of fiction for which "[f]iction and nonfiction are two poles of an analog continuum, and there is no definite, stable boundary between the two."[105] Instead, Ryan ascribes make-believe to "digital"[106] approaches, which serve well for a discussion of video games. "The digital model deals with hybrid phenomena by allowing texts to borrow elements from the other side of the border without being infected by these elements, because the reader makes separate judgments of fictionality on the local and global level."[107] "[T]he reader," Ryan continues, "will assume that

100 Iser, *Act* x; emphasis added.

101 Ibid., 53.

102 Meskin and Robson, "Fictional Worlds," 4.

103 Indeed, Domsch goes as far as to claim that "[t]here are almost no games in which there is not at least an element of fictionality." (Domsch, *Storyplaying*, 19). Yet the discussion of fiction in games might be more complex than he indicates. Things become complicated once the researcher includes every game genre in the equation, such as competitive multiplayer games or online worlds in which players together embark on (make-believe) adventures. Even racing simulations or city building games create problems. Especially in the latter case, it can be discussed whether their function is one of make-believe or whether these games primarily aim at the creation of belief or non-belief about certain real-world issues through exact simulations.

104 Ryan, *Avatars*, 53.

105 Ibid., 52.

106 Ibid., 53.

107 Ibid.

some statements are true in both the real and the fictional world, while others describe the fictional world only."[108] Such an understanding of fiction is especially beneficial when dealing with a hybrid phenomenon like the video game. Then, Aarseth's ontological distinction between *real*, *virtual* and *fictional* elements of a game becomes obsolete[109] in that both the player's *attitude* towards the game and, equally importantly, its *function* become of prime interest. This is especially important pertaining to the VGD, which generally holds the function of warning the player of troubling tendencies in her empirical present by having her make-believe the fictional storyworld to be true (thus creating the necessary credibility). Function thus becomes of utmost importance and assumes an aesthetic quality.

As a result, let me give a preliminary conclusion. Fiction in the Waltonian sense is best seen as a communication between the work of art and the appreciator who engages in the latter in a *playful manner* and with an open-minded (aesthetic) *attitude*. This dialectic is of a special kind, as is the nature of fiction, and the appreciator assumes a vital role in the participation process—a fact that Walton repeatedly stresses. "The basic appreciative role consists, in a word, in *participating* in a game of make-believe in which the appreciated work is a prop."[110] As a consequence of her involvement, the appreciator is willingly sucked into the game and becomes absorbed by the all-engulfing space known as fiction.[111] Fiction thus exerts an irresistible fascination for the appreciator. It not only draws her into lively games of make-believe but creates a fictional space that extends into the real world, surrounding the appreciator. This occurs, for instance, when the museum-goer becomes involved in a painting[112] or when she engages with a sculpture from different angles and distances.[113] If this is so, it follows that in video games even the player's extradiegetic "*play space*, meaning space of play, which includes

108 Ibid.

109 This is similar to a theatre play, where the materiality and ontological dimension of the stage is of no great importance to the spectators. Rather, what counts are the imaginative-evocative qualities of these stage props, which allow the spectators to become immersed in a fictional world.

110 Walton, *Mimesis*, 190.

111 Ibid., 190ff., 215-216.

112 Walton exemplifies this claim with the example of Willem Van der Velde the Younger's *The Shore at Scheveningen* (ca.1820-30). Here, "it seems to be fictional not only that there are several sailing ships offshore but also that Stephen [the museum-goer] sees them. His looking at the picture makes this fictional of himself." (Ibid., 215).

113 Ibid., 215, 338.

the player and the video game hardware,"[114] becomes surrounded by the irresistible veil of fiction.[115]

Yet Walton explicitly warns us not to conflate what is fictional in the games appreciators play with what is fictional in the work of art itself.[116] Therefore, to continue the investigation and to determine the nature of the relation between work of art and the participant, it becomes necessary to take a closer look at the specific form of communication that occurs between both parties and to explore if or how these insights hold true for video games. Moreover, although the appreciator's experience of non-ergodic artwork cannot be entirely equated with the player's experience of a video game, Walton's notion of fiction may help to shed light onto the player's *imaginative interaction* with a gameworld—with the notion of the *prop* being of specific interest here, which I will later regard in the Iserian sense as a perspective on the game and its world.

4.2.1 Work Worlds and Game Worlds

To underline the appreciator's decisive role in the communication process, Walton differentiates between "work worlds and game worlds, between the worlds of novels, pictures, and plays and the worlds of games of make-believe in which these works are props. Appreciators belong to the latter."[117] Such a statement necessarily endows Walton's theory with "ludic aesthetics,"[118] and to explain the function of props and the *playful imaginings* these trigger, he resorts to a discussion of Georges Seurat's painting *Un dimanche après-midi à l'Île de la Grande Jatte* (1884-1886)—amongst other examples.

Like other representations, *La Grande Jatte* functions as a prop in the appreciator's game of make-believe, and, by doing so, evokes a rudimentary storyworld

---

114 Nitsche, *Game Spaces*, 16.

115 This is particularly so when using the extradiegetic play space in a kinetic manner such as fictionally playing tennis with a Wii Remote or other similar games with Microsoft's Kinect.

116 Walton, *Mimesis*, 58-59.

117 Ibid., 215.

118 Doležel, *Heterocosmica*, 11; Klevjer similarly discerns Walton's theory as "a play-based theory of the nature of representation" and as a "'phenomenology' of art appreciation" that focuses on "*imaginative play* … as the central model for understanding representation in arts." (Klevjer, "Avatar," 23; emphasis added).

to interact with.[119] Unlike stumps or cloud formations, however, which serve as "ad hoc props … for a single game … on a single occasion,"[120] representations are generally more akin to toy trucks and dolls.[121] They are "designed" props for "games of certain kinds, ones in which they generate certain sorts of fictional truths."[122] These observations lead Walton to the following questions: what sorts of games does the storyworld of *La Grande Jatte* allow? And, in playing those, will the storyworld's "objective integrity"[123] be maintained?[124] To answer them, Walton distinguishes between *two sorts of imaginings*: those that conform to the rules (the principles of generation) of *La Grande Jatte*'s storyworld—like imagining a couple strolling in the park—and those which he claims of are a "misuse"[125] of the painting—for instance, imagining a number of hippopotamuses enjoying themselves in a mud hole.[126] Thus, Walton concludes:

> It is not the function of *La Grande Jatte* to be a prop in games in which fictionally hippos are wallowing in a mud hole, no matter what games people actually play with it. The hippopotamus game is inappropriate for the painting, *unauthorized* … to play it is to misuse the work. This is why it is not *La Grande Jatte*-fictional that hippos are wallowing in a mud hole.[127]

119 Walton, Mimesis, 60. In this respect, "figurative paintings" or artworks that "'point beyond' themselves", such as *La Grande Jatte*, in that they depict "people and objects distinct from the painting itself" and thus evoke a storyworld, can be distinguished from those which do not. (Ibid., 57). These "nonfigurative" (Ibid., 54) representations focus on abstract objects and only portray their "own elements in a certain manner" without evoking a greater storyworld. (Ibid., 57; cf. 54-57).

120 Ibid., 51.

121 Sicart describes the potential for toys (and also for video games) to evoke certain kind of reactions in the player: joyful or unsettling, etc. (Sicart, *Beyond*, 83-88, 93). Thereby, his explanations come close to Walton's on props in representational artworks—leading to the hypothesis that gameworlds may function in the same way.

122 Walton, *Mimesis*, 51.

123 Ibid., 67.

124 Ibid., 59-60.

125 Ibid., 60.

126 Ibid., 59-60.

127 Ibid., 60.

On the other hand, "authorized games, games it is the function of the work to serve in,"[128] organically work within the bounds of *La Grande Jatte*'s storyworld (or work world). According to Walton, then, it is both "*La Grande Jatte*-fictional that a couple is strolling in the park" and that an appreciator "sees a couple strolling in the park, for [such a game] (let's assume) is in accordance with the painting's function."[129]

Figure 12: Georges Seurat's 'La Grande Jatte' involves the appreciator in imaginative games of make-believe.

Georges Seurat, *Un dimanche après-midi à l'Île de la Grande Jatte* (1884-1886).

Arguably, the line Walton treads is slim, but with it he wants to direct attention to the fact that representations involve the appreciator in vivid games of make-believe, yet that these games are regulated by certain rules (principles of generation) the appreciator has to follow in order to experience a work's function. Such a directing effort through props will become of importance to the genre of the VGD, for it generally aims to evoke in the player a certain response to her experience in virtuality.

128 Ibid.

129 Ibid.

To sum up, props "prescribe imaginings"[130] and therefore "insulate fictional worlds from what people do and think."[131] Hence it follows that becoming involved in representations, appreciators generally follow certain guidelines or "rules"[132] they have to respect, which the work of art has *outlined* for them. If this is so, fictionality can be seen "in terms of prescriptions to imagine,"[133] or "what is fictional in a work is what appreciators of it … *are* to imagine."[134] What this does not mean is that the appreciator's part is devaluated or neglected. Rather, work worlds function as *dynamic frameworks* that guide the appreciator's participation and involve her in expressive games of make-believe. These games are most diverse, and primarily two sorts can be observed: *games of proximity* and *games of distance*.

The importance of participation is beyond dispute, and Walton highlights the appreciator's imaginative and psychological involvement in representational works of art throughout his entire work. There are many things appreciators *fictionally* do when engaging with representations: seeing,[135] fearing,[136] feeling, worrying, sympathising, enjoying, hoping, wanting, knowing, having "certain beliefs, expectations, suspicions, hunches," being "ignorant or uncertain,"[137] and so on.[138] Given the diversity of these actions, it follows that appreciators cannot be reduced to "mere spectators of work worlds, observers from the outside … That leaves out our *participation*."[139] Rather, they are wilful participants "blatantly playing along with the fiction."[140]

Yet Walton's observations go further and do not fail to recognise the appreciator's involvement on the level of concept. Besides her imaginative and psychological "involvement in the worlds of our games,"[141] critical observation remains a substantial aspect of her experience. "The appreciator's perspective is a dual one.

130 Ibid., 51.

131 Ibid., 67.

132 Ibid., 60.

133 Ibid., 58.

134 Ibid., 60-61; emphasis added.

135 Ibid., 215.

136 Ibid., 241ff.

137 Ibid., 259.

138 Ibid., 258-259.

139 Ibid., 208.

140 Ibid., 246.

141 Ibid., 272.

He observes fictional worlds as well as living in them [sic]."[142] Hence, to engage with representational works of art (paintings, literature, plays, and so on), appreciators "simultaneously"[143] play two sorts of games. These are in themselves multifaceted and include: 1) participatory games that involve the appreciator on a basic level of entertainment and affective emotions (games in which she comes to know the fictional world, either an abstract world or a storyworld including characters and their relations—that is, the developing plot), and 2) emancipatory games that allow for the close examination and reflection of props.[144] Consequently, the appreciator's game is best described as one of *proximity* and *distance*, as she constantly oscillates between the poles of inhabiting and observing a fictional world.[145]

It is clear that Walton does not stand alone with these observations, as they chime in with those of other scholars, such as Rancière or Iser, who describe the *phenomenology of art experience as an active participation process* in which the reader or emancipated spectator engages in a dialectical communication with the literary work or play. As Rancière holds: "Why identify gaze and passivity, unless on the presupposition that to view means to take pleasure in images and appearances while ignoring the truth behind the image and the reality outside the theatre?"[146] With this statement, Rancière opposes two misconceptions that reduce spectators to "passive voyeurs:"[147] for one, the didactic mindset that "viewing is

142 Ibid., 273.

143 Ibid., 285.

144 It is first and foremost works that show a certain amount of aesthetic quality and diversity which extend the appreciator's participation into "a long ... psychologically rich game of make-believe" that continues after she has closed the book or stepped out of the museum (Ibid., 254) while listening to a Blink record.

145 Ibid., 285. Ryan refers to the reader's involvement in narratives as follows: "Participating in the plot is a compromise between identification with the character and distanced observation. We simulate mentally the inner life of these characters, we transport ourselves in imagination into their minds, but we remain at the same time conscious of being external witnesses." (Ryan, *Avatars*, 124-125).

146 Rancière, *Emancipated*, 12.

147 Ibid., 4.

the opposite of knowing,"[148] that during the observation of a play knowledge undergoes a "straight, uniform transmission"[149] from "schoolmaster" to "ignoramus."[150] For another, that viewing "is the opposite of acting," that "the spectator remains immobile in her seat, passive."[151] Against these claims, Rancière holds the notion of *emancipation* and underscores the spectator's imaginative and interpretive participation in the spectacle.

At the heart of his argument thus lies the call for emancipation from certain oppositions, specifically from the "poles of distanced investigation and vital participation."[152]

> Emancipation begins when we challenge the opposition between viewing and acting; ... It begins when we understand that viewing is also an action that confirms or transforms this distribution of positions. The Spectator acts, like the pupil or scholar. She observes, relates, selects, compares, interprets. She links what she sees to a host of other things that she has seen on stages, in other kinds of places.[153]

Given these observations, spectators are far from passive and actively participate in plays on both an imaginative and interpretive level. To do so, they mobilise their world knowledge and relate the actions on stage to the empirical world they live in.[154]

Being a spectator thus means to enjoy *and* understand—and, indeed, it appears that to engage with representational art means to perform a courtship play between distance and proximity; to play between the poles of what Rancière has called "distanced investigation and vital participation"[155] and to what Iser refers to as "[t]he ability to perceive oneself during the process of participation [which] is an essential quality of the aesthetic experience; the observer finds himself in a

148 Ibid., 2.

149 Ibid., 14.

150 Ibid., 9.

151 Ibid., 2.

152 Ibid., 5.

153 Ibid., 13.

154 Ibid., 22.

155 Ibid., 5.

strange, halfway position: he is involved, and he watches himself being involved."[156] The result is a life-giving tension between critical reception and illusory immersion, and it is only through this tension that the appreciator's engagement with fiction comes to full fruition.

Considering these facts, one can start seeing work worlds—whether they are those of literary fictions, theatre plays, or video games—as *incomplete constructs* that necessitate the appreciator's involvement—and it is only through the mutual interaction between the two parties that the work of art may come to life.

> *Every* work of art, even though it is produced by following an explicit or implicit poetics of necessity, is effectively open to a virtually unlimited range of possible readings, each of which causes the work to acquire new vitality in terms of one particular taste, or perspective, or personal *performance*.[157]

In other words, representations (in the sense of fictions) involve the appreciator in complex participation processes, and work worlds thereby assume a vital part. For they are dynamic frameworks that outline the appreciator's imaginative involvement (in the case of non-ergodic fictions) and ergodic involvement in them (to anticipate the additional plane for hypertexts, video games, and so on) by using props to guide her imaginings, emotions, and ergodic actions (in children's games of make-believe or virtuality). *Work worlds*, one could argue, *imply* appreciators and allow participation to occur in the first place. The resulting *game worlds* are fascinating. They not only allow access to the work worlds but, in doing so, function as "expansion[s]" of them.[158] The result is the creation of something *new*, something brought about only through the act of engaging with representational art. As Rancière maintains: "from the schoolmaster the pupil learns something that the schoolmaster does not know himself. She learns it as an effect of the mastery that forces her to search and verifies this research. But she does not learn the schoolmaster's knowledge."[159]

The deliberations so far evoke the following questions: can the player's experience of a video game be compared to that of the reader, spectator, or appreciator? If so, in what aspects do their experiences coincide and how do they differ (which was implied above)? Video games are easily thought of as being fundamentally

---

156 Iser, *Act*, 134.

157 Umberto Eco, *The Open Work*, trans. Anna Cancogni (Cambridge, Mass.: Harvard UP, 1989), 21.

158 Walton, *Mimesis*, 216.

159 Rancière, *Emancipated*, 14.

different from non-ergodic representations, and there is a certain amount of truth to this claim. One has to bear in mind Aarseth's observation that in ergodic media, such as the video game, "nontrivial effort is required to allow the reader [/player] to traverse the text."[160] By doing so, both the game system and the player are responsible for producing semiotic sequences.[161] In addition, Walton recalls the fact that with different media, there are different kinds of games to be played[162]—and that the appreciator's games are restricted compared to children's games of make-believe, which are "active, physical, and involved."[163] Consequently, and because of the appreciator's more distanced involvement, her game is "more reflective, more contemplative. The restrictions on physical participation shift the emphasis to psychological participation."[164]

Now, one may think that video games show strong similarities to children's games of make-believe in that players have direct influence on the gameworld through what Gordon Calleja calls "[k]inesthetic involvement."[165] This influence is effectuated through the player's physical manipulation of some sort of input device: a controller, mouse and keyboard, or an even more physically demanding input source such as a Wii Remote or a Kinect camera. The logical conclusion to be drawn would be that through the player's physical input, the focus of her game world (similar to children's games) is shifted away from a more distanced reflection of what is being played towards a more involved one, towards the frenzy of spectacle ludic encounters so often show. Luckily, in many video games such a conclusion is not so easily drawn. To be precise, however, and to do justice to the player's multifaceted involvement in the virtual worlds of the video game, a brief excursion into Calleja's take on the phenomenon may be beneficial. His perspective is fruitful, as it sheds light on the intricate processes that emerge during play.

4.2.2 Becoming Involved in the Virtual Worlds of the Video Game

"Fictional worlds have always been meticulously designed to allure us into inhabiting them. With the advent of networked digital technologies, we now have the

160 Aarseth, *Cybertext*, 1.

161 Aarseth, "Ontology," 487.

162 Walton, *Mimesis*, 220.

163 Ibid., 224.

164 Ibid., 228.

165 Calleja, *In-Game*, 71.

ability to simulate these worlds and share them across the globe instantly."[166] It is widely understood that becoming involved in the virtual worlds of the video game differs in certain aspects from the appreciator's involvement in non-ergodic fictions—and some would claim that it is an entirely different phenomenon (as explained before). Calleja recognised this particular aesthetic, and in his influential work *In-Game: From Immersion to Incorporation* (2011) he makes keen observations on the player's extended involvement as a form of *inhabiting* "virtual spaces not just through our imagination, but also through the cybernetic circuit between player and machine."[167] He then continues to develop a model that does justice to the "multidimensional phenomenon"[168] of the video game and establishes six distinct but interwoven categories: kinesthetic, spatial, shared, narrative, affective, and ludic involvement.[169] In addition, Calleja goes on to distinguish between micro and macro involvement: the "moment-to-moment involvement within the respective dimension"[170] and "the ongoing motivation to interact with the game and the off-line thinking that fuels it."[171] These observations lead Calleja to the conviction that the term *incorporation* best describes the player's involvement in the gameworld, a term that transcends the concepts of immersion and presence. To Calleja, incorporation works on basically two levels: on the first level, the player incorporates the virtual environment she navigates and interacts with into her "mind as part of immediate surroundings,"[172] while on the second level, the direction is reversed. Here, it is the player herself who becomes part of that same virtual environment through her PC.[173]

To reach such a conclusion, Calleja postulates a close connection between the virtual and empirical world and argues that we make sense of virtual environments with the help of "*experiential gestalts* that inform being in everyday life."[174] In this line of thinking, the boundaries between *what is virtual* and *what is real* seem to blur, as there is "no longer need to draw a strict line of demarcation between stimuli emerging from the virtual environment and stimuli emerging from the

166 Ibid., 185.

167 Ibid., 167; cf. 181-182.

168 Ibid., 31.

169 Ibid., 37-38.

170 Ibid., 4.

171 Ibid., 37.

172 Ibid., 169.

173 Ibid., 169.

174 Ibid., 167-168; emphasis added.

physical world, for the emphasis is placed in the internally constructed consciousness of the individual."[175] Incorporation, he continues, therefore "allows us to move beyond the notion of virtual environments as *experientially separate other-worlds* and to treat them instead as domains continuous with the media-saturated reality of everyday life."[176]

Calleja is generally right about the blurring of boundaries between the virtual and empirical world, but is wrong about the unhindered communication this notion of connectedness implies, specifically when dealing with fictional game-worlds—for the access to fiction is even further away than that for virtual and empirical realms.[177] This is so because with fiction there can be no direct connection to empirical reality, and this results in a hindered and aggravated form of communication through both *world* and *agency* in which the player has to exert effort in connecting the dots and building up a situational context out of an estranged gameworld.[178] Although fictional worlds are also made accessible through experiential "gestalt groupings,"[179] the connection to the empirical world is dependent on a *secondary gestalt*, which is not explicit in the text … [and] brings out something which is not stated by the linguistic signs."[180] I will come back to this matter in chapter V, but for now it suffices to point out that, even with the complication of fictionality, the potential this form of communication entails—between the virtual and empirical world—is beyond dispute. The merging of realities brought about by the bi-directional communication between player and the game consequently shows the strong potential to influence the player in a lasting manner.

Ergodic media (such as hypertexts or video games) are thus different and, at the same time, very much connected to forms of non-ergodic media (such as the novel, the non-participatory play, or the film). The difference rests in the fact that

175 Ibid., 179.

176 Ibid.; emphasis added.

177 In fact, Calleja implies that ultimately, the processing of the virtual and empirical world occurs through the same psychological *gestalts*, so never in a direct, immediate way.

178 Indeed, one could argue that many aspects of the empirical world seem strange as well to the onlooker. But in fiction, the strategies of estrangement are employed for creative use and to exorcise the appreciator's inner demons for a cathartic purge.

179 Iser, *Act*, 120.

180 Ibid., 121.

in ergodic media the player's presence in the gameworld is directly "acknowledged by the system itself,"[181] whereas in non-ergodic media one can 'only' imagine oneself being present in the fictional world, "but the world does not recognize"[182] the reader or viewer.[183] These, Calleja concludes, are very "different forms of involvement"[184]—and while true, there is no denying the fact that they display yet another game of proximity and distance (indeed, a very much closer one). The connection, then, lies primarily in two forms of involvement Calleja and many others[185] have neglected in their descriptions of video games, which are that of the player's *imaginative* and often *emancipated involvement* in the diegesis.[186]

Calleja implicitly recognises these two forms of involvement in his *narrative involvement* but, surprisingly, has not integrated them as a separate category.[187] Therefore, and in order to do full justice to the player's spectrum of pleasures in a video game, I wish to underline the category of imaginative involvement and propose that of emancipated involvement, which are fundamentally related. Whereas imaginative involvement occurs on a basic level when the player engages cognitively with the game- and storyworld, filling in its indeterminacies and combining its perspectives, emancipated involvement goes further to designate a quasi-transcendental viewpoint in which the player not only becomes involved in the occurring gameworld events but, at the same time, occupies a detached, observing perspective on it.[188] Consequently, the player, as Daniel Vella eloquently puts it, finds

181 Calleja, *In-Game*, 22.

182 Ibid., 29.

183 Ibid., 22, 29.

184 Ibid., 29.

185 A similar mistake is committed by (Britta Neitzel, "Medienrezeption und Spiel," in *Game Over!? Perspektiven des Computerspiels*, ed. Jochen Distelmeyer, Christine Hank, and Dieter Mersch [Bielefeld: Transcript, 2008], 102-103).

186 Naturally, the connectedness also lies in the player's *psychological involvement* in fictions, and one could argue that there is a form of spatial and shared involvement (watching a movie together, for example) in any kind of fiction—although these are admittedly very different in video games.

187 Calleja, *In-Game*, 113-134.

188 This *dual performance of participation and observation* comes close to Walton's distinction between imagining *de se* and imagining *de re*. *De se* imaginings (of which the most prominent variant is *from the inside*) represent "a form of self-imagining characteristically described as imagining *doing* or *experiencing* something (or *being* a certain way), as opposed to imagining merely *that* one does or experiences something or possesses a certain property [*de re* imaginings]." (Walton, *Mimesis*, 29). Imagining

herself exposed to "two dimensions of subjectivity … by which the player inhabits both a perspective internal to the gameworld as the ludic subject, and a perspective external to the gameworld, which frames the ludic subject as an object of perception."[189] For my notion of emancipated involvement this means that it occurs when the player steps beyond the pleasures of entertainment and affective emotions (although these are vital parts of her experience) and reaches for the levels of significance/concept by contemplating on the events experienced and enacted. It is only then that the act of play can lead to a partial restructuring of the player's habitual dispositions.

Consequently (as chapter V and the rest of this study will further clarify), although the player's involvement in the virtual worlds of the video game is extended beyond that of the non-ergodic media participant, it would be a serious mistake to think that the newfound level of ergodic participation comes at the detriment of the player's imaginative, psychological, and emancipated involvement in these worlds. Rather, and following the observations above, the player of a video game shall be seen as *amalgam of both creatures*, living on the fragile border between *inhabiting* and *observing* a fictional gameworld—and, therefore, holds a liminal position between "identity and difference, proximity and distance, selfhood and otherhood – that play out across the gap between the player outside the game and her ludic subjectivity in the game."[190] The imaginative and emancipated games she plays, thereby, go naturally hand in hand with her ergodic (or physical) games.

There are a number of questions that remain to be answered. Work worlds in video games seem different to the ones appreciators have come to know in non-ergodic fictions, as the player's involvement in them is extended, allowing for

---

oneself to be the player-character and to experience the gameworld from the inside—while still retaining the critical distance of *de re* imaginings, when props prescribe imaginings about gameworld objects—thus holds the potential of affecting the player on a personal level. It is as Walton describes of the appreciator: "It is chiefly by imagining ourselves facing certain situations, engaging in certain activities, observing certain events, experiencing or expressing certain feelings or attitudes that we come to terms with our feelings—that we discover them, learn to accept them, *purge ourselves of them*." (Ibid., 34; emphasis added). This is how we understand fictional characters, "when I imagine *myself* in another's shoes … my imagination helps to understand *him*." (Ibid., 34; cf. 29, 33, 34, 106). Of course, enacting fictional characters in video games certainly does not hinder this process of comprehension but further enhances it.

189 Vella, "Ludic Subject," Abstract.

190 Ibid., 17.

more variety in participation. Then again: 1) What kind of work worlds are players actually interacting with? 2) Do they prescribe imaginings such as other forms of representational art do, and if so how? 3) In addition, do players play according to the rules of a storyworld? 4) Most intriguingly, does the player's extended game world (her imaginative and ergodic involvement in the game) alter the work world in a nontrivial manner or does the latter remain stable? Maybe it is even the creation of something new that emerges through the interactions of play? I will postpone questions two and three and tackle the remaining ones now. For this reason, a closer look at two perspectives on the video game medium is helpful.

4.2.3 Video Games as Objects and Processes

Approaches to the study of video games often focus on one of two interlinking perspectives (or points of view) on the medium: games as *objects* and games as *processes*. "Games are both objects and processes (a combination of states not dissimilar to the duality of language: langue/parole, paradigm/syntagm etc.)."[191] As such, if one follows the object perspective on a game, the focus inevitably shifts towards two formal aspects: 1) the game's rules, which are inscribed into its code, and 2) the representational sign system of the gameworld.[192] In combination, these two interlocking aspects formulate a semi-open and dynamic "*framework*"[193] that allows for play in the first place. Furthermore, Calleja argues, "[f]om the game-as-object perspective, the player is conceived as an ideal, or implied, player," not as an "actual, active player."[194] This is different in the games-as-processes perspective, where the player is referred to as an empirical being influenced by social and cultural contexts.[195] Here, the focus rests on the processes of play, whereby "[t]he term *processual* refers to the potential for variation in a game's enactment at every engagement and favors a dynamic and recursive view on games."[196] That is, each time a player interacts, or different players interact, with the same game framework, the result of their engagement will differ[197]—on both imaginative and ergodic levels, I might add.

191 Aarseth, "Ontology," 484.

192 Calleja, *In-Game*, 10ff.

193 Ibid., 12; emphasis added.

194 Ibid., 11.

195 Ibid.

196 Ibid., 10.

197 Ibid., 12.

It is no coincidence that such observations dovetail well with that of other scholars, such as Domsch or Tavinor. Domsch, for instance, distinguishes between 1) the game's *architecture*, or "the overall structure of the text, containing its rules, its nodal situations (e.g. tree or network), possible entry and exit points etc.;" 2) the *individual run*, which "is the concrete realization of one possible reading/playing;" and 3) the resulting *protocol*, "the perceptible, recorded result."[198] Tavinor, on the other hand, differentiates between "work *type*" (or multiples), for example the movie *Star Wars: Episode IV – A New Hope* (George Lucas, 1977)—or more appropriately, a play such as *Hamlet* (William Shakespeare, 1599-1602)—and "work *tokens*, [instances] which are comprised of individual screenings [or performances] of the world."[199] From this he concludes that "[w]here a film is reproduced by a screening, a video game is reproduced through its various *playings*, which are dual acts of performance and interpretation."[200] If this is so, in films or plays the work world the spectator engages with becomes synonymous with the work token, that is, the individual screening. From there, the spectator's imaginative and psychological game world merges with that of the work world to produce something new. Players, conversely, as Tavinor notes, produce tokens themselves as they interact with bare work types, whereas spectators only rarely do (for example in participatory plays or interactive films).[201] Hence, in games the player's dialectic with the work world starts a level earlier and her game world (which is created through interaction) comprises both her ergodic as well as imaginative, psychological interaction with the game.[202]

So what are work worlds in video games, then? What is the result of their interaction with the player's game worlds? If one follows Tavinor on this matter, "work worlds and game worlds play out differently than it does with traditional fictions," and "[w]hen appreciators interact with videogame fictions, the game world effectively *projects* into the work world of the fiction because the work is only rendered after the game has been played."[203] As a result, both worlds seem to

198 Domsch, *Storyplaying*, 48.

199 Tavinor, "Art," 64. Meskin and Robson claim that "[t]o each production (and arguably each performance [of a play]) there may correspond a distinct work world." Meskin and Robson, "Fictional Worlds," 17-18.

200 Tavinor, "Art," 64.

201 Ibid.

202 It could be argued that different versions of the same game, such as THE LAST OF US on PS3 and the remastered edition on PS4, represent different work tokens and thus different work worlds (the framework the player engages with).

203 Tavinor, *Art of*, 57.

fuse, and, therefore, video games may only "contain the *bones or possibility of a work world*"[204]—in other words, a framework for interaction on both an ergodic and imaginative level. In stark contrast to these claims are Meskin and Robson's observations on the phenomenon. For them, video games still retain a stable distinction between work world and game world, even though they allow "for actions by agents other than their creators to directly determine the nature of relevant work worlds"[205]—with work worlds, crucially, being equated to the instance of an individual playthrough.[206] They justify their claim with several examples, in which emotions (such as shame and guilt) are specific to the player's game world, or where there are divergences in what the player fictionally sees and what the PC sees, or in their respective states of knowledge about the fictional world, and so on.[207]

Frankly, it is true that one can equate exclusive emotions, states of knowledge, or what the player fictionally sees with her game world. But Meskin and Robson's implication that the player's game world may only start *after* the creation of a particular instance (or work world) is not remotely adequate. For one simply cannot reduce the player's involvement (that is, her game world) to imaginative and psychological interactions that occur *only after* the instance of a certain playthrough was created. This is so because a great deal of the player's personality—her values that drive the play experience[208]—has found direct entrance into the co-creation of the gameworld and the resulting narrative, in ergodic, psychological, and imaginative ways. Consider, for example, a playing of FALLOUT 4 (Bethesda Game Studios, 2015) where a player who in real life likes animals (and maybe has pets of her own) encounters a supposedly animal-friendly woman in the Commonwealth. She lives in a small shed together with several cats, but when the player chooses to barter with her, he encounters a shocking truth: the women sells cat meat. In a frenzy of potential fictional anger, now, which feeds back into FALLOUT 4's work world, the player may choose to deal with the situation in various ways. Since it is the virtual post-apocalypse, free of the constraints of the empirical world, he may choose to blow off the women's head. A player who dislikes animals may act differently, of course. But the point is that in both cases the player's game world—finding her real life experience partially reflected in the game—is

204 Ibid., 58.

205 Meskin and Robson, "Fictional Worlds," 20.

206 Ibid., 1, 18ff., 29.

207 Ibid., 20-27.

208 Sicart, *Beyond*, 15.

guided directly because of who she is,[209] and this self will inevitably project back into the work world through ergodic and imaginative action. What Meskin and Robson suggest, however, is a distinction between ergodic interaction, on the one hand (creating the instance of the playthrough), and imaginative and psychological involvement on the other (the player's game world)—a conclusion I find highly problematic.

Now, to approach the issue from an alternative vantage point, one may recall Tavinor's claim that the player directly affects the work world by means of her game world—and this fusion, so I claim, will not only result in the creation of another work world (if at all[210]) but in the creation of something *new*, something more personal, which comes into the world through the act of play. It begins with the game designers' creation of a dynamically incomplete framework: a work world for the player to complement (or fill in) through her personal game world (her emotions, feelings, ergodic and imaginative actions, etc.). Logically, the result is not simply another work world, or merely a resulting protocol (for this would in a sense devalue the player's efforts), but, as described above, something entirely new that brings forth an *aesthetic effect* as described by Iser—and this effect does not refer to something already in existence, yet brings forth a meaning new to the world.[211]

Consequently, in order to describe this all-pervasive effect, the intricate dialectic that evokes it becomes of interest: a dialectic between the *implied player* (the dynamically incomplete framework of the work world) and the *empirical player* (the player's game world on various levels). For this purpose, it first becomes necessary to inspect different empirical player types and how they are implied by a game's structure. In a second step, a critical scrutiny of Wolfgang Iser's original concept of the implied reader will be conducted, which informs the notion of the implied player as it has been used in video game studies (particularly by Aarseth) and that I will describe as a system of perspectives. Third, the creative dialectic between empirical and implied player becomes the focus of attention (chapter V)—and, most importantly, the *aesthetic response* the process of play triggers within the player, thus creating something new, which is neither to be found exclusively in the work world nor in the game world.

---

209 Of course, the player can also role-play and try out certain perspectives he might normally decline: this is not the point here.

210 Indeed, why should one call the result of the act of play a work world, if so much of the player's self and personality has found entrance into its co-creation?

211 Iser, *Act*, 22.

4.3 THE IMPLIED PLAYER

It has become clear that participating in representational works of art entails several restrictions to the appreciator's games. Props dictate imaginings of a certain kind and the appreciator generally plays according to the rules of a particular storyworld or abstract representation. This is so because the appreciator willingly agrees to an informal but binding contract with fiction, which is that of make-believe. Now, it is these restrictions in involvement that further link video games to other kinds of representations, but in video games the binding instructions are of course extended to include the player's ergodic interactions with them.

The role of rules is widely discussed in video game studies, and they are usually understood as prescribing the function of game objects and the player's ergodic participation in a game. However, it is not only the rules of a game that outline the player's involvement in it but also the props (or perspectives) the gameworld is composed of which guide her imagination.[212] Having both observations in mind is of utmost importance and is a vital starting point for further deliberations on the issue. The *rules of the gameworld* can therefore be seen to prescribe both the player's ergodic and imaginative interaction[213] with it—and to better illustrate the structure that affords the player's involvement, I will start with the hypothesis that *the implied player describes a structural construct and a dynamic work world that outlines the empirical player's participation on all levels of involvement (offering her various roles to perform)*. To reach such a conclusion and to expand on it, let me start with elaborations on how the implied player is used and defined in video game studies to then move back to the concept's origins found in Iser's observations on the phenomenology of reading. Both perspectives are fruitful, and in combination they will inform the definition and use of the implied player I am proposing here.

---

212 Tavinor explains this through the example of RED DEAD REDEMPTION (Rockstar San Diego, 2010), where the player is "guided by the depictions of a fictive prop, imagines that a man named Marston exists and that he has the various features ascribed to him in that fiction." (Grant Tavinor, "Fiction," in *The Routledge Companion to Video Game Studies*, ed. Mark J. P. Wolf and Bernard Perron [New York: Routledge, 2014], 437).

213 However, players often try to break the rules of a game, be it the rules outlining their ergodic participation or those of the storyworld's integrity. This form of play, which Aarseth has called *transgressive play*, should not be underestimated, as it is a common form of engaging with games. (Aarseth, "Implied.").

4.3.1 The Multi-Layered Qualities of the Implied Player: Popular Cultural Player Types and the Emancipated Player

When Espen Aarseth transferred Wolfgang Iser's concept of the *implied reader* to the study of video games, he defined the phenomenon as follows: The *implied player* "can be seen as a *role* made for the player by the game, a set of *expectations* that the player must fulfil for the game to 'exercise its *effect*'."[214] Although appropriately formulated, one cannot help but wonder as to what *effect* he is talking about, *because there are many*. Before coming back to Aarseth's take on the implied player (and tinkering with it and expanding on it), let me first dive into the multifarious effects a game can have on a player and the many roles she may assume.

The reason for the multifacetedness of games can easily be explained. Many video games these days are mass market productions (AAA games) that try to reach an audience as diverse as possible in order to maximise profitability—and this is also true for some indie game productions. They are designed to cater to a wide variety of different tastes and mindsets, and it comes as no surprise that an analysis of player types and their specific needs and tastes is an important aspect of the game development process.[215] Few games, however, can reach all potential player preferences, and hence it follows that certain kinds of games imply certain types of players—or, to be more precise, playing styles that cater to specific genres (whereby I mean both ludic and narrative genres).

Domsch, for instance, describes the empirical player of chess as somebody who "is implied in the game's structure" and who "wants to win" the game.[216] Similarly, the player of QUAKE III ARENA (Id Software, 1999) fulfils her role implied by the game's structure in that she participates in rapid ludic action against other players, which takes place in multiplayer arenas. Research has identified this type of player who enjoys *ludic pleasures* on a level of entertainment and affective emotions. It is common to Bateman and Boon's "Conqueror,"[217] John Kim's

214 Ibid., 132; emphasis added.

215 Chris Bateman and Richard Boon, *21st Century Game Design* (Boston: Course Technology, 2006), 49ff.

216 Domsch, *Storyplaying*, 10.

217 Bateman and Boon, *21st*.

"gamist"[218] attitude, or Richard Bartle's "Achievers"[219] that their agenda lies in "winning and 'beating the game'."[220] They enjoy "a fair challenge ... which may be tactical combat, intellectual mysteries, politics, or anything else"[221] and "regard points-gathering and rising in levels as their main goal."[222] Consequently, as a structural construct, QUAKE's implied player (or that of other ludically focused games) outlines a specific style of interaction for the player, letting her fulfil a certain role.

Besides ludically oriented player types, there are a plethora of others: for instance, players who are inclined towards *world exploration* and the *experience of a narrative*. While these still work on a basic level of entertainment and plot, the focus shifts towards the engagement with a virtualised storyworld. There is, for instance, the player of DEAR ESTHER (The Chinese Room, 2012) who engages in the dreamlike gameworld, not only to explore its bounds but also to decipher the story behind this world, and that of Esther. This intimate experience of playing a story may result in an emotional attachment on the player's side and implies types of players such as Bateman and Boon's "wanderer" and "participant,"[223] Kim's "dramatist,"[224] Craig Lindley's "performer" and "immersionist,"[225] or Karen and Joshua Tanenbaum's narrative-oriented player who "is concerned with participating in a fictional world where her decisions and actions are incorporated meaningfully into that fiction."[226]

In addition to these, there are "killers" and "socializers"[227] who play against or in tandem with other players (this is especially so in competitive multiplayer

---

218 John Kim, "The Threefold Model FAQ," *Darkshire*, accessed June 28, 2017, http://www.darkshire.net/jhkim/rpg/theory/threefold/faq_v1.html

219 Richard Bartle, "Hearts, Clubs, Diamonds, Spades: Players who Suit Muds," *Mud*, accessed March, 21, 2016, http://mud.co.uk/richard/hcds.htm#1

220 Bateman and Boon, *21st*, 58.

221 Kim, "Threefold Model."

222 Bartle, "Hearts, Clubs."

223 Bateman and Boon, *21st*.

224 Kim, "Threefold Model."

225 Craig A. Lindley, "Story and Narrative Structures in Computer Games," in *Developing Interactive Narrative Content: sagas/sagasnet reader*, ed. Brunhild Bushoff (München: High Text, 2005).

226 Karen and Joshua Tanenbaum, "Commitment to Meaning: A Reframing of Agency in Games," *Proceedings of Digital Arts and Culture Conference* (2009), http://escholarship.org/uc/item/6f49r74n?query=tanenbaum#page-1

227 Bartle, "Hearts, Clubs."

games); or "managers"[228] and "simulationists"[229] who are respectively concerned with "the mastery of the game" and "how to play well"[230] as well as "resolving in-game events based solely on game-world considerations, without allowing any meta-game concerns affect the decision."[231]

It is common, then, to all these types of players discussed above that they primarily work on a basic level of entertainment and affective emotions—and all address a specific aspect of the player's experience only. This is so because a discussion of player types inevitably mixes up *implied* and *empirical players*—and the above-mentioned players are, of course, to be seen as constructs, for empirical players are not confined to exhibiting solely one of the described attitudes. Similarly, there is a distinction to be made between games that primarily imply one type of player (which are mostly ludically oriented and unilinear in focus—such as many smartphone games) and those that cater to a diverse array of player preferences (for example, open world games, which generally allow for a variety of playing styles).

Now, it is especially the latter variant—which shows aesthetic complexity and allows for a variety of playthroughs, imaginings, and interpretations—that often implies a further group of players. These *aesthetically-oriented player types* step beyond the basic pleasures of entertainment and affective emotions (although these are certainly a vital part of their experience), and start pondering the meaning of these games for their lives. Video games, it is clear, "can disrupt and change fundamental attitudes and beliefs about the world, leading to potentially significant long-term social change."[232] Yet it is not only a game itself that is responsible for that change but, similarly, the player's attitude. In this regard, McKenzie Wark talks about a player type who is "playing with style to understand the game as a form," who "trifle[s] with the game to understand the nature of gamespace as a world ... and to discover in what way gamespace falls short of its self-proclaimed perfection."[233] In addition, Miguel Sicart addresses an *ethical player* who "voluntarily steps out of the pleasures of instrumentality and incorporates play as a way of understanding the world including experiencing and exploring ethical and political thinking."[234] However, although Sicart's player moves in the realms of the

228 Bateman and Boon, *21st*.

229 Kim, "Threefold Model."

230 Bateman and Boon, *21st*, 62.

231 Kim, "Threefold Model."

232 Bogost, *Persuasive*, ix.

233 McKenzie Wark, *Gamer Theory* (Cambridge, Mass: Harvard UP, 2007), pa. 21.

234 Sicart, *Beyond*, 78.

aesthetic, his focus on playing in an ethically correct manner does not suffice for the purpose of my claims.

Consequently, and in order to fully appreciate a game's manifold experiences, an emancipated player becomes necessary to further discussion.[235] The emancipated player slumbers in all empirical players and is (primarily) interested in experiencing play's aesthetic effect. She is critical about her involvement in the game- and storyworld and confronts it with an open attitude and the necessary respect. Compare, for instance, the 'standard' player of BIOSHOCK (2KGames 2007) who engages solely for entertainment purposes with the attentive player of this critical dystopia who continues to ponder the larger significance of her actions within the virtual diegesis. While the former's involvement remains caught up on the level of the plot and affective emotions (to recall Iser's distinction), the latter's goes on to establish links between the virtual and empirical world, thus aiming for levels of significance. This transforms the emancipated player neither into an ideal nor a model player but into a real-world player type who engages in a creative dialectic with the intersubjective structure of the implied player.

This being said, it is natural that for emancipated play to occur, high demands are imposed not only on the player—as she frees herself from the confining opposition between inhabiting and reflecting on a gameworld, letting both forms of involvement occur—but also on the video game (dystopia) at hand. For it is only when a game shows a certain degree of aesthetic complexity that the demands of the emancipated player are met and satisfied (certain conditions that can be found in a game's structure; having classified many video games according to what Eco has called multi-layered artefacts). To put it simply: for emancipated play to occur, the emancipated player has to be *implied* in a game's structure, and it is only then that the *preconditions* for experiencing play's aesthetic effect are given. To analyse these may be one way to answer Tavinor's question why some video games may be considered art while others may not[236]—and for this purpose, a closer look at the intersubjective structure of the implied player becomes necessary. Before embarking on this venture, however, let me formulate *five hypotheses* on the emancipated player that I will explicate and use in the further course of this study.

---

235 Although showing certain overlap, the emancipated player differs from Sicart's ethical player in that the former does not necessarily need to play in an ethically correct way. What is more important for the category of the emancipated player is that she tries to see through the occurrences of the gameworld while, naturally, being involved on an entertaining level. She thus retains enough critical distance to reflect on the enacted events.

236 Tavinor, "Art," 61.

It is a concept that expands on Rancière's notion of emancipation and Walton's and Iser's deliberations on the imaginatively active appreciator and reader.

1. *The emancipated player enjoys and understands. She refrains from accepting a languid attitude towards representational art and participates to her fullest potential in the video game (narrative). This means it will not satisfy her to be merely involved on a basic level of entertainment and affective emotions (that is, on a purely ludic or plot level), but only the playful thrills of significance will suffice. Emancipated play may thus only occur through the player's combined efforts of inhabiting and reflecting on the gameworld.*
2. *The category of the emancipated player is closely tied to an aesthetic complexity of video games, and it is only when this quality is given (that is, inscribed into the implied player) that the preconditions for experiencing play's aesthetic effect are given.*
3. *The emancipated player slumbers in all empirical players. However, it is clear that the affordance and appeal structure of the implied player can better be read (or filled in) the more knowledgeable (or informed) the player is and the more life experience she draws on. Consequently, an intellectual richness of playthroughs, imaginings, and interpretations becomes possible. This sort of emancipated (aesthetic) involvement necessarily distinguishes the emancipated player from popular culture player types.*
4. *The emancipated player expresses herself through play as she engages in a creative dialectic with the implied player (a playful trial action). While doing so, she resembles a scientific investigator who employs her world knowledge to establish links and associations. The emancipated player participates, observes, selects, interprets, and acts upon her deliberations. Not only does she imprint herself in the gameworld, but she constantly relates the diegetic events to facts about her empirical present or other works of art she has previously encountered. Consequently, and in her quest for truth and self-reflection, the emancipated player accepts the role(s) the implied player has offered her (otherwise the game could have no effect on her) but, at the same time, subjects them to meticulous scrutiny in an act of emancipation.*
5. *The emancipated player frees herself from a confining perception and interpretation of video games. Instead of solely analysing a particular aspect of the video game (its procedural rhetoric or semiotic layer, for example), the emancipated player tries to see the video game (narrative) in its entirety and in a holistic manner. Hence, a variety of different perspectives on the gameworld appear, the combination of which may create the most interesting blanks to close.*

It is easy to discern that the emancipated player represents an aesthetically oriented type of player[237] who delights in beauty and confronts the gameworld with the necessary respect and a critical attitude. Given the diversity of player types discussed above—most of which belong to the category of popular culture players—and the multitude of different playing styles a game may afford, the emancipated player becomes necessary to further the discussion on the VGD (or any aesthetically complex video game). This is so because her priorities do *not only* rest in winning the game, becoming immersed in ludic encounters, exploring the gameworld, or participating in the creation of plot, etc. (although these aspects necessarily formulate part of the emancipated player's experience). Instead, the emancipated player's involvement in the gameworld reaches further in that she tries to engage with the implied player on a higher level of sophistication—motivating herself to attain a quasi-transcendental viewpoint that is nonetheless never reached—and thus enters a creative dialectic with it (see chapter V). The result is the experience of the aesthetic effect, something that happens *naturally* if one does not play in an outright depreciating manner. The emancipated player should therefore *not only* be seen as a specific type of real-world player but can, through her interaction with the implied player, be used as a *method for analysis* that works especially well for VGNs, thus describing a specific phenomenology of play.

Until now, I have repeatedly mentioned a so-called *aesthetic effect* that deeply and lastingly affects the appreciator or player. To describe this phenomenon and

237 However, the emancipated player is not an elitist type of player, which would reduce her to a specific intentional community. To clarify: the emancipated player slumbers in all empirical players and is a very inclusive concept, designating the meaning-seeking animal human beings are. Emancipated play hence begins the moment the player allows a game to exercise its effect on her, that she becomes affected by it. Yet for this effect to be experienced, a certain amount of effort is necessary. I am referring to languid players who care less about the meaning of their actions or those who solely engage for entertainment purposes—for example, in frantic ludic encounters without taking into consideration the bigger picture. This may happen in the BORDERLANDS series when players go on scavenger hunts for hidden treasures while neglecting the storyworld, which would give them a different perspective on the selfish and brutal actions they commit. Such play comes close to superficial perceptions in everyday life, where human beings take pleasures in spectacle while—as Rancière would say—"ignoring the truth behind the image and the reality outside the theatre [or game]." (Rancière, *Emancipated*, 12). Emancipated play thus begins on a basic level, but becoming affected by art can be addictive and exposes the individual to pleasures unknown to her before the first encounter.

the player's experience of it, it is time to delve into the work of Wolfgang Iser, who convincingly explains not only the aesthetic experience of engaging with literature but, on a bigger scale, that of engaging with representational art.

4.3.2 On the Phenomenology of Reading and the Aesthetic Effect of Art

"As readers, we exercise a power over narrative texts that is arguably as great as their power over us. After all, without our willing collaboration, the narrative does not come to life."[238] In his groundbreaking work *The Act of Reading: A Theory of Aesthetic Response* (1978), Iser sets out to describe the reader's imaginative and emancipated involvement in a literary text (without using the term emancipation) and the potential effects the reading process has on her.[239] To illustrate this process and its consequences, the literary theorist explains the reader's communication with the literary text in terms of a tripartite *dialectic* between text, reader, and culture:

> [T]he art of our times has created a new situation: in place of the Platonic correspondence between idea and appearance, the focal point now is the interaction between the text and, on the one hand, the social and historical norms of its environment and, on the other, the potential disposition of the reader.[240]

Iser's basic premise thus rests on a rejection "of the nineteenth-century concept of literature"[241] in which it was common for critics to embark on a quest for "the hidden meaning"[242] of a text. Such an attitude and the frequent question, "what does it mean,"[243] are, according to Iser, highly detrimental to the literary text. For if meaning is reduced to "a *thing* which can be subtracted from the work … the work is then used up – [and] through interpretation, literature is turned into an item for consumption."[244] Instead, Iser suggests a different attitude.

---

238 H. Porter Abbott, *The Cambridge Introduction to Narrative, 2nd ed.* (Cambridge: Cambridge UP, 2008), 86.

239 Iser, *Act*, ix, 18-19.

240 Ibid., 13-14.

241 Ibid., 5.

242 Ibid., 4.

243 Ibid., 11.

244 Ibid., 4; emphasis added.

The phenomenology of reading is based on the reader's interaction with the text, yet for this process to occur, the literary work itself cannot be a closed and finished object. Instead, the solution Iser proposes is to situate the literary work in between the opposites of author and reader and to view it as basically "*virtual* [in the sense of *indeterminate*] in character, as it cannot be reduced to the reality of the text or the subjectivity of the reader, and it is from this virtuality that it derives its dynamism."[245] Now, if the text's work world remains dynamically incomplete awaiting the reader to fill in its particulars, it follows that the literary text may only come alive through the process of *actualisation* and the interaction between reader and text.[246] This interaction is of a special kind and differs from other forms of communication in that the fictional quality of the literary work hinders the reader's comprehension. The argument rests on the assumption that in literary communication the *validity* of familiar real-world experiences suffers, "[a]nd it is precisely this loss of validity which leads to the communication of *something new*."[247] I will come to this matter soon and look at it in detail (chapter V), but for now I would like to follow up on the question of what kind of *newness* the communication with fiction actually produces.

In this respect, Iser proposes a most interesting conclusion. To him, the newness being unveiled through the act of reading rest in "*what comes through it into the world*" (emphasis added), and herein lies the nature of the *aesthetic effect*.[248] "It is characteristic of [the] aesthetic effect that it cannot be pinned to something existing, and, indeed, the very word 'aesthetic' is an embarrassment to referential langue, for it designates a gap in the defining qualities of language rather than a definition."[249] Meaning, in other words, cannot be reduced to a thing (it cannot be grasped, defined, or extracted), but is "imagistic in character,"[250] always in motion, and should rather be compared to an "experience" and "a dynamic happening."[251] This is chiefly so because "[t]he aesthetic effect is robbed of this unique quality the moment one tries to define what is meant in terms of other meanings that one knows."[252]

245 Ibid., 21; emphasis added; cf. 20-21.

246 Ibid., 19, 21, 66.

247 Ibid., 83; emphasis added; cf. 83.

248 Ibid., 22.

249 Ibid., 21-22.

250 Ibid., 8.

251 Ibid., 22.

252 Ibid.

Imagine reading a book or playing a game—say, JOURNEY (Thatgamecompany, 2012, 2015)—and think about how the experience results in a personal response to the game and the establishment of unexpected connections to the empirical world. In JOURNEY the player is sent on a spiritual quest towards an enigmatic mountaintop and on a journey of self-discovery. The meaning-making process is thereby in a continuous flux, for the image of the mountain and the journey towards it (as floating signifiers, so to say) are in constant renegotiation. With every stage of the route—from the initiation in the vast desert, to the perilous underworld ruins, and to the final steps towards the mountaintop, which are at first tortuous then pleasurable—the player finds herself in *constant renegotiation of meaning*, incorporating newly found perspectives (and those she has helped create) into the horizon of past ones and aligning this experience with her real world knowledge.

Figure 13: The beautiful post-apocalypse of JOURNEY, *and the player's venture towards an enigmatic mountaintop.*

JOURNEY (Thatgamecompany, 2012, 2015)

Consequently, the moment the player tries to define the journey's meaning, it eludes her grasp, and only fragments of the experience remain—those unexpected connections established between the virtual and the real world, between a fictionally enacted perspective and the player's self. JOURNEY may thus be described as an *experiential epiphany* that is different for each and every player (but may as well fail to impress the languid player) and which stands in the tradition of fictions like Samuel Beckett's play *Waiting for Godot* (1952), which, as Martin Middeke correctly claims, reflects upon the insufficient validity of all systems of the production of meaning by a radically open formal structure in which readers of all backgrounds find their respective interpretive consciousness activated yet at the same time frustrated.[253]

So far, it has become clear that the process of ideation and the ongoing flux of images (to which I will dedicate the entirety of chapter V) are undeniably important to the appreciator's or player's involvement in representational art, and Iser is well aware of this. He therefore argues that both the literary text and art in general resist "translation into referential meaning,"[254] and this generally conforms to the characteristics of images. "The image cannot be related to any … frame [of reference], for it does not represent something that exists; on the contrary, it brings into existence something that is to be found neither outside the book nor on its printed pages."[255] Images, so it seems, are a vital aspect of *fictional communication*, and in order to explain the mechanisms behind this process, the investigation finally leads to the implied reader and player.

For this purpose, let me begin with a simple explanation. If the aesthetic effect is created through the interaction between reader, culture, and text, and the text exhibits virtual (indeterminate) qualities, then it must also be structured in a certain manner in order to guide the participation process. For there is a sense in which art—as my previous observations have shown—guides, or even prescribes, the appreciator's involvement in it. In this regard, the video game is similar to the literary text, whose "full potential"[256] can never be exhausted. Consequently, while the literary critic "elucidate[s] the potential meanings of a text,"[257] the video game scholar illuminates both the player's ergodic and imaginative possibilities of play, which she does by scrutinising the structure that affords them. This is an important insight, as one has to bear in mind that "[a] theory of aesthetic response [Wirkungstheorie[258]] has its roots in the text" (or game) and does not arise "from a history of readers' [or players'] judgments."[259] Consequently, what is important

253 Martin Middeke, "Reception Theory," in *English and American Studies: Theory and Practice*, ed. Martin Middeke, Timo Müller, Christina Wald, and Hubert Zapf (Stuttgart: J.B. Metzler, 2012), 194.

254 Iser, *Act*, 11.

255 Ibid., 9.

256 Ibid., 22.

257 Ibid.

258 The "[t]he German term 'Wirkung' comprises both effect and response." (Iser, *Act*, ix).

259 Ibid., x.

is the question "what happens to us through these texts"[260] and games, which inevitably lays the focus of the investigation on "the structure of 'performance':"[261] *on the bare work world that awaits the appreciator for complementation*, and not its "result."[262] The implied reader or player, therefore, is by no means to be confounded with any empirical being but rather resembles a *structure* that outlines the player's participation in a work world.[263]

4.3.3 The Intersubjective Framework of the Implied Player

Aesthetic response is … to be analyzed in terms of a *dialectic relationship between text, reader, and their interaction.* It is called aesthetic response because, although it is brought about by the text, it brings into play the imaginative and perceptive faculties of the reader, in order to make him adjust and even differentiate his focus.[264]

In his analysis of the reader's *aesthetic response* to a literary text (or fiction), Iser maintains that the meaning-making process primarily depends on two faculties: the text and the reader, who draws on her cultural knowledge to understand the text. Dissatisfied with previous concepts of readers, on which he nonetheless builds, Iser continues to develop his own model, which he names the *implied reader.*[265] As a structural concept firmly anchored in the text, which involves the empirical reader in a creative dialectic because of its incompleteness, the implied reader consists of primarily *two interlocking aspects*: 1) "the reader's role as a textual structure" and 2) "the reader's role as structured act."[266]

While the first part of the implied reader sheds light on the *strategies* (and the *repertoire* from which they draw) by which a text guides the empirical reader's participation in it (ascribing a certain role to her), the second part clarifies how the empirical reader becomes affected by the text, that is to say, how the text "induces structured acts in the reader"[267] and thereby exerts a gradual influence on her self.[268] To give Iser's full definition:

260 Ibid.

261 Ibid., 27.

262 Ibid.

263 Ibid., 27-28, 34; Aarseth, "Implied," 130-131.

264 Iser, *Act*, x; emphasis added.

265 Ibid., 27-34.

266 Ibid., 35.

267 Ibid., 36.

268 Ibid., cf. 35-36, 85.

> the implied reader … embodies all those predispositions necessary for a literary work to exercise its effect – predispositions laid down, not by an empirical outside reality, but by the text itself. Consequently, the implied reader as a concept has his roots firmly planted in the structure of a text; he is a construct and in no way to be identified with any real reader.[269]

It is clear that the effect Iser is alluding to is the *aesthetic effect* explained earlier, and this effect has its predispositions outlined by the literary text. Structure, therefore, becomes of utmost importance, for as Iser reminds us, "the role described by the text will be stronger" than "the reader's own disposition," which, in turn, will "form the background to and a frame of reference for the act of grasping and comprehending."[270] Iser's focus on *the determining nature of the text* has often been disputed by other scholars, for example by Strasen, who argues that the implied reader is a theoretical emergency brake, since Iser does not respect the unbounded diversity of meaning creation inherent to the reader's communication with the literary text which his theory implies.[271] Yet it is true that only when the work of art assumes a superior role in the communication process (or one sufficient enough), it may have a lasting effect on the participant, paving the way to incorporating "new experiences into our own store of knowledge."[272] Naturally, the empirical reader remains a vital part of the communication process, and Iser is well aware of that.

To Iser, the implied reader designates *a network of structured indeterminacies*—an appeal or "textual structure anticipating the presence of a recipient"[273]—and therefore fulfils a dual function. Because of the text's virtuality (indeterminacy) and the reader's consequent ability to be involved, it "must already contain certain conditions of actualization."[274] From this Iser concludes that one can start seeing the implied reader as "a network of response-inviting structures, which impel the reader to grasp the text."[275] During this process "the reader is … offered a particular role to play"[276] and this role is twofold, as I have alluded earlier—for otherwise the implied reader is reduced to a "structured prefigurement," which

---

269 Ibid., 34.

270 Ibid., 37.

271 Sven Strasen, *Rezeptionstheorien: Literatur-, sprach- und kulturwissenschaftliche Ansätze und kulturelle Modelle* (Trier: Wissenschaftlicher Verlag Trier, 2008), 67.

272 Iser, *Act*, 37.

273 Ibid., 34.

274 Ibid.

275 Ibid.

276 Ibid.

implies a uni-directional communication between both parties, and that the text would "imprint" itself on the reader.[277]

Consequently, and in order to underscore the inherent dynamism between text and reader, the first half of the implied reader constitutes the *textual strategies* that organise a particular viewpoint on the world. This viewpoint includes and, at the same time, transcends the author's point of view, since the text "constructs a world of its own out of the material available to it" and "*bring*[*s*] *about* a standpoint from which the reader will be able to view things that would have never come into focus."[278] It does so by organising the literary text in a framework that consists of "a system of perspectives," of which Iser postulates four basic ones: "the narrator, the characters, the plot, and the fictitious reader."[279] These "provide guidelines originating from different starting points," and it is only in their convergence achieved through the reader's acts of ideation that the locus of "the meaning of the text" may come to the fore.[280]

While this first part of the implied reader lays the emphasis on the text as structure that "offers guidance as to what is to be produced,"[281] the second part aims to underscore the reader's importance in the process. For one must not neglect the reader's involvement and "the extent to which this text can *activate* the individual reader's faculties of perceiving and processing"—thus triggering structured acts in her.[282] This is further reinforced by the fact that reading is a *selective process* in which the reader has to make decisions as to which possibilities are to be imaginatively actualised. In other words, "there are many different interpretations of a single text ...[different] attempts to optimize the same structure,"[283] which leads Iser to the conclusion that the implied reader should be seen as an "intersubjective structure"[284] and a "textual pattern"[285] that outlines the reader's involvement in a text, giving "rise to ... many different subjective realizations."[286]

In video games, the selective process is extended to the player's ergodic involvement in the game, and Aarseth recognised this potential for the implied

277 Ibid., 107; cf. 107.

278 Ibid., 35; cf. 35, 96, 107.

279 Ibid., 35.

280 Ibid.; cf. 35.

281 Ibid., 107.

282 Ibid.; emphasis added; cf. 107.

283 Ibid., 118.

284 Ibid.

285 Ibid., 9.

286 Ibid., 118; cf. 118, 123-124.

player. According to his working definition, "[g]*ames are facilitators that structure player behavior … and whose main purpose is enjoyment*."[287] In this line of thinking, empirical players have to follow strict guidelines. They subject themselves "to the rules and structures of the game" for the process of play to occur and to complement the dynamically incomplete work world.[288] Having learned and accepted these rules, players are now able to manoeuvre within the confines of a system—which closely aligns Aarseth's notion of player to that of Salen and Zimmerman, who conceive of play as a "free movement within a more rigid structure. The particular flavor of a game's play is a direct result of the game's rules."[289] These rules, they continue, "guide and shape the game play experience."[290] Still, Aarseth goes one step further in that he compares the empirical player's experience of a game to "the prison-house of regulated play."[291] Combining the implied player with "[Hans-Georg] Gadamer's notion of the unfree player subject," he concludes that "we can start to see the implied player as a boundary imposed on the player-subject by the game, a limitation of the playing person's freedom of movement and choice."[292]

Against this notion of regulated (or confining) play, in which players play according to a game's rules and representational aspects (by which he means visual aspects, not fiction), Aarseth holds the "counterweight"[293] position of "transgressive play."[294] This he describes as "a struggle against the game's ideal player" and the "active, creative, and subversive"[295] "rebellion against the tyranny of the game."[296] Although in essence a true observation (and one that in a modified version will be useful for describing the player's involvement in dystopia[297]),

287 Aarseth, "Implied," 130.

288 Ibid., 130; cf. 132.

289 Salen and Zimmerman, *Rules of Play*, 310.

290 Ibid.

291 Aarseth, "Implied," 133.

292 Ibid., 132.

293 Ibid., 133.

294 Ibid., 130.

295 Ibid.

296 Ibid., 132.

297 This perspective (though too narrow for games in general) is extremely beneficial when it comes to describing the VGD and the constricting rule system of these games. Their fictional societies are in disarray and show confining processes from which there seems to be no escape. Consequently, the player assumes *a transgressive role* in that

Aarseth's conception of the implied player remains narrow when compared to Iser's original phenomenon. Of course, it is possible to conceive of the implied player as a framework for play that outlines the player's ergodic participation in the game—but this is only part of it. Although Aarseth recognises the potential for a broader concept in that he ascribes a vital part in the creation of "ludic meaning"[298] to the implied player, this aspect remains underdeveloped. Indeed, one only needs to think about the gameworld as a system of props (or perspectives) that, besides obvious ludic functions, guides the player's imagination—or other aspects, for example: "characterization, themes, or even expressly narratological elements such as point of view,"[299] which David Ciccoricco also ascribes to the implied player.

Consequently, it is more than adequate to describe the implied player as "the game's formal structure" "for the player to inhabit," and as "the standpoint the game establishes for the player as an individual outside the gameworld."[300] Yet there are also alternative ways of setting things up that do not consider this structure as confining as Aarseth (and to a degree Vella) describes it to be. For the *implied player offers an intersubjective and (potentially) multi-layered framework of play that enables the empirical player to subversively engage in its structure and in a fruitful dialectic—delighting in the elegance of the form, but, at the same time, negotiating its contents and exposing it to meticulous scrutiny through play. It can thus be seen as the affordance and appeal structure of the game that offers the player various roles to perform and functions as a road map to catharsis and the aesthetic effect (harbouring all necessary predispositions).* As such, and to further explore this conceptual framework of play, it is beneficial to refer back to Iser's original notion of the implied reader, which is composed of an *entire system of perspectives.* This intricate structure offers the reader a participatory framework that grants her access to a work world and guides her involvement in it—an interaction that will eventually result in the creation of the aesthetic object.

---

she tries to escape the prison of her situation and to actualise the utopian horizon hidden within the system. But this role is already *inscribed into the implied player* (it is part of the VGD's strategies) and thus differs from Aarseth's notion of *transgressive play*.

298 Ibid., 131.

299 David Ciccoricco, "'Play, Memory': *Shadow of the Colossus* and Cognitive Workouts," *Dichtung-Digital: Journal für Kunst und Kultur digitaler Medien* (2007), http://dichtung-digital.de/2007/Ciccoricco/ciccoricco.htm

300 Vella, "Ludic Subject," 24.

4.4 THE GAME(WORLD) AS A SYSTEM OF PERSPECTIVES

To better understand this important aspect—and to describe the perspectival system of a game that comprises the implied player—consider again JOURNEY, whose virtual desert offers the empirical player *a semi-open framework for play and an indeterminate space for creative expression and interpretation—which is guided by structure but negotiated by the player*.[301] Stepping into the virtualised storyworld (gameworld), the player enjoys the liberty to inscribe herself into it and to express who she is through play. This means to play in a specific manner by trying out certain roles and playing styles: to travel with companions or not, to be gentle to them, or ignore them, and to imaginatively link the enacted to her own life experience. By doing so, the player has already entered a *creative dialectic* with the game's implied player and allowed it to exercise its effect on her—an experience that may have a lasting influence on the player's habitual dispositions and self-awareness by rearranging her stock of knowledge. This aesthetic effect is hard to explain, and one has to recall Iser's observation that as soon as one tries to define it—to explain why a work of art moves the appreciator in a particular way—it has already eluded one's grasp.

Such a conclusion is frustrating, if one strives to explain the reasons behind the creation of meaning. However, there might be an approximation of this issue. In order to approach the preconditions of the aesthetic effect, two interlocking aspects need to be addressed: 1) the *structural peculiarities* of the game framework (the implied player), which allows the player's participation in a work world (this will be conducted in the following by detailing the perspectival system of a game) and 2) the *interactions* between this framework and the empirical player (an aspect I will postpone to chapter V, which furthermore links my deliberations to the genre of dystopia). The moment the empirical player commences the act of play, she enters a creative dialectic with the intersubjective framework of the implied player, whose roles she interprets, performs, and scrutinises—and it is through this playful trial action that she will experience the beauty of the aesthetic effect.

Consequently, in order to analyse the game structure of JOURNEY, one has to take a closer look at its gameworld (and generally at that of any other VGN), which is framed as *a whole system of signs and perspectives.* These perspectives are

---

301 Fahlenbrach and Schröter similarly regard JOURNEY's gameworld in terms of a *structural openness* that revolves around its implicit backstory, the potential of interaction with the gameworld, and the emergent social interaction between two players. (Fahlenbrach and Schröter, "Rezeptionsästhetik," 201).

composed of both structural elements that formulate the game's discourse as well as gameworld items that oscillate in function between *virtual game objects* with which the player can interact and *props* that evoke specific imaginings—to expand on Klevjer's distinction between *dynamic reflexive props* and *perceptual props*. In other words, the perspectives of the game not only afford the player's ergodic involvement—by offering her the possibility of exploring the gameworld and acting within its bounds—but also, as Domsch claims, guide her understanding of the abstract gamespace, its objects and rules through imagining a world.[302] Hence it follows that becoming involved in a game, both the player's ergodic and imaginative faculties are at work (and, of course, her psychological/affective ones). As such, it is only through these combined efforts and by inserting herself into the game—her feelings, attitudes, and stock of knowledge—that the player may experience play's aesthetic effect.

Thereby, the term perspective is used in a *specific way* and should be understood similarly to what John Sharp has called "the rhetorical perspective[s] embedded in a game's design."[303] Likewise, in the Iserian sense, "perspective … implies a channelled view (*from the standpoint of* narrator, characters, etc.)"[304]—that is to say, "the different views and patterns" through which "the reader passes"[305] and that the player helps create.

Hence, the moment the player sets foot in the estranged gameworld of JOURNEY, she encounters a panoply of these perspectives: the vast desert where her journey begins, its ruins and tombstones, the characters that inhabit it, and so on. These trigger imaginings of times long forgotten, which are complemented by the question of how the characters managed to survive in this world. In addition, the ever-present mountain looms pompously in the background. This is where the journey inevitably leads—and the mountain therefore holds a dual function: as a point of orientation for the player's navigation of the gameworld (leading her through its labyrinthine structures) and her imaginative interaction with it (triggering diverse imaginings in her because of its indeterminate nature).[306]

302 Domsch, *Storyplaying*, 19.

303 John Sharp, "Perspective," in *The Routledge Companion to Video Game Studies*, ed. Mark J. P. Wolf and Bernard Perron (New York: Routledge, 2014), 107.

304 Iser, *Act*, 113; emphasis added.

305 Ibid., 21.

306 In essence, the mountain can assume a host of meanings for the player. From being regarded as a final destination towards which life inevitably leads or a life goal one aims to surmount, to more creative interpretations such as the finding of happiness or

Being a video game, however, the player's experience of the gameworld does not halt at this initial level of *discovery* but is complement by that of *creation*. For, in contrast to non-ergodic forms of art, where the appreciator can also decipher a multitude of perspectives that constitute the storyworld and its discourse (its mode of transmission), the player's involvement in the world of JOURNEY extends beyond that. This is explained through a game's systemic nature that enables the simulation of a world in which every occurrence, process, or event creates a perspective on the gameworld, which includes the player's actions. Consequently, setting foot into the virtual worlds of the VGN, the player enters a space of disclosure and creation, of perceiving the gameworld and acting within it—or reacting to it. This experience is similar to how human beings generally perceive the real world (to recall Calleja's observations). Indeed, Iser mentions that he has derived his theory of aesthetic response from "a basic rule of human perception, as our views of the world are always of a perspective nature"[307]—which, nonetheless, is more intricate, because the reader needs to compose a secondary *gestalt* to establish the link between fictional and empirical world (chapter V).

Clarifying why these insights apply so well to VGNs and VGDs will be undertaken in the remainder of this study by describing a phenomenological experience in which the player perceives the gameworld as a *perspectival system of discovery and creation.*[308] The individual perspectives (or perspective segments) of the VGN, then, comprise:

1. **Sensorial perspective***:* *the player's sensorial (visual, auditory, haptic) perspective on the world of the game, which grants her access to it in the first place, but that is also guided by the movements of a virtual camera*.
2. **World perspective:** *the gameworld including its settings, objects, architecture, and labyrinthine structures; the sounds and music of this world; and the characters who inhabit it (what they say and do).*
3. **Plot perspective:** *the plot developments that are outlined according to a narrative framework which structures the gameworld in organising its 'loose' elements by giving them a purpose and role within the overarching plot.*

---

sexual pleasures, which culminate in the orgasmic experience of the mountain's upper regions in which the player ecstatically flies towards its peak.

307 Iser, *Act*, 38.

308 To view games as a perspectival system is not confined to VGNs. Games that do not fit in this genre feature several perspectives as well, though one has to subtract the plot perspective and, in rare cases, the world perspective (for instance, in very abstract games).

4. **System perspective:** *processes, playing styles, and player actions that are outlined by the game's dynamic system and rules. These structure the gameworld on a basic and profound level and grant the player agency within it.*

Consequently, whereas the first of these perspectives primarily, though not exclusively, resides in the player, the rest of them are used to *incorporate* the player into the world of the game by involving her in the act of play. In combination, they thus constitute the *general framework of the game*,[309] a dynamic structure that implies various types of players and playing styles and that enables the player's participation in a work world on both an ergodic and imaginative level. Being confronted with this *larger framework of the implied player*, the empirical player assumes a particular role that is both informed by who she is yet, at the same time, is guided by the structural finesse of the implied player and its system of perspectives.

To participate in a game, therefore, means to engage in a creative dialectic with its *work world*: to encounter a panoply of perspectives, to perceive them, to interact with them, and to scrutinise their appearance, but also—and this aspect is specific to ergodic media—to aid in their creation. Engaging on such a personal level with a form of representation means to experience a *game world* (the player's interaction with the game) that is more intimate than that of the non-ergodic media participant and potentially more subversive, because the player inserts part of herself into it. What is similar, though, to other forms of representations is that the creation of the *aesthetic object* (which is experienced in and through the act of play) depends on the player's acts of ideation and on the consequent convergence of the perspectives.

Before coming to this matter and the interaction of perspectives in chapter V, it is necessary to elucidate *the nature of the perspectives*. This starting point requires one to bear in mind the player's intimate involvement in a game that is due to the creation of a private *fifth perspective* that aims at a quasi-transcendental view point on the game. This *player perspective* oscillates between the poles of proximity and distance and offers the player a participatory yet reflective window into the world of the game—in other words, a dual position from which the player can imagine the gameworld and act within it and one from which she can observe

309 Markus Engelns devises a different categorisation of *narrative affordances* that direct the player: "Narrem, Historisierung, narrative Mitte, Konsequenz, Achse der Handlung, topischer Pool, Isotopien, Achse der Topoi, narrativer Ursprung." (Markus Engelns, *Spielen und Erzählen: Computerspiele und die Ebenen ihrer Realisierung* [Heidelberg: Wissenschaftlicher Verlag der Autoren Synchron Publishers, 2014], 393).

it from afar.[310] To explain this communication model for the act of play, the following scheme frames and gives an introduction to the forthcoming deliberations.

Figure 14: A Communication Model for Video Game Narratives

The communication model describes the tripartite dialectic between culture (world), player, and game. Thereby, the empirical player draws from her cultural knowledge to play a game and engages in a communication with the implied player. The dashed lines imply the fluidity of boundaries between game, player, and empirical world (similar to the fictional being part of the actual). The sensorial perspective lies at the threshold to a game, for it grants the player partial entrance into its world. The two-sided arrows describe the mutual communication between all parts of the model. This is to say, by becoming involved in a game, the player's cultural knowledge also experiences a restructuring as a result of play. Consequently, her values and habitual dispositions may change.

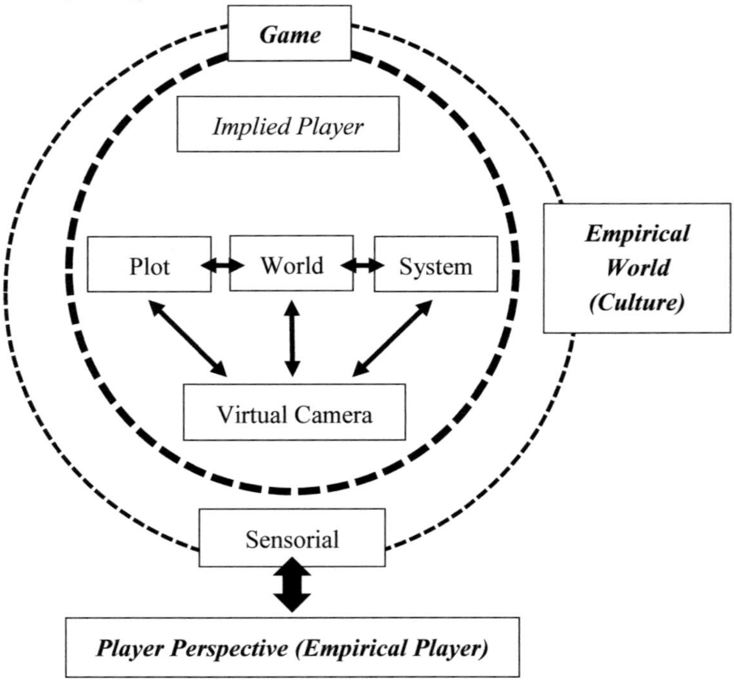

310 Neitzel would describe this as the oscillation between "*Point of Action*" and "*Point of View*" (Neitzel, "Medienrezeption," 100) and as a game of *proximity and distance* ("Nähe und Distanz"). (Ibid. 102).

4.4.1 The Player's Sensorial Perspective

To begin with the description of the player's sensorial perspective on a game—which includes her *visual*, *auditory*, and *haptic involvement* (through the controller or gamepad)—one has to recapitulate Vella's claim that the PC, or playable figure (as he describes it), can be regarded as an "*ontological duality* ... between *self* and *other* ... and the *phenomenological duality* ... that the figure is both a *subject*, with a perspective upon the gameworld that the player adopts, and an *object*, being itself perceived."[311] As such, the player's sensorial perspective is to a degree always filtered through her PC (if that is the case) and situated on the threshold between the game and the player's empirical world.

In this regard, Klevjer has observed that the avatar is, in fact, "a prosthetic extension of the body-in-the-world,"[312] and Keogh claims that during the act of play, the game and player "come together ... to form particular modes of embodiment through which a videogame work is both *interpreted* and *perceived*."[313] This entanglement leads to the player's augmentation through "the phenomenon of videogame play to perceive, sense, and ultimately feel a liminal presence within a virtual world"[314]—"touching the controller in their hands, looking at the moving images on the screen, listening to the music and sound from the speakers."[315] However, Keogh continues, "[a]s the player acts and engages with and makes choices about how to perform or enact the videogame, so too is the videogame already constricting, affording, and shaping the player's habit and perceptions in some way," which leads to a "irreducibility of player and videogame."[316] With this statement, Keogh lays emphasis on the importance of both the empirical and implied player in the act of play, who engage in a dialectical communication with one another.

This directs attention to the fact that the player's *sensorial perspective* cannot be exclusively constituted by the empirical being living outside of the diegesis—who engages with the gameworld—but also by the game's discursive strategies that outline this participation. Two of these are the PC, as a being living in the gameworld (which I will describe in the gameworld perspectives), and the player's *visual viewpoint* on the gameworld. A useful description of the latter is formulated

311 Vella, "Ludic Subject,"11.

312 Klevjer, "Avatar," 93.

313 Keogh, "Play of Bodies," 15; emphasis added.

314 Ibid., 17.

315 Ibid., 15.

316 Ibid., 28.

by Michael Nitsche, who describes a "virtual camera"[317] that is positioned in the game environment and through which the player *sees* the events of the gameworld, through which they *are conveyed* to him.

> It is the nature of a camera (virtual or real) to select, frame, and interpret. Through this selection, the moving image infuses the virtual world with a perspective. It narrates the space to the player … [and] constitutes a particular perspective that uses a specific range and features a genuine narrative force.[318]

The virtual camera is thus akin to Sharp's first type out of five perspectives that is concerned with the "means of constructing images with the illusion of dimensionality."[319] Yet it would be a mistake to reduce the virtual camera to an external discourse device only, and instead of exclusively conveying something to the player, its *performativity is shared* between the game's dynamics system and the player—for it is mostly she who is able to steer its movements, although in a more or less restricted form.[320] Consequently, a structural analysis of the gameworld implies a close observation of in-game objects and their perspectival arrangements (through both system and player) that will foreground certain aspects of the world (such as the mountain in JOURNEY) or create other visual illusions. In any case, the virtual camera offers the player a perceptual and participatory entrance into the gameworld, with the player enjoying the possibility to (figuratively) step beyond the *fourth wall*—which supposedly separates the realms of fiction from those of reality.[321]

4.4.2 Items of Setting and Characters

Right at the outset of JOURNEY, the player encounters a magnificent place whose excellence she thus experiences in a haptic (feeling the sticks, buttons, and rumble of the controller), auditory (hearing the sounds in the environment), and visual form (by moving both the PC and the virtual camera around the environment).

---

317 Nitsche, *Game Spaces*, 77. During the development of a game, the virtual camera (which is depicted through a camera icon in game engines such as Unity or Unreal) is placed into the gamespace by the designer, and the player's viewing angles can be determined through a variety of settings.

318 Ibid., 77.

319 Sharp, "Perspective," 107.

320 Nitsche, *Game Spaces*, 112-113.

321 Murray, *Hamlet*, 113ff.

However, it is not only this sensorial exercise that requires the player's attention (although it will always feed into her experience), for the gameworld she encounters holds many mysteries, and wandering through its spaces, the player not only tries to assemble its past and figure out the roles of its inhabitants,[322] but, most importantly, aims to decipher her own role within this process.

Consequently, being confronted with the gameworld as a confusing *network of indeterminacies*, the *existents* (the setting, its objects, and characters)[323] are amongst the most obvious perspectives of the game and illustrate guidelines to support the player's process of comprehension. These include apparent perspectives, such as *the game environment, its locations, signs, sounds, and architecture*; but also more obscure ones, for example: its *musical score* (diegetic or extradiegetic) or the *underlying labyrinthine (topological) structure* of the world—which in the case of JOURNEY follows a linear corridor that moves from level break to level break and includes larger areas for exploration and task fulfilment. In addition, there are the important perspectives of *in-game characters* (NPCs or additional human players) that inhabit the gameworld and what they say and do. These are sometimes (but not in JOURNEY) complemented by a reliable or unreliable *narrator*, be it a *homodiegetic one* (who belongs to the diegetic gameworld) or a *heterodiegetic one* (who does not belong to it)—a distinction originally proposed by Gérard Genette for the literary text.[324]

As such, these initial perspectives do not only constitute parts of the gameworld but, at the same time, they convey (discourse/narrate) something to the player and guide her participation in a decisive manner (in both ergodic and imaginative terms). I will explicate environmental storytelling techniques in chapter V, but for now I wish to direct attention to a fundamentally important perspective I have outlined before and which could also be ascribed to the sensorial perspectives. However, as the *player-character* belongs to the diegetic gameworld, his relation to the player is best discussed here. Confusion is hardwired, as the PC shares his point of view and other commonalities with the player herself, who takes control of him and not only executes most of his actions but, as a consequence for her agency, cognitively transforms into him.[325] This may result in *a*

322 Fahlenbrach and Schröter, "Rezeptionsästhetik," 195.

323 This distinction is informed by Chatman's observations on the ingredients of narrative. (Seymour Chatman, *Story and Discourse in Fiction and Film,* 2nd ed. [Ithaca: Cornell UP, 1980]).

324 Gérard Genette. *Die Erzählung. 2nd ed.* trans. Jane E. Lewin (Paderborn: Fink, 1998), 244-245.

325 Murray, *Hamlet*, 170.

conflation of PC and player perspective (through the player's sensorial involvement), yet there are also important differences that favour a separation, and for this reason a brief investigation into the issue of *focalisation* becomes necessary.

The concept of focalisation was first introduced by Genette who describes it as the point of view through which the reader gains access to a story, the perspective through which it is filtered to her. Genette thereby discerns three subtypes of focalisation: 1) the now rare "*zero focalization*,"[326] which grants the reader access to the storyworld through an omniscient point of view; 2) "*internal focalization*,"[327] in which the story is filtered through the perception of a) one specific character ("*fixed*"), b) of alternating characters ("*variable*"),[328] or c) of various characters ("*multiple*"),[329] each of which sheds light on the same event (or chain of events) from a different vantage point. Finally, there is 3) "*external focalization*,"[330] "in which the hero performs in front of us without our ever being allowed to know his thoughts or feelings."[331] Transferring Genette's observation to video games can be difficult, however, and requires medium-specific deliberation. In the following, I will thus further develop Allison Fraser's premise that there are primarily *three different aspects* that affect focalisation in games: the "audiovisual presentation, its selection and restriction of private knowledge, and its ludic affordances."[332]

To begin with, from a purely *visual standpoint,* the transition works surprisingly flawlessly—Sharp's "view perspective."[333] Games that move the virtual camera to a great distance from the action and allow the player to assume a 'god-like' view from above—such as TETRIS (Nintendo, 1989), BLACK & WHITE (Lionhead Studious, 2001), or strategy games that are depicted from an isometric point of view—could be compared to Genette's zero focalisation, because they offer the player a rather *omniscient* point of view on the events.[334] This perspective

326 Genette, *Die Erzählung*, 189.

327 Ibid.

328 Ibid.

329 Ibid., 190.

330 Ibid.

331 Ibid; cf. 189-190.

332 Allison Fraser, "Whose Mind is the Signal? Focalization in Video Game Narratives," *Proceedings of the 2015 DiGRA International Conference* 12 (2015): 1, http://www.digra.org/digital-library/publications/whose-mind-is-the-signal-focalization-in-video-game-narratives/

333 Sharp, "Perspective," 111.

334 Ibid.

Neitzel has called "the *objective* point of view."[335] Conversely, games that simulate the viewpoint of a character through a first-person perspective and "through the character's eyes"[336] are akin to Genette's internal focalisation, which offers the player an *internal* point of view of the events. Neitzel calls this "subjective perspective."[337] Lastly, there are games that depict the action from a third-person perspective, which creates an *external* point of view of the game events.[338] Neitzel calls this the "*semisubjective*" perspective.[339]

Problems of strict classification arise rather quickly, however, for example in games such as FALLOUT 3 (Bethesda Game Studios, 2008) where the player can switch between a visual first-person and third-person perspective. What complicates matters further is when analysing the problem from a literary standpoint as originally intended by Genette. Here, it is rather the *information conveyed through the fictional character* and the *established closeness or distance* to the reader that are of interest. Hence if the subject is to be analysed more thoroughly, one has to take into account the player's *relative proximity or distance* to the PC (her level of immersion and identification), which is not only based on the visual view perspective. This is important, since it prevents a premature conflation of PC and player perspective—which although sharing certain points, are to be separated nonetheless.

> The point of view of the player character is always different from that of the player, even in games where this difference is minimised as much as possible. A player, no matter how good the immersive experience, is always aware that he sits in front of a screen using some kind of interface device. The character always inhabits the game world.[340]

In order to explain the differences, let me first begin with the similarities. In many cases it would seem reasonable to suggest that games using a first-person perspective come close to a form of internal focalisation that literary fictions are not able to produce. For these games leave the visual perspective to the player as they set

---

335 Britta Neitzel, "Narrativity in Computer Games," in *Handbook of Computer Game Studies*, ed. Joost Raessens and Jeffrey Goldstein (Cambridge, Mass.: MIT Press, 2005), 238.

336 Sharp, "Perspective," 111.

337 Neitzel, "Narrativity," 238.

338 Sharp, "Perspective," 112.

339 Neitzel, "Narrativity," 238.

340 Steve Ince, *Writing for Video Games* (London: A & C Black, 2009), 65.

her into the PC's skull, thus enabling him to become "the lens through which we see [and act with] the [game]world, rather than through a narrator's recounting."[341]

> There is no distinction between what someone accomplishes as a character and as a player. While this is true of all styles of videogames, FPSs are the only genre where you and the character are 'virtually synonymous'. And in many ways, you aren't in the character's head so much as the character is you.[342]

This fusion of player and PC perspective is reinforced by Fraser's claim that a game increases the subjective feeling of internal focalisation through gameplay elements, for instance: *diegetic markers* in the game environment that help the player manoeuvre through it—such as the colour *red* in MIRRORS EDGE (DICE, 2008) that marks certain objects in the otherwise white gameworld. While obviously fulfilling a ludic function (navigation and pathfinding), this strategy succeeds in linking the player's consciousness to that of her PC, Faith, in that the former executes her abilities. For in the fictional world of the game, the ability to perceive effective ways to traverse the environment is explained through the so-called Runner Vision.[343]

Peak conflation between player and PC perspective is reached in first-person games with a silent-player-character—who, as the name suggests, refrains from talking and is often void of personality.[344] This creates an interesting situation for the player, for it is solely her perceptions, imaginings, and actions that constitute the PC perspective—for example in DEAR ESTHER or in HALF-LIFE 1 and 2. Still, as soon as these characters show hints of personality, the internal focalisation can never be complete. Gordon Freeman, for example, is a scientist and has a history that evokes the desire to role-play in the player.[345] Also, there are fully fledged personalities with many lines of dialogue such as Booker DeWitt in BIOSHOCK INFINITE or William J. Blazkowicz in WOLFENSTEIN: THE NEW ORDER. To talk in these cases of a complete conflation between PC and player perspective would be misleading, for several blanks arise between player and PC due to different opinions on gameworld aspects and so on.

---

341 Sharp, "Perspective," 109.

342 Lucien Soulban and Haris Orkin, "Writing for First-Person Shooters," in *Writing for Video Game Genres: From FPS to RPG*, ed. Wendy Despain (Boca Raton: A K Peters, 2009), 62.

343 Fraser, "Focalization," 7-9.

344 Soulban and Orkin, "Writing," 62.

345 Sharp, "Perspective," 113.

Similarly, it is not always easy to speak of external focalisation in the case of a third-person perspective. While it is true that in most of these games, the player enjoys an external perspective on her PC—from both a visual and informational standpoint (for example, in THE LAST OF US, the player does not know Joel and he rarely discloses his thoughts)—there are other instances, such as HEAVY RAIN (Quantic Dream, 2010), in which this is not the case. Here, three out of the four characters[346] give the player an insight into their feelings and thoughts by the use of *interior monologue*, which can be triggered by the push of a button. Yet this moves the focalisation decisively towards an internal one, since the interior monologue is one of the latter's most complete forms[347] (which is only trumped by stream-of-consciousness techniques).

Whereas such a state of affairs moves the focalisation away from an external one, this may also occur on a more general note, which can be explained through the player's relative proximity to her PC. Even in games with a third-person perspective, where players are able to control the virtual camera above and around the PC (and thus witness the gameworld from a slightly different vantage point than his), they are nonetheless "*viewing-with*" him, as their "point of view is connected to the movements of the avatar."[348] Such a closeness to the PC is further increased by player decisions that determine the PC's self—for example, choosing between dialogue options for the PC to utter.[349]

What is surprising, however, is that the contrary is also the case. Players role-play fictional personalities and are *only* able to actualise certain types of behaviour the game system allows them to perform and they are thus closely linked to the PC's self.[350] This is so because

> ludic affordances [that outline the actions the player may perform] … convey a great deal about the character's nature, goals, and mental models, as well as their abilities. In doing so, the player's own perspective and way of thinking is shaped according to what is required to operate the video game.[351]

346 Except for Scott Shelby, who is the Origami Killer—which the player does not know. Although Shelby gives the player insights into his thoughts, his true nature remains hidden until the game's later events.

347 Genette, *Die Erzählung*, 193.

348 Neitzel, "Narrativity," 238; cf. 238.

349 Fraser, "Focalization," 4.

350 Ibid., 11-12.

351 Ibid., (13-14)

Having these facets in mind, any attempts at external focalisation in games may be disrupted—and, on a more general note, they illustrate how difficult it can be to translate literary theory to the study of video games. For my present observations, this does not pose a huge problem, however, since I have embarked on this excursion to prove a certain point. Although showing overlaps, *the PC perspective cannot be equated entirely to that of the player* and the *relative closeness or distance* between the two depends not only on the visual viewpoint from which the gameworld is perceived. What is also of importance is both the roundness of the PC (the conflation being stronger with a *flat* PC and looser with a *round* PC) and how well the ludic affordances tie into the PC perspective. If done well, these have the player perform a certain role—types of behaviours and mental models—and offer her insight into the PC's self. What this does not mean is that the player has to agree with every kind of behaviour the PC affords (such as morally ambiguous situations in which the PC has to kill certain characters for the greater good or simply for the game to continue). Consequently, friction arises between the player[352] and PC perspective—between the former's beliefs and emotions and the perspective of a fictionally enacted character—a state of affairs that will give rise to the most interesting blanks to close (see chapter V and Part III).

Thus far I have scarcely scratched the surface of the gameworld perspectives, and in order to devote more attention to their diversity, I wish to address the *perspectives created by a game's topological structures* to then summarise and expand my findings in a table—which categorises the gameworld perspectives according to the player's potential interaction with them: *tending* either towards imaginative interaction or a mixture between imaginative and ergodic interaction. Before doing so, a vital way of outlining the player's participation in a gameworld can be done by employing certain *labyrinthine structures* that organise the gameworld and direct the player's movement within it. In general, there are "five clearly different topological structures."[353] These may occur in several combinations and variations.[354]

352 My notion of player perspective is similar to Sharp's (but does not exhaust itself in it), which is built up from "the way the player is represented [through her PC], what the player can do … ; the micro- and macro- goals assigned to the player," and the emotions she develops during play. (Sharp, "Perspective," 113; cf. 113-114).

353 Aarseth, "Narrative," 131.

354 Ibid., 131.

1. "[T]he linear corridor"[355] or unicursal labyrinth, "where there is only one path, winding and turning, usually toward a center."[356] This structure sets the player on a mostly linear route through physically confined spaces and occurs in primarily two forms of progression: a) in a purely "linear,"[357] where the game progression is structured according to consecutive levels that are interrupted by a pause screen, cutscene, or organic forms of these, like a blocked pathway (EA Montreal's 2010 ARMY OF TWO: THE 40TH DAY or THE LAST OF US); b) in a "continuous" form, where the unicursal labyrinth is not interrupted by level breaks and allows the player to revisit previous areas of the gameworld (BIOSHOCK, HALF-LIFE 2).[358]
2. "[T]he multicursal or hub-shaped labyrinth,"[359] "where the maze wanderer faces a series of critical choices, or bivia"[360] (THE STANLEY PARABLE, DEUS EX: HUMAN REVOLUTION). This labyrinth tends towards an open-world space but is still confined to smaller or bigger areas and mazes that add up to the gameworld. Due to its confusing structures, the multicursal labyrinth aggravates the player's pathfinding and can be employed in the form of maze-like areas to heighten the intensity of combat (by having the player struggle to find the maze's exit in hostile situations).[361]
3. Akin to the multicursal labyrinth but used for different purposes is the hubworld—if one "separate[s] hub from multicourse."[362] In this structure—which Boon refers to as "domain structure"[363]—the player's access to the gameworld is channelled with the help of a central hub that connects the individual parts of the gameworld (which may then employ any of the five labyrinthine structures available) (DEUS EX: HUMAN REVOLUTION).[364] Due to their centrality in

355 Ibid., 131.

356 Aarseth, *Cybertext*, 6.

357 Richard Boon, "Writing for Games," in *Game Writing: Narrative Skills for Videogames*, ed. Chris Bateman (Boston, Mass.: Charles River Media, 2007), 59.

358 Ibid., 59; cf. 59-60.

359 Aarseth, "Narrative," 131.

360 Aarseth, *Cybertext*, 6.

361 Another application of the multicursal labyrinth is the simulation of city spaces in a fairly realist fashion. These convey the sensation of being lost in a city and the enticing experience of navigating through its mazes (similar to a tourist's experience of Venice).

362 Aarseth, "Narrative," 131.

363 Boon, "Writing," 60.

364 Ibid.

the gameworld, hub spaces are regularly visited by the player and become familiar locations for her in which she can rest from the strains of travelling (Princess Peach's castle in Nintendo's 1998 SUPER MARIO 64 or the Normandy in Bioware's MASS EFFECT trilogy (2007, 2010, 2012), a spaceship that offers the player a base of operations with which she travels the galaxy). Hub spaces create the sensation of *being at home* in the gameworld and represent places of sanctuary.

4. Moving one step further towards gameworld realism is "the open world"[365] structure in which the player can almost freely roam the environment (Bethesda Softworks' FALLOUT 4 or the 2011 THE ELDER SCROLLS V: SKYRIM). This "contiguous structure attempts to create the illusion of a complete, explorable world" in that it virtualises a realistically open and interconnected space with various locations the player can visit.[366] Games that use an open world format are also referred to as *sandbox games*. These are said to create a malleable space that can be shaped by the player in many ways by offering her a great degree of freedom and creativity.[367] There may, however, be some restrictions to the player's movement within the open world such as mountain formations, larger multicursal (city) spaces, or confining indoor spaces integrated within the open world.
5. Lastly, there is the rare "one-room game"[368] (PAPERS, PLEASE), which, as the name suggests, is confined to a particular (small) space in which the action occurs. Because the rest of the gameworld is visually and physically inaccessible to the player, she has to imagine most of it.

---

365 Aarseth, "Narrative," 131.

366 Boon, "Writing," 60.

367 Ahmet Saad, "Writing for Sandbox Games," in *Writing for Video Game Genres: From FPS to RPG*, ed. Wendy Despain (Boca Raton: A K Peters, 2009), 137.

368 Aarseth, "Narrative," 131.

Table 5: Gameworld perspectives and elements of discourse[369]

Visuals, Objects, and Signs *imaginative interaction*	The visuals, objects, and signs of the gameworld invoke in the player the desire for what Chatman has called "*reading out.*"[370] This means extracting meanings from what is depicted and giving it causality by connecting its elements and interpreting genre conventions.[371] Consequently, the settings of a story (including any objects or signs) are an important discursive vehicle that reinforces the narrative, its characters, and the gameworld's depth.[372]
Topological Structures and Labyrinths *imaginative and ergodic interaction*	**Linear corridor:** Directs the player through a unicursal labyrinth towards a certain destination or goal. **Multicursal labyrinth:** Sets the player in a multi-branched maze through which she needs to manoeuvre. Such a structure is beneficial for combat encounters or simulating intricate cave structures or city spaces. **Hub world:** Creates a central home space for the player from which she can explore the gameworld. **Open world:** Creates a fairly realistic open space for the player to discover at her leisure. Open world games often give the player vast possibilities for interaction and creative play. **One-room:** confines the player to a single, small room.
In-Game Artefacts *imaginative interaction*	In-game artefacts are gameworld objects that contribute to the plot and give the player various information.[373] They include: textual writings (letters, diaries, notes, emails) or visual and auditory forms (answering machines, tape-re-

369 See Domsch for a similar segmentation and description of a VGN's discourse. (Domsch, *Storyplaying*, 31-47).

370 Chatman, *Story and Discourse*, 41.

371 Ibid., 36-42.

372 John Feil, "Writing for Action-Adventure Games," in *Writing for Video Game Genres: From FPS to RPG*, ed. Wendy Despain (Boca Raton: A K Peters, 2009), 31, 33.

373 Richard Dansky, "Introduction to Game Narrative," in *Game Writing: Narrative Skills for Videogames*, ed. Chris Bateman (Boston, Mass.: Charles River Media, 2007), 4.

	cordings, mobile phones, television and radio transmissions).[374] In-game artefacts are mostly optional for the player to find, collect, read, or view,[375] but they add to her understanding of the gameworld.
Music and sounds *imaginative interaction*	The sounds and noises of the gameworld make the player aware of certain events (the proximity of enemies, animals, or characters), while the musical score (intra- or extradiegetic) contributes to the emotional impact of the game by bestowing additional lustre to the gameworld. This may sometimes create perspectives on it: for example, by having songs or radio transmissions critically comment on gameworld events.
Characters and Dialogue *imaginative and ergodic interaction*	Characters (NPCs and PCs) are one of the most effective methods to convey story. "By what they say and do, you'll expose the beats of the story, reveal personalities of the characters, and unveil your plot."[376] The player may interact with characters by entering into a conversation with them, often via so-called dialogue trees. These give her a certain amount of freedom in that she may choose from several options to initiate a conversation or respond to NPCs.[377] Again, the player is confronted with two kind of perspectives: those she perceives from an external POV (the NPCs she is talking to) and those in which creation she helps (that of her PC), thus co-determining the PC's self from an internal view point.
Focalisation: Player and Player-Character *imaginative and ergodic interaction*	Focalisation in games depends on the player's relative proximity to or distance from her PC, which may vary according to three aspects: 1) the player's sensorial perspective on the gameworld through which she perceives and interacts with it; 2) the level of informational proximity to the PC, which varies according to the roundness of these characters and how much insight is allowed into them (their thoughts and feelings); 3) ludic affordances illustrate the

374 Boon, Writing," 49-50; Dansky, "Game Narrative," 4; Soulban and Orkin, "Writing," 54.

375 Feil, "Writing," 31.

376 Ibid., 32.

377 Domsch, *Storyplaying*, 38.

	PC's self and shape the player's role of performing him. Although sharing overlaps, the PC and player perspective are not to be confused, and blanks may emerge between them.
Narrators *imaginative and ergodic interaction*	Narrators that are not also NPCs are rare. Such narrators comment on the gameworld and its events (reliably or unreliably), and in some cases (Supergiant Games' 2011 BASTION; THE STANLEY PARABLE) they interact with the player by addressing her upfront or responding to her actions.[378]

4.4.3 Plot Developments and Narrative Framework

So far the investigation of the gameworld perspectives has revealed more obvious ones—such as the game environment, its objects or characters, which are quickly perceived and understood by the player—but also more obscure perspectives that shape the player experience and her access to the gameworld in a hidden fashion (the labyrinthine structures). Into these, one can also include the *developments of plot* that arise out of the interacting gameworld elements. These are framed in a specific manner for the dramatic action to occur, and I have pointed out in Part I the importance of such a *narrative framework* to the VGD (the clash between official narrative and counter-narrative that structures the game and its events).

Now, it is easy to see that any VGN is shaped according to a certain plot framework, and this framework adds considerably to the clarity of the gameworld events by giving them a focus and lending purpose to its existents[379]—be it in a looser form (Hello Games' 2016 NO MAN'S SKY, JOURNEY) or a stricter form (THE LAST OF US, BIOSHOCK INFINITE). For, as Nitsche argues, "[w]ithout narrative elements, a 3D video game would be in danger of disintegrating into singular unconnected splinters of momentary interaction."[380] This may be a blunt statement (and perhaps

378 See Froschauer for a specific contribution of narration in video games. (Adrian Froschauer, "Der Kampf um die Erzählhoheit: Voice-over-Narration im Computerspiel," in *Spielzeichen: Theorien, Analysen und Kontexte des zeitgenössischen Computerspiels*, ed. Martin Hennig and Hans Krah (Boizenburg: Werner Hülsbusch, 2016), 117-126).

379 Nitsche, *Game Spaces*, 64-65; Egenfeldt-Nielsen Simon, Jonas H. Smith, and Susana P. Tosca, *Understanding Video Games: The Essential Introduction*, 3rd ed. (London: Routledge, 2015), 201.

380 Nitsche, *Game Spaces*, 64-65.

not very accurate to describe all types of games), yet Nitsche has a point when it comes to the VGN. Consequently, only when such a narrative framework is given will I talk about a VGN—because otherwise, even the most basic worlds and abstract actions would qualify, such as that of TETRIS (Nintendo, 1989) or THE MARRIAGE (Rod Humble, 2006). This is a crucial aspect, for if the framework is integrated well, the individual parts of the gameworld are united and held together by a strong *theme* and a *trajectory* for the player to follow—for instance, in the open world game THE WITCHER 3: WILD HUNT (CD Projekt Red, 2015), whose gameworld is given causality through the theme of war and suffering.

Before going into further detail here, let me first clarify my use of the terms *story*, *discourse*, and *plot*, which together make up what is usually called *narrative*—some of which I have employed above without further discussing their application. These terms are often mixed up by scholars and the general public alike and, to avoid this mistake, they should be strictly separated from one another. Consequently, a first distinction can be made between *story* and *discourse*, with which I follow general narratological theory which claims that "[t]his analytically powerful distinction between story and its representation is, arguably, the founding insight of the field of narratology."[381] Similarly, Seymour Chatman delimits both terms as follows:

> Structuralist theory argues that each narrative has two parts: a story (*histoire*), the content or chain of events (actions, happening), plus what may be called existents (characters, items of setting); and a discourse (*discours*), that is, the expression, the means by which the content is communicated. In simple terms, the story is the *what* in a narrative that is depicted, discourse the *how*.[382]

When discussing VGNs (from both a game studies and game writing perspective), the distinction between story and discourse seems to withstand the transfer to this medium, although one has to make adjustments to what these terms refer to. For this purpose, consider narrative designer Richard Dansky's claim that in video games "story is a launching point for the narrative, not the narrative in toto."[383] This statement rest on the assumption that "story is what happens, the flow of the game that can be separated from the game mechanics and retold as a narrative,"[384]

381 H. Porter Abbott, "Story, Plot, and Narration," in *The Cambridge Companion to Narrative*, ed. David Herman (Cambridge: Cambridge UP, 2007), 40.

382 Chatman, *Story and Discourse*, 19.

383 Dansky, "Game Narrative," 2.

384 Ibid., 2.

whereas he defines "narrative as the methods by which the story materials are communicated to the audience."[385] These are most diverse and (as previously explained) include "cut scenes, characters, dialogue, and more."[386] Clearly, Dansky interchanges the term narrative and discourse here, and for reasons of clarity, I will stick to the latter term when it comes to how a story is conveyed. What is interesting, though, is that he refers to some *basic building blocks* (launching points) out of which a narrative may be constructed.

This observation chimes with those of narrative and game studies scholars who discern *story* as basically a *virtual* construct—thus implicitly aligning the concept of fiction with that of virtuality. In this line of thinking, *narrative* "is the textual actualisation of story, while story is narrative in virtual form"[387] and, therefore, only exists at some "abstract level … quite separate from any particular kind of manifestation."[388] If this is so, narrative is automatically decoupled from any kind of medium, for story is then conceived as "a mental image, a cognitive construct that concerns certain types of entities and relations between these entities"[389]—and this construct may be discoursed (or actualised) by a variety of media[390] (such as film, theatre, poetry, literature, music) or plain language itself.[391] Consequently, from this post-structuralist perspective, which embraces "the complexity of narrative across modes, media, and genres," "games can be studies from a narrative standpoint by examining how they renew, complicate, or transform our understanding of what a narrative is, and of how narration can operate."[392] These are vital questions when it comes to the VGN, and to answer them one has to first

385 Ibid., 1.

386 Ibid., 2.

387 Ryan, *Avatars*, 7.

388 Chatman, *Story and Discourse*, 44.

389 Ryan, *Avatars*, 7.

390 The concept of narrative I am using can be seen as a *broad approach* to narrative—which includes video games. Such an approach includes "anything that 'tells a story'" and assumes "that narrative is primarily characterised by the representation of noteworthy events." (Birgit Neumann and Ansgar Nünning, *An Introduction to the Study of Narrative Fiction* [Stuttgart: Klett Lernen und Wissen, 2008], 11). As such, it stands in contrast to *narrow approaches* to narrative which restrict the concept "to verbally narrated texts" and require the telling of a narrator. Ibid. From this point of view, many video games (if not all) would be excluded.

391 Ryan, *Avatars*, xx; Neumann and Nünning, *An Introduction*, 8-9.

392 Arsenault, "Narratology," 477.

observe the abstract elements of story, "the raw material"[393] out of which a (video game) narrative is composed.

In this respect, Aarseth has discerned some "common denominators"[394] between games and stories. These include "*four independent, ontic dimensions: WORLD, OBJECTS, AGENTS, and EVENTS*. Every game (and every story) contains these four elements, but they configure them differently."[395] These elements, to which Chatman has referred as *existents* and *events*, constitute the basic material of how narratives (in games) can be constructed and serve well as a starting point. However, the existence of such elements in a game does not necessarily mean it can be included in the genre of the VGN. Consider, for example, DR. MARIO (Nintendo, 1990), a puzzle game that features some rudimentary world and characters that allude to the bigger Mario-universe, or MINECRAFT (Mojang, 2009), which lets the player construct an entire gameworld in the form of a Lego building set. Aarseth rightfully describes MINECRAFT as a pure game and not a story[396]—and what both examples lack is a *unifying plot framework* that artfully outlines the player's participation in the resulting plot.[397] It is as Dominic Arsenault holds: what is of importance are not so much the *extrinsic* elements of a story (its basic building blocks) but rather "the means by which they are put into play by the unique properties of the video game."[398] Consequently, it is only through an analysis of these *intrinsic* narratological methods, which include both game system and player —"when the algorithm ... orders the events and relays the effects of

---

393 Ryan, *Avatars*, 7.

394 Aarseth, "Narrative," 130.

395 Ibid. Ryan similarly argues that "computer games present all the basic ingredients of narrative: characters, events, setting, and trajectories leading from a beginning state to an end state." (Ryan, *Avatars*, 182).

396 Aarseth, "Narrative," 132.

397 Indeed, one could argue that anything that includes some rudimentary world, characters, and events (actions); and that is represented in the one or other way in some medium could be called a narrative. (Abbott, *Narrative*, 19, 23; Marie-Laure Ryan, "Towards a Definition of Narrative," in *The Cambridge Companion to Narrative*, ed. David Herman [Cambridge: Cambridge UP, 2007], 23-24, 28-31). But, then, almost anything (and almost any virtual world) would classify as such, and the concept runs the risk of losing its validity. Consequently, I will restrict the VGN to those games that besides showing the above-mentioned elements, feature a *plot framework* that structures the gameworld and the player's involvement in it in a dramatic form—however loose and rudimentary this framework may be.

398 Arsenault, "Narratology," 479.

actions and current state of the fictional world through visual semiotics"—that one may understand the specific nature of the VGN.[399]

In order to integrate "play within a narrative and fictional framework,"[400] then—which strings together the loose story elements and endows them with potential artful arrangements—additional aspects have to be considered. For this reason, consider Dansky's statement:

> In many ways, creating a game story is about creating opportunities and effects. The opportunities are for gameplay, moments in the story where the player takes heroic control and succeeds in action. The effects are chiefly those experienced by the player: moments of emotional intensity. The story, then, must be created with more than its artistic component in mind. It also needs to serve as a framework for gameplay to be hung upon, and a road map to reward and catharsis.[401]

What Dansky describes, in other words, is a *malleable framework* that not only contributes to structuring the game's progression with the help of a "*story arc*,"[402] which paces the action and leads to some sort of "climax" (14), but also the player's involvement in it.[403] As a result, the player participates in this plot framework and experiences both the *effects of discourse* as well as showing the ability *to act* within the bounds of the framework.[404] This leads to the conclusion that story in VGNs is actualised/conveyed/discoursed through both the game's discursive strategies (some of which I have detailed above) as well as the player's decisions and actions, which are also to be viewed in terms of discourse.[405] So, whereas in literature story is primarily actualised and conveyed through a narrator's descriptions and the arrangement of events (this is what gives the story its substance and form), video games explore this aspect in a radically different way. Here, the *discourse of the story* is divided between game and player[406]—it is co-discoursed between a dynamic work world and the player's game world—and both constitute a collaboration which actualises and conveys the virtual construct of the story (in

399 Ibid., 482; cf. 479-482.

400 Ryan, *Avatars*, 182.

401 Dansky, "Game Narrative," 13; emphasis added.

402 Ibid., 13.

403 Ibid., 13-14.

404 Calleja similarly distinguishes between "*scripted narrative*" and "*alterbiography*." (Calleja, *In-Game*, 115) (see chapter V).

405 Aarseth, *Cybertext*, 5; Klevjer, "Avatar," 44.

406 Nitsche, *Game Spaces*, 54-56.

other words, they decide *what* happens and *how* it happens). The result of this interaction is not only a specific effect on the player—an often cathartic experience previously described as the aesthetic effect—but also the creation of a certain *plot* that was guided by the framework Dansky describes.

As a consequence, the second important distinction to be made is the one between story and plot. Whereas *story* referred to *the mere agglomeration of existents and potential events* existing in virtual form (in JOURNEY these included the derelict but beautiful gameworld, its characters, and the panoply of potential events that may occur), *plot* exists at a more pragmatic level.[407] Here, the individual building blocks of story have become actualised and artfully arranged into a coherent whole by the discourse, and this adds *causality* to them.[408] The argument thereby runs as follows: whereas all stories move from an initial state towards an end state (in JOURNEY, for example, the story begins long before the player enters the gameworld, with a proud population that has now vanished into oblivion), the selective work of the discourse fashions the bare story elements into a graspable whole, which in the Aristotelian sense includes "a beginning, middle, and end"[409] (in JOURNEY these would be the playable parts of the game, beginning in the desert and ending on the mountaintop). In this sense, plot designates "a type of story"[410] that can depart from the chronological order of the story in that it chooses to depict (or have the player enact) specific events while leaving out others (which happens frequently in Campo Santo's 2016 FIREWATCH or Naughty Dog's 2009 UNCHARTED 2: AMONG THIEVES) as well as reorganising the temporal structure of the story, for instance, the plot may begin at the story's end and move towards its beginning, or in other fashions[411] (a drastic rearrangement of the story chronology occurs in Quantic Dream's 2013 BEYOND: TWO SOULS, in which Jodie's important life events are played in an almost random order).

Now, while in some cases the arrangement of story events into a plot may undermine and confuse the reader's understanding of it (creating suspense in detective fiction, or confusing the reader in fragmentary postmodern fiction), the structuring usually contributes to clarity of the work. This is so because a specific arrangement of the events endows the separate story elements with causality, and this can be strengthened by explicit or implicit elaborations. Chatman illustrates this point through E. M. Forster's example: "'The king died and then the queen

407 Abbott, *Narrative*, 19.

408 Abbott, "Story," 43; Chatman, *Story and Discourse*, 43.

409 Abbott, "Story," 43.

410 Ibid.

411 Ibid.; Chatman, *Story and Discourse*, 43.

dies' is only a 'story' … [whereas] 'The king died and then the queen died of grief' is a 'plot'."[412] Whereas such explicitness helps the reader to grasp the causality of a story (in games this can be done through a variety of discursive features), Chatman does not fail to acknowledge the reader's involvement in the creation of plot. To him, human beings invariably seek meaning and structure, and they do so by closing the blanks between the elements of a story, thus adding causation to them. In other words, readers infer that the queen died out of grief, even if this is not explicitly stated.[413] This is an important insight, because it places emphasis on the appreciator's imaginative involvement in representations as *co-creators of plot* (a fact that I will elaborate in the following chapter when discussing what Iser calls the *process of synthesis*).

Considering these deliberations, a preliminary conclusion concerning the VGN's plot framework can be formulated. In structuring a game in a decisive manner, leading to closure and catharsis, the plot framework adds to the overall structure of the game by outlining an indeterminate but framed story space for the player to interact with on both an ergodic and imaginative level (to recall what Domsch has called the *architecture* of the game). In this virtual space, the building blocks of narrative lie dormant in a state of superposition awaiting the discourse to give them shape and a perceivable form—with discourse including both the game system and the player (out of which the system assumes the dominant role in the process). These two select, actualise, and arrange the game events (the *individual run*), and out of this interaction one of many plots is created (the resulting *protocol*).[414] It is as O. B. Hardison puts it: "Each arrangement [of the story events] produces a different plot, and a great many plots can be made from the same story."[415]

---

412 Ibid., 45.

413 Ibid., 45-46.

414 See Nitsche for a similar conception of how story is conveyed in VGNs, but who reduces the creation of plot to a "cognitive process," "the order and connections between events as understood by the reader" or player. (Nitsche, *Game Spaces*, 50; cf. 49-56). In my conception, *plot* is the result of an interaction process (the *discourse*) in which both the system and the player's ergodic and imaginative faculties are at work. Here, different plots stemming from the same virtualised *story* prototype may be created.

415 O. B. Jr. Hardison, "A Commentary on Aristotle's Poetics," in *Aristotle's Poetics: a Translation and Commentary for Students of Literature*, ed. O. B. Hardison (New Jersey: Prentice-Hall, 1968), 123.

To illustrate these points, let me again resort to JOURNEY, whose game structure is meticulously outlined with the help of Joseph Campbell's[416] famous plot framework of the Hero's Journey.[417] Many of its steps are also discernible in the game and primarily outline the player's *creation* and *comprehension* of plot. This is possible due to "[t]he flexibility of the Hero's Journey"[418] as a plot framework, as it "provides a scalable and adjustable matrix" which allows "a form of quest that comes to live in the player's comprehension and his or her interaction with the game space."[419] This flexibility is discernible in JOURNEY, and specifically in one of the framework's aspects I will come to later: the player's agency to find a virtual companion. Beginning with the player's *departure* (including the *mentor* she meets and the first *threshold* she passes), the road continues as the player faces a series of *obstacles* and *enemies*, but also encounters *friendly companions*. Having found her *death* in the torturous ascension of the snowy mountain, the player is *resurrected* and relishes the final ascent towards the mountaintop.

All these steps endow the journey with structure and causation. They formulate *milestones* the player will pass through and contribute to the formation of the plot, which is structured by the system and negotiated by the player. In Chatman's terms, these steps of a story are called *kernel events.*[420] They are of indispensable necessity to the logic of a story (that is, they make a certain story that story and not a different one) and cannot be deleted without severe consequences[421]—without Ganondorf capturing princess Zelda at the beginning of THE LEGEND OF ZELDA: OCARINA OF TIME (Nintendo, 1998), a completely different story would have been actualised and turned into a plot; without the PC's resurrection in

416 Joseph Campbell, *The Hero with a Thousand Faces. Bollingen Series XVII*, 3rd ed. (Novato, California: New World Library, 2008).

417 See Jacobs for a general application of the Hero's Journey to video games. (Stephen Jacobs, "The Basics of Narrative," in *Game Writing: Narrative Skills for Videogames*, ed. Chris Bateman [Boston, Mass.: Charles River Media, 2007], 28-30); and Fahlenbrach and Schröter who argue that JOURNEY is structured according to Campbell's monomyth. (Fahlenbrach and Schröter, "Rezeptionsästhetik, 190).

418 Nitsche, *Game Spaces*, 63.

419 Ibid., 64.

420 Aarseth suggests a similar segmentation of game events into *kernels* and *satellites*, remarking that "[t]hese two concepts, kernels and satellites, allow us to say something about the ways games can contain one or several potential stories." (Aarseth, "Narrative," 131; cf. 130-132).

421 Chatman, *Story and Discourse*, 32, 53-55.

JOURNEY, the game would come to an early end.[422] In JOURNEY, the linear succession of kernel events (the steps of the Hero's Journey) is facilitated through the game's structure of progression, which follows the trajectory of a unicursal labyrinth with level breaks between the different sections. Consequently, the player witnesses the game's major events in a mostly linear fashion, since "level order is fixed", [kernel] event order is also fixed, and story material may be planned exactly."[423]

What this does not mean, however, is that the player is robbed entirely of her freedom to explore and influence this world, and her agency primarily rests in actualising certain *satellite events*. These are, according to Chatman, of minor importance to a story (they may influence the plot but not the deeper story level) and can be omitted without disrupting its logic. In JOURNEY, these include exploring different areas of the gameworld in more detail, taking different routes through it, or savouring its vistas instead of rushing through the game, etc. Although not altering the main story in a vital fashion (what turns JOURNEY into a mainly *linear story*), the player's experience and creation of satellite events should not be underestimated, for they aesthetically enrich the resulting plot through diversity and precision. In other words, they formulate "the flesh on the skeleton" of the kernels.[424]

Besides these interactions on a relatively minor level, JOURNEY also allows the player to participate in the creation of one of its kernel events—which is where the most intriguing pleasures of participatory narratives lie. Connecting the PS4 console to the Internet, the player is able to share her journey with a fellow human being and to engage in the pleasures and dangers of this world cooperatively. This not only includes the choice of taking the route together, but also of performing various actions such as communicating via music, waiting for the additional player, or helping him out in dangerous parts of the gameworld—a fact that Fahlenbrach and Schröter attribute to the social dimension of the game, which affords a feeling of solidarity between the players.[425] As such, these potential ker-

422 However, as Abbott remarks, it is sometimes difficult to judge whether a certain event can be endowed with the status of a kernel, as the process is a subjective one. (Abbott, "Story," 41). Nonetheless, such a distinction can be of importance to VGNs, as kernels offer the player branching points that may lead the story in a different direction (a fact of essential importance to the critical dystopia variant II).

423 Boon, "Writing," 59.

424 Chatman, *Story and Discourse*, 54; cf. 32, 53-55.

425 Fahlenbrach and Schröter, "Rezeptionsästhetik, 198-200.

nel events represent a vital change not only to the game's plot but also to its underlying story, for the journey differs considerably (as does its interpretation) once the second player has joined (which I will illustrate in the following).[426]

As Chatman puts it: "Kernels are narrative moments that give rise to cruxes in the direction taken by events. They are nodes and hinges in the structure, branching points which force movement into one or two (or more) possible paths."[427] Consequently, the degree of the player's influence on (or creation of) kernel events contributes highly to the amount of *agency* she has in the gameworld: "the capacity to effect meaningful changes" in it, "or at least the illusion that the player has this capacity."[428] This power to influence the gameworld in a decisive manner is of course dependent on the gameworld's level of indeterminacy (which the player may fill in through ergodic action) and also on the specific plot structure employed. These may differ from one another,[429] and range from fairly *linear stories* (JOURNEY, THE LAST OF US) in which the player's potential to actualise a certain plot is restrained by the number of possibilities the virtual story space offers, to structures that allow for several endings in which a variety of branches may be actualised and turned into plot. These are *interactive stories* such as HEAVY RAIN and FALLOUT 4.

---

426 One could object at this point, for even though the second player joins the journey, its steps remain the same, and only the indeterminate space between them is filled differently (consequently reducing the interactions with the second player to satellite events). This points to the inherent difficulty of determining a kernel event (which remains subjective to a degree).

427 Chatman, *Story and Discourse*, 53.

428 Boon, "Writing," 63.

429 See Ryan or DeMarle for a description of interactive plot structures. (Marie-Laure Ryan, *Narrative as Virtual Reality 2: Revisiting Immersion and Interactivity in Literature and Electronic Media* [Baltimore: John Hopkins UP, 2015], 160-185; Mary DeMarle, "Nonlinear Game Narrative," in *Game Writing: Narrative Skills for Videogames*, ed. Chris Bateman [Boston, Mass.: Charles River Media, 2007], 71-84).

Table 6: The perspectives of the plot framework as elements of discourse

Plot framework and its various steps *imaginative and ergodic interaction*	The plot framework is the narrative skeleton of the gameworld that endows its elements with structure and causation and sends the player on a journey to closure and enlightenment. It does so by organising the gameworld through various discursive ways (see chapter V) and by the strategic distribution of kernel events the player will witness or actualise. Their distribution differs according to the game structure employed (a fairly linear distribution in case of unicursal labyrinth and a non-linear distribution in case of an open world). Either way, kernel events offer cruxes in the storyline and will eventually determine whether the player is dealing with a form of *interactive story* (kernel events offer more than one choice to the player) or *interactive plot* (only satellite events offer more than one choice).
Cutscenes *imaginative interaction*	Kernel events can be portrayed in different ways, and the cutscene is a traditional method of doing so. A cutscene is a movie clip that plays between sections of gameplay. Most often, important events are depicted through this storytelling technique.[430] This is because cutscenes deprive the player of agency and allow the game designers to take control over the action. They thus enjoy the possibility to structure the story in a careful manner without the player interfering, and reward her by offering cinematic visuals.[431] However, there are instances in which the choice between several options results in different cutscenes, which makes them slightly more interactive.
Quick-Time Events *imaginative and ergodic interaction*	QTEs are akin to cutscenes in that they are presented cinematically but differ from them in that they allow a minimal amount of player agency. They do so by interspersing the cinematic action with moments in which the player may intervene by the push of a button. This will either ensure the continuation of the action (if the button push occurred within a limited time span) or dramatically change

430 Dansky, "Game Narrative," 4.

431 Boon, "Writing," 54-55.

	the course of events, thus actualising a new event branch in the ongoing plot.[432]
Static Images *imaginative interaction*	Static images are akin to cutscenes in that they represent non-interactive pieces of visual information—such as paintings, drawings, or comic strips—that can be used to tell the story. They may also occur in dynamic form with the help of camera zooms or pans and are often given context by the words of a narrator.[433] In addition, static images can be found as objects that are organically integrated into the gameworld such as paintings, graffiti, or drawings.

4.4.4 Processes, Playing Styles, and Player Actions

In 1997, when Janet Murray elaborated on the *four essential properties of digital environments*—which are "procedural, participatory, spatial, and encyclopedic"[434]—she set a milestone for researchers. I have already addressed the participatory aspect of playing games (the player's ergodic and imaginative interaction) and touched on the remaining ones. Amongst these, the *procedural aspects* of video games and their underlying *rule system* have been fervently discussed by game scholars. They are often designated "the deep structure of a game from which all real-world instances of the game's play are derived"[435] and, thus, its "formal identity."[436]

Rules, in other words, structure the gameworld and its underlying system on a basic but profound level and are diverse in their respective areas of application. Although some researchers (Aarseth or Salen and Zimmerman) consider rules to be restrictions on the player's freedom to interact with the gameworld, there are others—such as Jesper Juul—who regard rules in a more creative way as both "*limitations* and *affordances*."[437] Such a view is especially interesting pertaining to the VGN and the semi-open framework of the implied player I am proposing. To elaborate on this aspect, Domsch's narrative-oriented classification of game rules becomes beneficial, for which he discerns basically two sorts: "rules that state the game's existents, and rules that define the valorisation of options and

432 Domsch, *Storyplaying*, 35-37.

433 Boon, "Writing," 53.

434 Murray, *Hamlet*, 71.

435 Salen and Zimmerman, *Rules of Play*, 120.

436 Ibid., 121.

437 Juul, *Half-Real*, 58.

outcomes."[438] Consequently, while the former type of rule is responsible for *determining aspects of the gamespace and its mechanisms*—the virtual space in which the game takes place (its size and variations), the agents to be encountered (NPCs and their behaviours), and the range of potential player actions and their consequences—the latter type describes *the values at play in the gameworld.* These also determine the player's goals and objectives and specify which options or paths in the gameworld are considered desirable.[439]

Now, what becomes of interest when analysing a particular VGN or VGD are the choices that create *frictions* between the structuralist level of the game system (system of rules) and that of the gameworld's semiotic aspects (system of props)—thus playing with the values at hand. This may happen when a player decides to take a choice simply for the game to proceed (or to succeed in the game at any cost) without having in mind the consequences for the storyworld (blowing up a door that will kill NPCs instead of taking the longer route to avoid this confrontation). Acting according to such "gameplay rationality"[440] is more common than Domsch suggests (the gamist player-type, for example, for whom the pleasures of *ludus* stand in the foreground). But the literary theorist has a point when he claims that players (at least the narrative-oriented type) inevitably endow their choices with meaning and significance—especially in the midst of a fictional storyworld.[441] This is also why game *processes* in themselves (which are a direct result of the game's algorithm and rules system) are meaningless when not aligned with the remaining perspectives of the gameworld.[442] In this case, however, they contribute to the significance of a game and the player's experience of meaning—which brings me to the following aspect.

---

438 Domsch, *Storyplaying*, 53. A distinction Domsch has developed based on Juul's observations of game rules. Juul, *Half-Real*, 55-120.

439 Domsch, *Storyplaying*, 16, 53, 61, 68, 150-151.

440 Ibid., 124.

441 Ibid., 124-125.

442 Aarseth compares two similar yet fundamentally distinct games: THE HOWARD DEAN FOR IOWA GAME (Frasca and Bogost, 2003) and KABOOM: THE SUICIDE BOMBING GAME (Fabulous999, 2002). While both games are identical from a game mechanics perspective—the player's goal is to bring as many people as possible into a certain area of the gamespace to gain points—the representational aspects differ in that in DEAN FOR IOWA, semiotic aspects of the gameworld visualise an electoral campaign in which the player's goal is to gather followers, while KABOOM revolves around the machinations of terrorism. (Aarseth, "Ontology," 489-490).

For Ian Bogost, the creation of meaning in video games (and, most specifically, the subgenre of persuasive games) is a result of what he calls *procedural rhetoric*, the "practice of using processes persuasively."[443]

> *Procedurality* refers to a way of creating, explaining, or understanding processes. And processes define the way things work: the methods, techniques, and logics that drive the operation of systems, from mechanical systems like engines to organizational systems like high schools or conceptual systems like religious faith. *Rhetoric* refers to effective and persuasive expression.[444]

This "art of persuasion through rule-based representations and interactions" can thus be seen as the predominant locus for the creation of meaning in games, "rather than the spoken word, writing, images, or moving pictures."[445] To explain his claims, Bogost resorts to examples such as THE MCDONALDS VIDEOGAME (Molleindustria, 2006). The game mounts "a procedural argument about the inherent problems in the fast food industry"[446] by involving the player in the inner mechanism that fuel it and having her control four aspects: "the third-world pasture … the slaughterhouse … the restaurant … and the corporate offices."[447] During the process of play, each "unit-operation" further elaborates how the grander system of McDonald's operates, and through a combination of these processes, the player comes to see beyond the capitalist strategy of the company.[448]

Bogost thus ascribes the authority in the meaning-making process in games (and thus a normative role) to the processes that occur during play. Such an argument runs contrary to many observations on representational art (see the entire previous discussion) and also meets strenuous opposition in game studies from some scholars.[449] For processes may only hold meaning if aligned with the remaining perspectives of the game(world), and to neglect these would be a fatal act considering their diversity. Yet if regarded in their context, the gameworld processes (and unit-operations) can provide a vital perspective for the player's acts of ideation. This aspect is of importance to the VGD, since there the player engages in a

443 Bogost, *Persuasive*, 3.

444 Ibid., 2-3.

445 Ibid., ix.

446 Ibid., 31.

447 Ibid., 29.

448 Ibid., 36; cf. 31, 36.

449 Miguel Sicart, "Against Procedurality," *Game Studies* 11, no. 3 (December 2011), http://gamestudies.org/1103/articles/sicart_ap

confining system of rules and resulting processes which she seeks to disrupt (see chapter V).

For now, however, it is important to state that a *game system,* its *code,*[450] *rules* and *mechanics* assume vital perspectives in the player's participation process.[451] They constitute a basic layer of perspectives in a VGN (and games in general), and can be located on the same level as the virtual story construct (to recapitulate Ryan's assertion that games are machines for generating stories) awaiting actualisation by the player, who through her actions breathes life into them. These ground-layered perspectives give rise to more pragmatic ones, which the player can perceive on a visual level: *the results of processes* and *gameworld events* and particular *playing styles* or *player actions.* All of them contribute to the formation of plot and create fictional truths[452] in the Waltonian sense. To further elaborate on this aspect, the discussion leads to the player's *agency* and her diverse interactions within the gameworld.

The issue of agency is heatedly discussed in video game studies, and for my current observations, those theories that align agency with the player's participation in a fictional storyworld become of prime interest. In this context, most are familiar with Janet Murray's definition of *agency* as "the satisfying power to take meaningful action and see the results of our decisions and choices."[453] This stands in strong contrast to pure *interaction*, which Murray reduces to the mere ability to move a joystick or to press a button. Player actions based on agency are thus *intentional* and of specific interest within a "narrative environment."[454] Of these, the

---

450 Ea Willumsen goes as far as to observe code itself, which can be regarded as a commentary (and perspective) inscribed into the game by the game designers (or programmers). (Ea Christina Willumsen, "Source Code and Formal Analysis: A Hermeneutic Reading of *Passage*" *Proceedings of the First International Joint Conference of DiGRA and FGD* 13, no. 1 [2016], http://www.digra.org/digital-library/publications/source-code-and-formal-analysis-a-hermeneutic-reading-of-passage/).

451 Sharp, "Perspective," 114-115.

452 Tavinor argues that "this account of fiction also means that the activities the player carries out in the game world, activities that constitute gameplay, are fictional." (Tavinor, "Fiction," 437). These actions are conducted through "the player's *fictional proxy* [her PC] in the game world." (Tavinor, *Art of*, 70).

453 Murray, *Hamlet*, 126.

454 Ibid.; cf. 128.

"constructivist pleasure is the highest form of narrative agency the medium allows"[455]—that is to say, the pleasures of filling in the gameworld's indeterminacies, of building objects within it, and of creating plot and altering story through expressive action.

Given these early observations, it is no coincidence that Murray's theory has often been taken up when discussing games in a narrative context. Karen and Joshua Tanenbaum, for instance, argue that *narrative agency* is experienced by players when their actions are in accordance with the internal logics of the plot—that is, make sense within the framework of a fictional world. Here, the "player is less concerned with limitless – but meaningless – freedom, and is instead interested in some systematic reification of the meanings which she is performing as an inhabitant of this world."[456] Agency in their sense is clearly *bounded* and adheres to the internal logics of a fictional storyworld. Another convincing theory in this respect is offered by Noah Wardrip-Fruin et al., where agency is connected to both the "player and game" and occurs "when the actions players desire are among those they can take (and vice versa) as supported by an underlying computational model."[457] In this regard, the gameworld and its imaginative-evocative qualities as a *prop* (or an agglomeration of props)[458] becomes strikingly important. For it prescribes imaginings about that world, which, in turn, will influence the player's desire to act within its bounds. Wardrip-Fruin et. al.'s notion of agency thus works in alignment with the internal logics of a fictional world and its plot structure.[459]

To create the phenomenon of player agency in relation to a fictional world it is necessary to suggest dramatically probable events, make material affordances available for taking those actions, and provide underlying system support for both the interpretation of those actions

455 Ibid., 149.

456 Tanenbaum and Tanenbaum, "Commitment to Meaning."

457 Noah Wardrip-Fruin, Michael Mateas, Steven Dow, and Serdar Sali, "Agency Reconsidered," *Proceedings of the 2009 DiGRA International Conference: Breaking New Ground: Innovation in Games, Play, Practice and Theory* 5 (2009): 1, http://www.digra.org/digital-library/publications/agency-reconsidered/

458 Klevjer builds on Walton's observations on props and distinguishes between "'world props' and other props" (Klevjer, "Avatar," 28)—such as "complex props – like computer games – that are, in a sense, both like tapestries and statues at the same time" and which are made up of several individual elements. (Ibid., 29).

459 Wardrip-Fruin et. al., "Agency Reconsidered," 1, 3, 4, 7.

and the perceivable system response to those actions … In other words, agency requires the construction of a playable software model of the domain of the fictional world.[460]

What follows from this model is that it attunes the rule framework (which affords the gameworld processes and player actions) to that of the world and plot framework. This is not necessarily the case in all video games or VGNs, but if done successfully, it results in the creation of participatory narratives that captivate players. Clarifying this collaboration of working forces (specifically for the VGD) will be attempted in the remainder of this study. By doing so, I will explore how the perspectival system of the game collaborates as *one framework*, consisting of a system of props[461] that prescribes imaginings about a fictional storyworld and a system of rules that affords processes and player actions within a virtual gamespace, the results of which also function as props. Before coming to this aspect and to how the perspectives coalesce in the player's act of ideation, there is one last issue that requires clarification.

Thus far, I have aligned video game theory with that of fiction and narratology, which has led to the description of the necessary requirements of the VGN. However, even the most reasoned argumentations are bound to run into minor bumps eventually, and there is one aspect I would like to address now, to then come to a nonetheless positive conclusion about the genre. In the beginning of this chapter, I posed the question of whether players of VGNs *play according to the rules and integrity of a fictional storyworld*—and the answer is both *yes* and *no*. This is because player types and preferences vary considerably, and not all of them are willing to play according to the function of a particular VGN—which directs attention to the multiple ways in which games can be played. None of this is surprising given the diversity of the medium, yet clarifying the boundaries of the VGN as a genre (and those of the VGD) is necessary.

For this purpose, let me consult Domsch's distinction between player choices that only affect the *game state* (the properties of game existents and the relations between them in a certain moment of play) and those that additionally have an influence on the *state of the fictional storyworld*. Only the latter choices can also be called "narrative" or "semantic choices."[462] As such, narrative choices should have "consequences on the internal development of a game's storyworld"—such

460 Ibid., 4.

461 These *props* can include "the graphical, auditory, and haptic elements of a video game display" (Tavinor, "Fiction," 438)—and I will use them as *perspective segments* in the Iserian sense.

462 Domsch, *Storyplaying*, 127.

as the choice for or against Megaton in FALLOUT *3*—instead of solely on its "external shape"—secondary aspects such as "the type of landscape, or choosing whether a protagonist is male or female."[463] These are of importance to a player's sense of *narrativity*—"the quality of being ... narrative, the set of properties characterising narratives"[464]—and involve her in the role of one or more PCs whose actions contribute to "the gameworld's narrative development."[465]

Domsch's observations, however, scarcely dent the problem, for how can one distinguish between these two types of choices and how can minor player acts, such as killing a random monster or picking up an item, be classified? To clarify the diversity of this matter, consider WATCH_DOGS 2 (Ubisoft Montréal, 2016) (or similar open world games) in which the player can engage in activities offered in virtual San Francisco besides following the main story line. These include, for example, dressing in different fashions, playing a tourist (a player type who wanders the gameworld and takes selfies or pictures of the environment), going on a sailing trip, or coming up with creative tasks such as swimming from Alcatraz to the shore, thereby role-playing an escapee. Many of these do not or barely have an effect on the status of the fictional storyworld or the main story, yet they constitute inherent aspects of playing this game.

To solve the issue, let me refer back to Walton's distinction between those games that organically work within the bounds of a storyworld (*authorised games* that are in accordance with the work's function) and those that are a misuse of it (*unauthorised* or *transgressive games* that disobey the rules/integrity of a storyworld). Such a distinction is more *inclusive* than the one Domsch proposes, and what becomes of interest is not so much whether certain activities influence the status of the storyworld, but whether or not they are *in accordance with its function and themes*. In this sense, even the supposedly random activities described above are both *virtually true*—they affect the game state—and *fictionally true*—they work within the bounds of the storyworld's function; in the case of WATCH_DOGS 2, they respect the game's theme as technocratic dystopia and the utopian enclave of an expressive lifestyle led by the main characters, which aims to disrupt the confines of a system in which every niche is controlled.[466] Such a dynamic fictional framework has the benefit of including a variety of actions (satellites) into

---

463 Ibid., 128.

464 Gerald Prince, "Narrativity," in *Routledge Encyclopedia of Narrative Theory*, ed. David Herman, Manfred Jahn, and Marie-Laure Ryan (London: Routledge, 2005), 387.

465 Domsch, *Storyplaying*, 127; cf. 126-128.

466 In this regard, even random activities, such as toying around with the hacking possibilities of San Francisco, are in accordance with the game's function, for they create

its possibility space and, thus, playing styles the player will comprehend as part of the overall plot—in a wide sense of the concept. Consequently, only those player actions that are nonsensical (running around in circles for hours in THE LAST OF US) and that, therefore, work outside the bounds of the game's function, are to be considered *virtually true only*.

All in all, my conclusions on the VGN as a genre are largely positive, and to bring the discussion on the perspectival system of the VGN to a close, it is beneficial to come back to JOURNEY. For this game represents an example in which most (if not all) player actions are both virtually and fictionally true and assume vital perspectives in the player's acts of ideation (in other words, the intricate games of the emancipated player). This is so because JOURNEY's rule system virtualises a dynamic space for creative expression and the performance of several roles (or playing styles)—all of which are implied by the greater structure of the implied player and are meaningfully integrated within the game's theme as a *life journey*. Such a journey can, of course, vary considerably depending on how the participant and life wanderer chooses to conduct it. Consequently, playing JOURNEY with consciousness of a *gamist attitude* (a *conqueror* and *achiever*), one may come to the creation of the following perspective: this player rushes towards the story's end and lays the focus on collecting pieces of scarf, which might distract her from savouring the gameworld's particulars and beauty. In an allegory of life, such a playing suggests a lifestyle focused on success and the fulfilment of duties. It neglects life's precious moments, while blindly rushing towards its end—thus creating a certain perspective on the game that works in accordance with its function.

A *wanderer* or *explorer*, conversely, may experience JOURNEY in a different manner. Here, the player is interested in the gameworld itself, in its intricacies and mechanisms. Such players have a lot in common with *narrative* player types and would stop once in a while to marvel at the gameworld's beauty—thus shutting their windows to the barren yet overloaded landscapes of their existence in order

a perspective on this world—and this is also the case for more VGNs. Consider, for example, GRAND THEFT AUTO V (Rockstar North, 2013) and its stunt driving throughout the city, random shooting of NPCs in broad daylight, exploitation of sex workers, or infamous pigeon hunts. Although these activities do not substantially influence the state of the storyworld (or at all), they are in accordance with the game's *function as satire*—for the GTA series can be seen as social critique on American society and the Western world. Consequently, all of the described actions are virtually *and* fictionally true.

to experience happiness. Finally, there arises an interesting difference when considering the playthrough of a *killer* compared to that of a *socialiser/ethical* player. While the *killer* represents a lone wanderer through life who may not be inclined to cooperate with NPCs or the additional player (and might even try to harm them), the socialiser (especially when showing an ethical attitude) will act differently and embark on the journey in companionship.

The list could be developed further, but it suffices to prove a certain point. Not only are these playing styles in accordance with the game's larger function as a participatory fictional artwork (being both virtually and fictionally true), but they also create important kernel events and perspectives on it. In playing JOURNEY, the player thus experiences *the highest form of narrative agency* a VGN can afford: the pleasures of creating a kernel event and altering the story in a decisive manner. Hence, with JOURNEY the player encounters an instance of an *interactive story* (specifically in multi-player) for which at least one kernel event or node has to offer the choice to actualise one out of two or more branches in the resulting plot—when "dynamic kernels" create multipath games.[467] Moreover, for this highest form of narrative agency to occur, the enacted event has to be an *action* conducted by the player, which Chatman defines as a "change of state brought about by an agent or that affects a patient. If the action is plot-significant, the agent or patient is called a character."[468] I thereby follow general *action theory* which claims that actions

> are construed as deliberate, planned behaviours within a larger context that also includes unplanned events ... or happenings; more or less durative processes that may have been triggered by an agent, but that then continue to unfold over time; and actual as well as possible state or conditions in the world, i.e., ways the world is before, after, or as a result of the performance (or non-performance).[469]

This stands in contrast to what can be called a *happening*, which also evokes a change of state but "entails a predication of which the character or other focused existent is narrative object: for example, *The storm cast Peter adrift*."[470] Consequently, actions that locate the player as an *agent* (as described by Herman) within

---

467 Aarseth, "Narrative," 132, cf. 132.

468 Chatman, *Story and Discourse*, 44.

469 David Herman, "Action Theory," in *Routledge Encyclopedia of Narrative Theory*, ed. David Herman, Manfred Jahn, and Marie-Laure Ryan (London: Routledge, 2005), 2.

470 Chatman, *Story and Discourse*, 44; cf. 44-45.

the gameworld and determine the nature of kernel events (without the game system interfering) are responsible for affording the highest pleasures of narrative agency. They can be contrasted with such events in which the player assumes the role of a *patient* such as QTEs, scripted events, forced choices, or cutscenes (as an extreme form). However, one should not underestimate the importance of satellite events, for such activities furnish the resulting plot with variety and personality. Consequently, if a game fails to offer the choice between kernel events but grants the player the possibility to choose between satellite events, one can talk about an *interactive plot*—what Aarseth calls a "playable story."[471]

To close the chapter, I wish to direct attention to the insight that narratology in the sense of *structuralism* is perfectly applicable to VGNs—in a creative way that critically rethinks the concept and takes it one step further. As such, I partially reject those claims that argue that narrative in games may only be understood through *cognitive narratology*—a branch of narrative theory that "overcomes the shortcomings of essentialist approaches" and is rather "understood as anything that is conductive to the user's mental linking of (at least) two events and the creation of a storyworld."[472] While this is certainly part of the player's experience of a VGN (in a phenomenological sense)—the "road", as Nitsche claims, "exists in the mind of the player and is constantly fueled by stimulants from the game"[473]—it does not exhaust itself in it. For the implied player as a dynamic framework of play has already done more than half of the job in structuring this road—and one must not forget the player's ergodic interaction with the gameworld that co-determines the ongoing plot. This is not to say that I deny the importance of the player's imaginative interaction with a game—nothing could be further from the truth. For this reason, I will now focus on the player's interaction (ergodic and imaginative) with the intersubjective structure of dystopia's implied player: the creative dialectic in which the empirical player engages. This playful trial action will not only result in the creation of a certain plot but, on a grander scale, of the *aesthetic object*.

471 Aarseth, "Narrative," 132.

472 Domsch, *Storyplaying*, 2.

473 Nitsche, *Game Spaces*, 43.

Table 7: The rule system, its resulting processes, playing styles, and player actions as perspectives and elements of discourse.

For all of these, player involvement oscillates between ergodic and imaginative interaction

System	The code, system, rules, and mechanics designate the deep-layered perspectives of the game. They are found in all forms of video games and can be situated on the same level as the virtual story construct and the plot framework. These perspectives will result in more pragmatic ones the player encounters and co-creates in the gameworld. As a whole, the system perspective is of importance, for it illustrates both how a system works independently and how it responds to player interference.
Processes	The underlying processes of a system result in pragmatic events and happenings that contribute to the player's understanding of the gameworld. They add to her knowledge of the gameworld as a system and may include: player independent behaviour such as the mechanisms and routines of a city and its inhabitants (WATCH_DOGS 2) or those that govern a wasteland (FALLOUT 4, MAD MAX), to uni-operations that concern sub-systems such as the behaviours of characters. During the act of play, the player will negotiate the functions and boundaries of this system through interacting with it and experience the results of her interference.
Emergent Events	Because of a game's dynamic system, it is not uncommon for emergent events to occur in the gameworld. These include events and happenings the game designers did not consider and emerge out of interlinking factors. They surprise the player and add to the feeling of participating in a gameworld.[474]
Scripted Events	An opportunity to merge gameplay and plot are scripted events. These do not interrupt the flow of the player action and leave her partially in control—thus adding to a

474 Salen and Zimmerman, *Rules of Play*, 151-168.

	player's sense of proximity to the gameworld.[475] Moreover, although scripted events grant the player some agency, they are happenings in which the player rather assumes the role of a patient (imagine a crumbling house the player tries to escape from).
Player Actions, Movements, and Agency	In a video game nothing will occur unless the player acts and moves her PC through a 2D or 3D environment.[476] Most often, it is the player who triggers certain events through her actions and movements. These include 1) "location-based" triggers where once the player steps over an invisible marker, a cutscene or other scripted events are triggered; 2) "event-based"[477] triggers that have the player fight enemies or solve tasks before further narrative material will be displayed; 3) NPCs triggers that occur once the player approaches a character.[478] Consequently, any action or movement the player undertakes can be seen as an event within an unfolding narrative[479]—be it as simple as turning the virtual camera to view a flock of birds.
Playing Styles	The rule system of a game and its mechanics may afford various playing styles. These constitute the sum of player actions and can be seen as perspectives on player behaviours and how she conducts herself within the larger framework of the implied player. A particular playing style is that of emancipated play in which the player assumes a quasi-transcendental role by trying to interpret the meaning of individual playing styles within a grander context (game and empirical world context).

475 Dansky, "Game Narrative," 4; Soulban and Orkin, "Writing," 61.

476 Egenfeldt-Nielsen et.al., *Understanding*, 201.

477 Boon, "Writing," 63.

478 Ibid., 63-64.

479 Domsch, *Storyplaying*, 35.

5 Estrangement Through World and Agency

Chapter IV has covered a lot of ground, from describing VGNs as a form of fiction in the semantic sense to detailing the framework of the implied player, which constitutes a dynamically incomplete work world and the affordance and appeal structure of a game that guides the player's participation (his game world) in all levels of involvement. This communication was described in terms of fictionality, as a dialectical communication between work of art, appreciator, and culture. Thereby, it was hinted that (similar to the appreciator) the player's acts of ideation and the constant renegotiation of images are responsible for creating the connection between both realities. Becoming involved in fictional gameworlds (and works of art) is thus different from how human beings perceive the empirical world or the purely virtual, for although the player (appreciator) makes sense of his experience through experiential *gestalt* groupings, these initial *gestalts* remain conceptually open and require a second filling to establish the connection between fictional and empirical world.

This process was described as being anchored in the openness of representations, which necessitate the appreciator's involvement in order to come to life in his imagination. As a consequence, the perspectival system of a game was illustrated in detail and included: the player's sensorial perspective, the items of setting and characters, the plot framework, and the system perspective. Furthermore, with the example of JOURNEY, the perspectives were described as oscillating in function between virtual game objects with which the player can interact and props that guide his imaginings—and, all together, as creating a vast network of indeterminacies for the player to fill in through both ergodic and imaginative action and by closing the blanks between the perspectives. Yet different players may negotiate these indeterminacies differently, and, consequently, the emancipated player was described as a necessity for a game to exercise its full (aesthetic) effect.

It is this process, now—*the player's experience of the aesthetic effect through acts of ideation*—that becomes the focus of attention in this chapter. For this purpose, I will integrate my findings into the realms of dystopia and describe the specific nature of this *playful trial action*—that is to say, how the communication between fictional gameworld and empirical world results in a form of play that is *precarious* and detrimental to the player's habitual dispositions in a positive, *regenerative* way. By granting him new insights into social totality, it reorganises his stock of knowledge and drives him to potential action. Such descriptions necessarily bring forth a conception of play that can be described as a *dialectical negotiation between guiding structure* and *playful expression,* between Iser's categories of *the fictive* and *the imaginary*, which take their inspiration from *the real*, yet, at the same time, negotiate its contents.[1] As such, the fictive permeates the gameworld. It models its realms after empirical reality (drawing from norms, conventions, social procedures) but disrupts reality through doublings and distortions, thus showing the player its result in a refracted mirror. In doing so, the fictive opens up a *space for play* for the imaginary to take a shape, which manifests itself in the player's psyche, his ergodic and imaginative interactions with the gameworld, and has him negotiate what the fictive has channelled. The player is thus invited to a trial action and a sort of boundary-crossing, vertigo, and transgression—and this form of involvement brings together the fictional and empirical reality in subversive ways.

In the VGD, the usual distortions/doublings of the fictive are thereby taken to an extreme and create nightmarish dreamworlds that are extrapolated from the empirical world and shown in a magnified form. They thus appear strange but, at the same time, familiar to the player, and to decipher these *games of estrangement through world and agency*, the player enters a *creative dialectic* with dystopia's implied player that has him close the *blanks* between the perspectives he encounters and co-creates. To describe this process, and the image creation in the act of play, will be the main purpose of this chapter. It will unite the strands from chapter IV, specifically how fiction in VGNs can be regarded from a phenomenological perspective. Such a perspective on the phenomenon does not contradict the gameworld's ontological status as a virtual object and seduces the emancipated player into a creative dialectic with the implied player, which is fuelled by the latter's strategies and the repertoire it employs, and eventually brings forth the aesthetic response and insight into the player's self and surroundings so important to dystopia. Of particular interest in this regard is the *function of the blank*, which is responsible for communication (play) to occur in the first place and induces the

1 Iser, *Imaginary*.

creation of *images of first and second order* in the player's mind. The second of these will align the fictional reality with the empirical world in a steady *feedback oscillation between* both worlds and the player's self. To better illustrate these theoretical manoeuvres, I will integrate an analysis of METRO 2033 into my deliberations.

5.1 PLAYFUL TRIAL ACTIONS IN DYSTOPIA

Fictional worlds invite appreciators into a sort of *trial space* that allows for imaginative explorations of estranged places and alternative imaginaries. This trial space—called make-believe—enables the appreciator to grasp these worlds, which are unknown to him, to explore them in detail, their curiosities and cultures, and to face situations that may prepare him for similar circumstances in real life. Consider a reader's exploration of the fictional Sprawl in *Neuromancer* (Gibson, 1984) or a player's of the virtual Japan in YAKUZA 0 (Sega, 2016), which introduces them to these respective cultures. Make-believe, in other words, allows the appreciator to perform alternative ways of living by endowing him with roles that are partially unfamiliar to him—and by practicing them, he gains insight into the lives of these fictional characters and his own self.[2] Such *explorations of alternatives in a playful manner* are far from the didactic lessons of the school teacher who imposes his wisdom on the pupil (to recall Rancière's observations). Instead, they offer emancipatory potential to attain a deep *understanding* of the imaginatively (or ergodically) enacted, and through this understanding to come to an insight into certain life situations and one's own self.[3]

Make-believe thus creates a harmless but, at the same time, subversive space for the appreciator. It allows him to fictionally explore perilous situations (like Captain Ahab's hunting of Moby Dick or the exploration of outer space in MASS EFFECT) within the relative safety of his home yet is simultaneously harmful on another level: the *psychological*.[4] This psychological infliction on the self—what is usually meant when the reader says that a story has moved him in a particular way, and by this I do not mean his interpretation of it, although these are necessarily intertwined processes—should by no means be seen pejoratively but as a positive ramification of fiction's enchanting spell. This *naturalness of aesthetic*

2 Walton, *Mimesis*, 12, 19, 272; Iser, *Act*, 19.

3 Walton, *Mimesis*, 34-35, 42, 228.

4 Ibid., 68, 241.

response triggers a *transformative process*. It restructures the reader's stock of knowledge and leads to a change of habitual dispositions.[5]

Such a subversive (regenerative) effect is of specific interest to the function of dystopia. For dystopia—similar to fiction in general, though with a reinforced effort—involves the appreciator in a *playful trial action* that sets him on a venture into a physically harmless but nonetheless precarious situation. By letting him playfully explore an estranged place, dystopia leads the appreciator to emancipation—in the sense of gaining knowledge, and through this knowledge to a radical break with the contemporary world system—and shows him in a creative manner what could be, what already is, and what can be done about it.[6] Dystopian fictions, in other words, invite the appreciator into a "complex dialogue"[7] with alternative imaginaries that are different from the rigid descriptions of nonfiction. This is because they produce reduced versions of social totality and "cognitive maps of the system,"[8] and through involving the appreciator in them, playfully question the norms and discourses of contemporary reality—thus revealing surprising perspectives that have thus far remained hidden in the social totality.[9]

Utopia, consequently—and fiction in general—turns into a "heuristic device ... an epistemological and not ontological entity"[10] that gains additional value in video game fiction because of the player's increased involvement in a game. When Baccolini and Moylan thus speak of the non-ergodic dystopia's potential (in the form of the genre of the critical dystopia) to enable "an exploration of agency"[11] that offers an "avenue for *the necessary tracking and testing* of these new maps of hell and *possible paths through them*,"[12] the VGD comes to mind. For the genre not only offers the player cognitive maps to imaginatively engage with but simulates these in a virtual gamespace, allowing "players to become [ergodic] participants in the flawed worlds they simulate."[13] It thus offers *a participatory space*

---

5 Iser, *Act* 18, 24.

6 Baccolini and Moylan, "Introduction," 11; Ferns, *Narrating Utopia*, x; Jameson, *Archaeologies*, 228; Levitas and Sargisson, "Utopia in Dark Times," 13-14, 17-18; Moylan, *Scraps*, 96-97; Seeber, *Selbstkritik der Utopie*, 33, 45-46, 106-107; Suvin, *Metamorphoses*, 52; Vieira, "Concept," 23.

7 Ferns, *Narrating Utopia*, x.

8 Moylan, *Scraps*, 106.

9 Seeber, *Selbstkritik der Utopie*, 33, 45-46.

10 Suvin, *Metamorphoses*, 52.

11 Baccolini and Moylan, "Introduction," 8.

12 Moylan, *Scraps*, 107; emphasis added.

13 Schulzke, "Virtual," 327.

and "*play-ground*"[14] for *creative expression* and *trial action*, a fact that is widely acknowledged amongst scholars in different regards.[15] Schulzke, for example, links it to "[t]he exploration of possible worlds" in general, "especially when the possible worlds challenge prevailing assumptions about how social and political institutions should be structured and what values people should live by,"[16] and Domsch foregrounds that playing dystopia can "make us reflect on what it means to act within a system."[17] However, in order to explain the pervasive effect of the VGD on the player, it is not enough to reduce the emphasis on the aspect of simulation (though it necessarily includes it). What is of bigger importance is how fiction exerts its seductive charm on the player by involving him in *games of estrangement*. These create an aggravated referentiality between the fictional and empirical world, and it is only through the player's acts of ideation and a minute comparison between the gameworld and the empirical world (from which the former inevitably draws) that he will make sense of the enacted.

To understand this peculiarity of fiction, Iser's perspective on the concept becomes of fundamental importance. Contrary to ontological definitions, he rejects the often expressed "basic and misleading assumption … that fiction is an antonym of reality"[18] and continues to describe the concept not as "the opposite, but the complement"[19] to it—and thus in terms of *communication*. Fiction, here, turns into a *functional strategy* "of telling us something about our reality."[20] Yet in order to comprehend and respond to what is communicated (that is, to experience fiction's aesthetic effect), one has to become fluent in the language it employs. As Doležel argues:

> Fantastic fiction provides us with numerous examples of fictional encyclopedias that contradict the actual-world counterpart, as any visitor to the non-natural or supernatural worlds quickly discovers. As he or she crosses from the natural into the non-natural world, his or

14 Walz, *Toward a Ludic Architecture*, 142; emphasis added.

15 Domsch, "Dystopian," 395-397, 408; Hall, "Trying to Believe," 70; Nyman and Teten, "Lost and Found," 1, 4, 12; Schulzke, "Virtual," 318, 321; "Bioethics," 56; Walz, *Toward a Ludic Architecture*, 139, 143; Wilder, "Synthetic Fear," 125.

16 Schulzke, "Virtual," 318.

17 Domsch, "Dystopian," 408.

18 Iser, *Act*, 53.

19 Ibid., 73.

20 Ibid., 53; cf. 53.

her encyclopedia has to be modified. The visitor has to learn the encyclopedia of the alien world.[21]

Consequently, to understand fictional worlds, the readers' actual-world encyclopaedia is not enough and needs to undergo change to have them "become cognitive residents of the fictional world they visit."[22]

Things become clearer if one regards fiction as a form of "simulated reality" that "has organized itself"[23] in a particular form by converting materials drawn from empirical reality into something new—be it the imaginary worlds of literature or the virtual ones of the VGN. As such, this newly manufactured 'reality' seems strange to the appreciator, and he has to decipher its underlying code and establish the absent "frame of reference"[24] in order to understand it. Such a frame is vital to all forms of communication, and Iser describes the normal illocutionary act as based on accepted "conventions, procedures, and guarantees of sincerity"[25] that are familiar to all participants and ensure successful communication between them.[26] However, these conventions and procedures are partially frustrated in fictional communication and require the establishment of a common code to ensure a successful dialectic.

Before coming to this matter, I wish to elaborate on how the *simulated reality of fiction* is created. This aspect will become essential when describing how the VGD exerts its effect on the player, and to explain it, a step back to the *perspectives that compose the reality of fiction* becomes necessary. Such a process can also be discerned for video game fiction, where, similar to the literary text, the gameworld is created in a "process of fiction-making,"[27] that takes elements from the empirical world and redistributes them as representations throughout the gameworld. Consider the game HALF-LIFE, which takes familiar objects, such as "vending machines," "lockers," "a crowbar," "physical traits and behaviours" of

21 Doležel, *Heterocosmica*, 178-179.

22 Ibid., 181.

23 Iser, *Act*, 181.

24 Ibid., 60.

25 Ibid.

26 Ibid., 53, 60.

27 Julian Kücklich, "The Playability of Texts vs. The Readability of Games: Towards a Holistic Theory of Fictionality," *Proceedings of the 2003 DiGRA International Conference: Level Up 2* (2003): 101, http://www.digra.org/digital-library/publications/the-playability-of-texts-vs-the-readability-of-games-towards-a-holistic-theory-of-fiction ality/

characters, "but also the interrelations between them,"[28] and sets them as virtual representations within the unfamiliar context of the game (in which an alien invasion functions as the imaginative novelty of this world).[29] None of this is surprising in the light of Jameson's "unknowability thesis,"[30] that one cannot imagine the radically Other without discerning familiar objects and relations.[31] As such, gameworlds are composed of both "realistic elements," which help foster the player's understanding of the world, and "exaggerations" of these elements that create excitement.[32]

While certainly a true observation, the problem lies in a different area—not so much in what kind of objects and relations are taken from the empirical world but in *how they are rearranged* as representations in the game to create *unforeseen connections*. It is the function of modern art "to reveal and perhaps balance the deficiencies resulting from prevailing systems,"[33] and this effort can be traced back to how conventions, values, and norms of the empirical world are distributed and rearranged in the textual perspectives to create unforeseen relations between them.

> The fictional text makes a selection from a variety of conventions to be found in the real world, and it puts them together as if they were interrelated. This is why we recognize in a novel, for instance, so many of the conventions that regulate our society and culture. But by *reorganizing them horizontally*, the fictional text brings them before us in *unexpected combinations*, so that they begin to be stripped of their validity. As a result, these conventions are taken out of their social contexts, deprived of their regulating function, and so become *subjects of scrutiny* in themselves.[34]

Literature, in other words, calls into question "conventional validity … not because it is without conventions (for then it could not call their validity into ques-

28 Ibid.

29 Ibid., 101-102.

30 Jameson, *Archaeologies*, 142.

31 Ibid., 106, 108, 114, 142.

32 Henry Jenkins, "The Art of Contested Spaces," Publications Henry Jenkins, 2002, accessed February 10, 2014, http://web.mit.edu/21fms/People/henry3/contested spaces.html

33 Iser, *Act*, 13.

34 Ibid., 61; emphasis added.

tion)—but because it disrupts this vertical structure and begins to reorganize conventions horizontally"[35] by redistributing familiar norms, values, and social conventions throughout the perspective of a text—in the characters, narrators, aspects of the world, and plot.[36] These conventions represent "the familiar territory within the text," its "repertoire."[37] They ensure that fictional worlds are understood by the reader because they link the newly created 'reality' to other works of art and to the social and cultural spectrum of the empirical world.[38] What remains to be shown for the VGD is whether this process of allocating norms, values, and conventions to the individual perspectives of a game holds true—which include the gameworld's underlying processes, player actions, and playing styles.

With the process of selection, the fiction-making is however not complete, and the estrangement the reader has come to know from fictional worlds is created in a second step. This is because during the selection process, norms, values, and conventions are "reduced or modified, as they have been removed from their original context and function,"[39] and this process is reinforced by the above-mentioned horizontal rearrangement of perspectives, which results in their depragmatisation. Consequently, the reader is set between *the poles of the familiar and unfamiliar*, between what is known from the empirical world and an estranged fictional world that draws its materials from the former. These materials are both *reduced* and *rearranged* in a creative manner and now enable the formation of unexpected connections while leaving the old ones partially intact to inform the reader's comprehension.[40]

---

35 Ibid.

36 Ibid., 61, 99. This stands in contrast to "rhetorical, didactic, and propagandist literature" (such as the anti-utopia) that "will generally take over intact the thought system already familiar to its readers. That is to say, it adopts the vertically stabilized validity of the thought system and does not reorganize its elements horizontally, as is always the case when norms are to be reassessed." (Ibid., 83). This is also where Iser separates *aesthetically complex literature* from those "of a more *trivial nature*." (Ibid., 77; emphasis added). While in the former prevailing systems of thought and conventions of the empirical world are undermined, the latter support "prevailing systems" by affirming "specific norms with a view to training the reader according to the moral or social code of the day." (Ibid).

37 Ibid., 69.

38 Ibid.

39 Ibid.

40 Ibid., 61, 69-70.

It is this movement, now, from the known into the unknown, the creative dialectic between a familiar background and "an unfamiliar direction," that grasp the reader's attention[41]—*where the aesthetic effect turns into aesthetic response*. The result is the creation of something *new* that endows the old conventions and norms with new significance and has them shine in a different light, thereby revealing aspects that were previously hidden from the observer.[42] Fiction, as such, assumes *aesthetic quality* with the function of destabilising prevailing "thought systems or models of reality"—not by copying these systems or deviating from them, but by *reacting* to those "it has chosen and incorporated into its own repertoire."[43] Consequently, it is more than adequate to compare *fiction to a system itself*, which "gives insight into the reality it simulates"[44] by revealing "those possibilities that have been neutralized or negated by that system" and "by drawing attention to its deficiencies."[45]

It follows that fiction is not so far from the *concept of simulation*, which, similarly, offers a *reduced and abstracted form* of reality[46] and a *trial space* that Aarseth describes as

> an effective pedagogical tool that privileges – actually, demands – active experimentation, rather than observation, of its subject material. It is also a way to explore the partly unknown, to test models and hypotheses, and thus to construct and acquire new knowledge in a way narrative never could.[47]

Consequently, what links these concepts is the notion of a *personal experience* and the *participant's efforts* that go into the process of experimentation (although these are extended to the level of the ergodic in games). Having to work through both forms of simulation (the imaginative and the ergodic) requires considerable

41 Ibid., 70; cf. 61, 69-70.

42 Ibid., 69-70.

43 Ibid., 72.

44 Ibid., 181.

45 Ibid., 72; cf. 71-72, 181.

46 Salen and Zimmerman, *Rules of Play*, 439, 457; Klevjer, "Avatar," 16, 25-29.

47 Espen J. Aarseth, "Allegories of Space – The Question of Spatiality in Computer Games," in *Cybertext Yearbook 2000*, ed. Markku Eskelinen and Raine Koskimaa (Jyväskylä: University of Jyväskylä, 2001), 154, http://cybertext.hum.jyu.fi/index.php?browsebook=4

effort from the participant. While the reader tries to bridge the "apparent inconsistencies"[48] which the dialectic between fictional and empirical world unfolds, the player experiences the same pleasures with the additional benefit of acting on his deliberations and the cognitive maps of his creation.[49] As a result, both forms of involvement resemble a personal experience as described by Iser that has aspects of the empirical world shine differently, when "the familiar is transcended or undermined"[50] and restructures the reader's knowledge.[51]

So far, the interaction process described by Iser, in which the reader engages in a creative dialectic with a fictional world, are most fruitful for the genre of dystopia. They are, in fact, similar to Suvin's deliberations on a *feedback oscillation* that marks the reader's participation in the genres of SF, utopia, and dystopia (to which I will come shortly) and describe a form of play that is common to all sorts of emancipated involvement in representations. This form of play I would like to call *precarious play*. It is detrimental to the participant's self in a specific sense and restructures his habitual dispositions by granting him a different view of empirical reality, which may be *regenerative*. Precarious play is thus of particular interest to playing dystopia. For what is to be lost in the participation process is not only the prospect of a virtual future, by failing to attain Utopia in the fictional reality of the game, but also its loss in the empirical world. To avert this, the seductive potential of fiction drives the participant towards emancipation, allowing him to free himself from languid attitudes concerning contemporary world situations and pushing him onwards towards a utopian horizon.

What is precarious, then, in playing dystopia in any medium is the inherent conflict between the fictional and empirical world, and the ethical responsibility this ascribes to the participant—because the transition towards Utopia is injurious in both worlds and requires a self-sacrificing stance.[52] As such, it is not only the safety of the participant within the protective veil of fiction that is questioned but also the distressful route of liberating oneself from those deep-seated parts of the self that have become acclimatised to the status quo and the complexity of social

48 Iser, *Act*, 17. Indeed, Iser argues that modern works of literature "are full of apparent inconsistencies—not because they are badly constructed, but because such breaks act as hindrances to comprehension, and so force us to reject our habitual orientations as inadequate." (Ibid. 18).

49 Jenkins, "Architecture," 126.

50 Iser, *Act*, 131-132.

51 Ibid., 17-18, 69, 131-133.

52 Ferns, *Narrating Utopia*, 202-203.

totality. This is "utopia's troubled, precarious, yet at the same time life-giving relationship with reality,"[53] which comes to full fruition in the video game medium.

5.1.1 Between Estranged Order and Creative Disorder: On the Notion of Precarious Play

To explain the notion of precarious play, it is beneficial to think back to 1950, when Johan Huizinga affirmed that *"[i]n play there is something 'at play'* which transcends the immediate needs of life and imparts meaning to the action."[54] Such a focus on play's creative faculties and the creation of meaning have inspired many followers (such as Caillois or Sicart), and I wish to expand on this by resorting to Wolfgang Iser's *The Fictive and the Imaginary: Charting Literary Anthropology* (1991).

This work is beneficial to my argument in that it attempts to explain the literary work's anthropological dimension—that is to say, why human beings are dependent on their involvement in fictional worlds and the playful interactions these evoke. For this purpose, Iser basically discerns "the text as a space for play,"[55] where *the act of play* finds itself exposed to "the interplay between the *fictive* and the *imaginary*"[56] yet, at the same time, is created and regulated through it. These two primordial forces defy any attempts at an ontological definition and are only graspable through play itself—for "[t]hrough play they provide a context for each other, since what is groundless can assume tangibility only by being contextually assessed."[57]

Consequently, Iser explains the *fictive* as

> an operational mode of consciousness that makes inroads into existing versions of the world. In this way, the fictive becomes an act of boundary-crossing which, nonetheless, keeps in view what has been overstepped. As a result, the fictive simultaneously disrupts and doubles the referential world.[58]

53 Ibid., 205.

54 Johan Huizinga, *Homo Ludens: A Study of the Play Element in Culture* (Boston: The Beacon Press, 1955), 1; emphasis added.

55 Iser, *Imaginary*, xiv.

56 Ibid.; emphasis added.

57 Ibid., 271; cf. xv, xvii, 271.

58 Ibid., xiv-xv.

The fictive thus works in the mode of "disguise" through wearing a "mask."[59] It enables the reader's acts of boundary crossing and "to act out either what they are denied in the socio-historical world … or what seems impossible."[60] This sort of trial action as a playful act of transgression has the advantage of creating a "double-meaning, with the manifest meaning not supposed to mean what it says in order *to make the disguise reveal what it hides*."[61] As such, the fictive is close to how Freud describes the night-time dream, "whose disclosure through disguise corresponds to a basic desire to overstep boundaries" and whose true meaning can only be revealed through interpretation.[62]

In this sense, Iser's later observations stand in the line of his earlier ones, as the fictive holds the *specific function* of revealing those possibilities the empirical system has denied or that lie hidden in the confusing social totality. It thus offers all the preconditions necessary for art to reveal its *aesthetic effect,* to uncover its newness, and to trigger a subversive response in the appreciator. This statement is underlined by Iser's description of how the fictive *opens up a space for play* and "compels the imaginary to take on a form at the same time that it acts as a medium for its manifestation."[63] For although the *imaginary* is similar to what might be called "fancy, fantasy, and imagination," it nonetheless transcends these, because they are "specific, context-bound demarcations of the imaginary whose potential eludes cognition."[64] The imaginary, consequently, is rather described as "a featureless and inactive potential" which "has to be brought into play from the outside itself by the subject … consciousness … or by the psyche."[65]

As such, the creative interplay of these primordial forces—the fictive, which transforms the work of art into a disguised and distorted dreamworld that allows for boundary crossing and trial action, and the imaginary, which manifests itself through the appreciator's acts of ideation as an experience—shows the potential to trigger an *aesthetic response* in him, of instigating him "to extend beyond habitual dispositions"[66] in an act of "self-cultivation" or "fashioning."[67] To explain the inner workings of the fictive and the imaginary in more detail, and because the

59 Ibid., xv.

60 Ibid.

61 Ibid.; emphasis added.

62 Ibid. cf. xv; Freud, *Dreams*, 33.

63 Iser, *Imaginary*, xvii.

64 Ibid., xvii.

65 Ibid.; cf, xvii.

66 Ibid., 17.

67 Ibid., 297; cf. 17, 297.

two forces are difficult to grasp, Iser resorts to Roger Caillois' *four types of play*—this formulates an important aspect that will constitute the link to the form of precarious play I am proposing.

To do so, Iser begins to critically rethink the concept of play for the literary text as it occurs between the poles of *instrumental play* and *free play*.[68] "Games in the text always pit free and instrumental play against each other,"[69] and through this interaction "they break all the bounds of everyday pragmatic needs that otherwise hold our attitudes in check."[70] As such, the reader's imaginative play with the literary text is seen in a subversive manner, oscillating between the forces of "free play," which undoes "the possible results of gaming, and instrumental play as a recuperation of what free play disperses."[71] These two opposites stand in close relation to what Caillois describes as *ludus* and *paidia*: games that are determined by a strict set of rules and those in which freer forms of involvement are possible (but between which a clear line of demarcation cannot be drawn).[72] Taking a closer look at Caillois' original description reveals their overlaps:

> At one extreme an almost indivisible principle, common to diversion, turbulence, free improvisation, and carefree gaiety is dominant. It manifests a kind of uncontrolled fantasy that can be designated by the term *paidia*. At the opposite extreme, this frolicsome and impulsive exuberance is almost entirely absorbed or disciplined by a complementary, and in some respects inverse, tendency to its anarchic and capricious nature … to bind it with arbitrary, imperative, and purposely tedious conventions … I call this second component *ludus*.[73]

Although Iser does not directly equate these two forms of games with his established forces of the fictive and the imaginary, they nonetheless show a certain family resemblance in that *paidia* comes close to a sort of *unbounded fantasy*, whereas *ludus* channels this involvement into *a more guided form*. What is more, the similarities between the two opposites can further be discerned in the way Iser describes Caillois' four forms of play—*agôn, mimicry, alea, ilinx*[74]—as emerging

68 Ibid., 247, 258-259.

69 Ibid., 258.

70 Ibid., 260.

71 Ibid., xviii.

72 Caillois, *Man, Play and Games*, 13, 27; Egenfeldt-Nielsen et.al., *Understanding*, 36-39.

73 Caillois, *Man, Play and Games*, 13.

74 Caillois categorises these four different yet related forms of play as follows: "One plays football, billiards, or chess (*agôn*); roulette or a lottery (*alea*); pirate, Nero, or Hamlet

out of the spectrum of *ludus* and *paidia* games and which are permeated by the fictive and the imaginary in certain ways.

To illustrate this point, Iser compares *the meaning-making process of traditional games* to that of the literary text, where the aesthetic object emerges out of the interrelation of different attitudes and perspectives that it is composed of. Similarly, "games always express attitudes that are not manifested in rigid forms but take on their individuality only through interactive relationships."[75] In this sense, Iser continues to describe the game of *agôn* (conflict) as one that "consolidates the opposing norms, values, feelings, thoughts, opinions, and so on into positions"[76] or perspectives the reader encounters in the literary text. Yet, at the same time, it "initiates the surmounting of what has congealed into positions through the conflict."[77] *Agôn*, in other words, "has to be played towards a result that will overcome the antagonisms"[78] created by the frictions of the perspectives—and to ease this tension, the reader will have to close blanks, the unstated connections, between them (an aspect which I will discuss shortly). Together with *mimicry*—an aspect of role-playing and doubling so familiar to playing games—*agôn* is permeated by the fictive, which has taken dominance in both games.[79]

As such, the fictive seems to assume a directive role in the participation process, because it offers the reader various riddles and conflicts to be solved. But "[a]lthough the fictive gives cognitive guidance to the imaginary, in so doing it unleashes the imaginary as something uncontrollable,"[80] and it is through "this uncontrollability" that play can emerge in the first place, "*for play proceeds by changing whatever is in play*."[81] This liberating, expansive and ever-changing process of meaning-creation Iser now explains through the remaining games of *alea* and *ilinx*. *Alea* (chance) works against the confining games of *agôn*. By breaking "open the semantic networks formed by the referential worlds and also by the recurrence of other texts," it intensifies the difference between the textual perspectives and prompts the reader to maximise his combinatorial efforts.[82] The result of

(*mimicry*); or one produces in oneself, by a rapid whirling or falling movement, a state of dizziness and disorder (*ilinx*)." (Ibid., 13).

75 Iser, *Imaginary*, 259.

76 Ibid., 260.

77 Ibid.

78 Ibid.

79 Ibid., 271.

80 Ibid.

81 Ibid.; emphasis added.

82 Ibid., 261.

such games and their effect on the reader can be summarised in the last variant Caillois mentions: the game of *ilinx*. In the sense Iser is using it, *ilinx* can be understood "as a game of subversion whose 'vertiginous' element consists in the carnivalization of all the positions [perspectives] assembled in the text."[83] It thus clearly shows "an anarchic tendency … and this not only liberates what has been suppressed; it also reintegrates what has been excluded."[84] In *ilinx*, therefore, "free play is at its most expansive,"[85] negotiating what the fictive has presented and, by doing so, exerting a lasting influence on the reader.

The four games of *agôn, mimicry, alea,* and *ilinx* may, of course, occur in various combinations in a text, with each game occurring in a more or less dominant form, depending on the text at hand.[86] What is of importance to the VGN and dystopia, is that, similar to the literary text and representational art, there emerges *a dialectic between guiding structure and playful expression* (see my discussion in chapter IV). It is as Iser remarks: "Free play [*paidia*] triumphs over instrumental play [*ludus*], but no matter how drastically the former rejects the latter, instrumental play will still be a necessary foil in order to prepare for the unexpected."[87]

In this sense, the fictive permeates the work world of a game and its implied player. It offers guidance by structuring a system of perspectives yet, simultaneously, disrupts the apparent order through doublings and distortions—taking its material from the real, but showing the player its result in the refracted mirror of the fictional world. This form of confusion and disarray is further complicated by the opposing structures of the perspectives. Although these complement each other to a degree (thus contributing to the player's understanding of the game-world), they are often in dire conflict, leading to increased player involvement, as he tries to solve their riddles. If this is not confusing enough, the guiding structure of the fictive opens up a space for play for the imaginary to unfold. This space invites the player to cross boundaries, to vertigo, and transgression; in which the imaginary runs wild by taking form in the player's psyche, imagination, and ergodic efforts. The result of this interaction between various forces is a form of *precarious play* that is detrimental to the player's habitual disposition but, at the same time, offers the opportunity for new experiences to invade his mind—an aspect that is of indisputable importance to playing dystopia.

83 Ibid., 262.

84 Ibid.

85 Ibid.

86 Ibid., 263.

87 Ibid., 261.

To give my hypotheses further substance, I now wish to find approval in video game studies before applying them. Since Huizinga, the *(aesthetic) effect of play* has been discussed from various perspectives, and I wish to begin in the same manner I started to discuss the concept of play in representational art. There is no denying the fact that, like non-ergodic forms of art, VGNs and dystopias allow for a *playful trial action* in which players may explore within the relative safety of their homes a seductive but often perilous gameworld in a variety of ways: "to try out and observe the effects of different behaviours ... test different strategies for problem solving ... live in danger, and experience strong feelings."[88] Imagine the player of GTA V and how he acts out all kinds of behaviours, from savouring the environment and going shopping, to robbing banks and exploiting prostitutes. Such acts of boundary-crossing, vertigo, and transgression not only reveal the player's *innermost self*[89]—because "through play, we explore the forbidden, the impossible, the things that we are not allowed to explore in our daily lives"[90]—but also *disrupts the safety of the player* on the level of the psychological. As Miguel Sicart argues: "[i]n this sense, fiction operates much like play,"[91] and it is my belief that because of this connection, playing dystopia (and VGNs in general) is essentially precarious, in a similar manner to reading a novel, watching a film or play, or attending an art exhibition.

Consequently, it is more than adequate to state that *the gameworld as a space for play* is organised by the structural finesse of the implied player and is permeated by the fictive. Such playgrounds are artificially designed virtual spaces (by

88 Perron, *Silent Hill*, 12; cf. 12.

89 Similarly, the fictional character William in the television series WESTWORLD (Nolan and Joy, 2016)—based on Michael Crichton's eponymous 1973 movie—exclaims: "I used to think this place was all about ... pandering to your baser instincts. Now I understand! It doesn't cater to your lower self, it reveals your deepest self. It shows you who you really are." (WESTWORLD, season 1, episode 7, "Tromp L'Oeil," directed by Jonathan Nolan and Lisa Joy, aired November 13, 2016, on HBO [Warner Bros. Television, 2016], min 28). Westworld is a theme park that was established to allow ordinary people to act out their deepest fantasies within the relative safety of a physically harmless environment in which robots are incapable of injuring humans and programmed to fulfil the visitors' fantasies. However, once the machines are updated to experience reveries, they start to gain consciousness and eventually rebel.

90 Sicart, *Beyond*, 15.

91 Ibid., 15.

the game developers) that facilitate and afford play.[92] Yet this is only the beginning. In many regards play can be pleasurable as well as liberating. It frees players from "moral conventions but makes them still present, so we are aware of their weight, presence, and importance."[93] As such, play grants players the *necessary distance to see values and norms from afar*, and by practicing them—specifically in the estranged context of a fictional world, I might add—they challenge their applicability for the real world.[94] "Play", as Colleen Maclin puts it in a personal conversation with Sicart, "lets us explore new roles, be transgressive, and push at the boundaries of both the game's rules and the rules of society."[95] It is where the imaginary takes hold of the player—in his imagination, emotions, and ergodic efforts—and negotiates the fictive, thus turning the experience into a perilous one by breaking open its semantic veil.

> Play is always dangerous, dabbling with risks, creating and destroying, and keeping a careful balance between both. Play is between the rational pleasures of order and creation and the sweeping euphoria of destruction and rebirth, between the Apollonian and the Dionysiac.[96]

It is this "struggle between order and chaos,"[97] between the "sobriety of the Apollonian" and the "passionate, irrational, and irreverent" of the "Dionysiac"[98] that makes the act of play precarious and that creates a dialectic, and the potential transition from safety to danger.

In dystopia, this transition to an uncertain and perilous situation transcends the bounds of fiction and extends into the empirical world—if players miss acting upon the lessons of the VGD, their own future may be at stake. Consequently, such an intrusion into the self should not be seen in a negative manner and, instead, offers a *regenerative space* for cultivating the individual and encouraging him to incorporate the newly found experiences into his stock of knowledge. The disruption, in other words, is a positive one. Play "breaks the state of affairs,"[99] and by

92 Sicart, *Play*, 7.

93 Ibid., 5.

94 Ibid.

95 Sicart, *Beyond*, 68.

96 Sicart, *Play*, 9

97 Ibid., 10.

98 Ibid., 9.

99 Ibid., 14.

doing so, "disruptively reveal[s] our conventions, assumptions, biases, and dislikes."[100] It aids the player in "finding expression," in "understanding the world, and through that understanding, challenging the establishment, leading for knowledge, and creating new ties or breaking old ones."[101] As such, play incorporates the disruptive function of fiction into its repertoire and makes it graspable for the reader, spectator, appreciator, and player—for the aesthetic response that is activated through play is a direct result of the interaction between the fictive and the imaginary.

Consequently, it comes as no surprise that, like Iser, Sicart comes to the conclusion that play is "the source of the aesthetic experience,"[102] "the message," which cannot be reduced to a game's individual parts.[103] Since it is through play that players will satisfy their desire "for beauty, appropriate a context and rules, make them their own, and live and feel them through their values," and through this experience come to "a deeper understanding of [them]selves."[104] Play, in other words, leaves its mark on the player, but in order to experience its aesthetic response, "we should demand not only entertainment [from it] but also something that moves us and leaves us changed."[105] Consequently, it is only when the player accepts this "invitation to reflection"[106] that he emancipates himself from ignorance and decay—which becomes of specific importance to playing dystopia, for what is at stake is the future itself.

5.1.2 The Incompleteness of Gameworlds as a Basis for Communication and Play

To further elaborate on how (precarious) play emerges, I wish to return to a fundamentally important aspect. In chapter IV, I fostered an understanding of *representational art as a sort of incomplete construct* that needs the appreciator's participation to come alive and develop its influence on him. Fictional worlds, in other words, are basically incomplete, and it is because of this incompleteness that

100 Ibid., 15.

101 Ibid., 18.

102 Sicart, *Beyond*, 18.

103 Ibid., 96; cf. 96.

104 Ibid., 21.

105 Ibid., 95.

106 Ibid., 78.

communication with them is possible in the first place.[107] This is also what separates the fictional world from the empirical world, as "it shows that the opposition completeness/incompleteness reflects the contrast actual/fictional,"[108] and that "denying incompleteness to fictional entities is tantamount to treating them as real entities."[109]

To explain this facet of fictional communication, it is beneficial to resort to the concept of *indeterminacy*. For there is a sense, as Iser argues, in which *gaps* and *blanks* (dynamically open aspects and vacancies in a conversation that hinder the participants' understanding of what is communicated) promote and "function as a basic inducement to communication" of all sorts.[110] Of course, should this aspect hold true for the VGN and dystopia, it will help to explain the player's communication with a game and with the dynamically open work world of the implied player. Hence, before coming to Iser's observations, I wish to begin by pinpointing the *form* of indeterminacy to be found in game fiction—for as Klevjer claims: "in principle, games are no less 'complete' as fictional worlds."[111] My initial hypothesis, thereby, revolves around the fact that although VGNs are similar to non-ergodic forms of art in this respect—the video game, as Sicart claims, can be seen as a continuation of Eco's open work[112]—they differ from them in the way they employ this *incompleteness, indeterminacy, or virtuality*.

To prove this point, consider Christoph Bode and Rainer Dietrich's claim that *future narratives* (which include the VGN/VGD) do not "only *thematize* openness, indeterminacy, virtuality, and the idea that every 'now' contains a multitude of possible continuations" but also involve the reader/player in a specific manner, where his decisions move the plot in different directions and where "what happens next" depends on the player's "actions … values and motivations."[113] With this observation, Bode and Dietrich touch upon the different forms of involvement the VGN offers (which I have labelled as oscillating between the player's ergodic, psychological, imaginative, and emancipated efforts) and link the concepts of indeterminacy and virtuality.

---

107 Walton, *Mimesis*, 64; Doležel, *Heterocosmica* 22, 169.

108 Ibid., 22.

109 Ibid., 23.

110 Iser, *Act*, 166; cf. 166-167, 169.

111 Klevjer, "Avatar," 33.

112 Sicart, *Beyond*, 29, 158.

113 Christoph Bode and Rainer Dietrich, *Future Narratives: Theory, Poetics, and Media-Historical Moment* (Berlin: De Gruyter, 2013), 1; cf.1.

Such a point is well worth pursing, and Domsch aligns their deliberations with the VGN. To him, "[t]he existence of nodes" (which are a fundamental aspect of future narratives) not only "heightens the ludic quality of narrative"[114] but results in a *specific incompleteness* of the gameworld.

Every fictional world also has blank spots, but they cannot be filled in without violating the integrity of the fictional world. In a game that creates a fictional world and … grants some degree of agency to the player, though, this world is *unfinished by design* and *can only be complemented through the active* [ergodic] *involvement of the player*.[115]

Although Domsch highly underestimates the complexity of what he calls "the usual activity of imagining it [the storyworld] and filling occasional gaps"[116] that mark the reader's involvement in the literary text, his observations point to the mutual filling in of indeterminacy through the player's ergodic and imaginative faculties. Consequently, it is not only the player's imaginative involvement that creates *narrative closure* (contemplating about the gameworld and using his world knowledge to establish links and connections) but also his ergodic one. This effort of actualising one of several paths results in a form of closure not available to the non-ergodic media participant[117]—and the creation of a perspective that is shared between the player and PC.[118]

So far, the observations on the incompleteness of gameworlds have revealed a form of play that, as Ryan has observed, "transposes the ideal of an endlessly self-renewable text from the level of the signified to the level of the signifier"[119] and

114 Domsch, *Storyplaying*, 2.

115 Ibid., 30; emphasis added.

116 Ibid.

117 Ibid., 30, 36.

118 Sicart offers a comparable description of the gameworld as a *possibility space* which is composed of "limited determinacies." (Sicart, *Beyond* 15). In this sense, players may fill in the semi-blank attributes of their PCs (they co-determine the PC's personality through actions) and the gameworld. (Ibid., 16). They are thereby presented "with a space of limited determinacies that are perceived as being open but that need to be closed and interpreted through play." Ibid. Hence, through "inserting the self into a system that is experienced as and through a world" players may narrow the possibility space of the story and its world. (Ibid., 27; cf. 27). Narrative closure is thus reliant on the player's ergodic and imaginative faculties, and Sicart argues that the "space of possibility" in games is both "physical and mental." (Ibid., 45).

119 Ryan, *Narrative*, 5.

oscillates between them. However, in order to describe the player's participation process in VGDs, a finer granularity of the topic becomes necessary, and a brief excursion into the concept of the *virtual as a form of indeterminacy*.

For this purpose, let me refer to Marie-Laure Ryan and her seminal work *Narrative as Virtual Reality: Immersion and Interactivity in Literature and Electronic Media* in which she distinguishes between primarily two forms of the virtual: "the virtual as fake" (Baudrillard) and "the virtual as potential" (Lévy).[120] Such a distinction echoes my discussion of fiction and is underlined by how Ryan describes both senses of the virtual: an "*optical sense*" which "carries the negative connotations of double and illusion ... the treacherous image" and a "*scholastic sense*" suggesting "productivity, openness, and diversity."[121] Given my current observations, it seems adequate to stick with the notion of the virtual as potential, because it will explain the possibility space of the VGN and dystopia, which players may fill in creative fashions. To underscore her claims, Ryan continues to establish a table of meanings of the virtual, the top part of which illustrates the scholastic use of the concept. Here, she enlists oppositions that stand in analogy to the overarching concepts of the *actual* and the *virtual* and that include: "enacted" and "potential," "closed" and "open," "complete" and "incomplete," "determinate" and "indeterminate."[122] Such a comparison not only aligns the concepts of virtuality and indeterminacy but further links the latter to my use of fiction.

Ryan, then, follows the concept of the virtual as potential. She emphasises Lévy's[123] remark that the virtual shall *not* be seen in opposition to the real and continues to describe the *dialectic* between the *virtual* and the *actual*. Here, "a phase of actualization" which offers a "concrete solution to answer a need" is complemented by a "phase of virtualization" that formulates "a return from the solution to the original problem."[124] This is an important aspect, because "reopening the field of problems that led to a certain solution"[125] after an apparent solution has been found (through actualisation) points to a *fluidity of meaning creation* and its constant rebuttal as described by Iser (for instance, when the player solves an issue in a certain manner, but its resolution creates further issues—such as in the

120 Ibid., 27.

121 Ibid.; emphasis added; cf. 27.

122 Ibid., 28.

123 Pierre Lévy, *Becoming Virtual: Reality in the Digital Age*, trans. Robert Bonono (New York: Plenum Trade, 1998), 16.

124 Ryan, *Narrative*, 36; cf. 35-37.

125 Ibid., 37.

Tenpenny Tower example mentioned previously). Ryan realises the similarity between such processes in literature and digital forms (hypertexts, virtual reality, and so on) and links her observations to theories of reader-response criticism.

She is thereby particularly interested in those of Roman Ingarden[126] and Wolfgang Iser, who both utilise the virtual (indeterminate) in the sense of potential.[127] Ryan thus begins by describing Ingarden's conception of the literary text "as an incomplete object that must be actualized by the reader into an aesthetic object" and which "requires of the reader a filling in of gaps and places of indeterminacy."[128] Each reader will fill in the gaps in his own way by resorting to his world knowledge, and, therefore, the literary text is best conceived as a *virtual object* (following Iser) that allows many fillings and the creation of possible worlds.[129] Such an insight is nothing new to my discussion (see chapter IV), but in order to expand on the usage of *virtuality-as-indeterminacy* to the digital era, Ryan underlines the importance of an "additional level" that fuels the hypertext's "textual machinery."[130] This additional level lies both in the fact that the hypertext is represented on the computer screen and that it offers the user the ability to actualise one of many potential manifestations of the text depending on his input.[131] One could argue that Ryan's conclusions cannot be directly transposed to the video game, since she only rarely mentions digital games, but this problem can be circumvented. For in video games the additional level is, of course, the *ergodic*—which Aarseth also uses in the sense of hypertexts[132]—which complements the player's *imaginative* participation in the game. The *virtual as potential*, then, and the concept of *indeterminacy* are undeniably close and will henceforth, and for my particular purposes, be viewed as synonymous.

Such an understanding of *indeterminacy-as-virtuality* is also beneficial in another way and helps to answer Egenfeldt-Nielsen et. al.'s question of whether one can consider the gameworld as "a prop for the player's imagination (in Walton's

126 Roman Ingarden, *The Literary Work of Art: An Investigation on the Boundaries of Ontology, Logic, and the Theory of Literature*, trans. George G. Grabowicz (Evanston: Northwestern UP, 1973).

127 Ryan, *Narrative*, 44.

128 Ibid., 44.

129 Ibid., 44-45. Walton differentiates between *incomplete worlds* (which he equates to fictional worlds that require the appreciator's filling) and *possible worlds* (which are complete constructs). (Walton, *Mimesis*, 64).

130 Ryan, *Narrative*, 46.

131 Ibid.

132 Aarseth, *Cybertext*.

sense), or is the world a virtual object or stage for the game to take place on?"[133] The only answer to this question, then, is both—for the *gameworld* constitutes an amalgam of basically two layers: the *virtual gamespace* and the *fictional storyworld*.[134] This approach bridges the incompatible quarrels about the gameworld's ontological status, because it regards the gamespace as the *virtual and ontological layer* (following Aarseth, in that the actions of the player are real; see chapter IV), which is nonetheless heaved into the realms of fiction from a *phenomenological point of view* (where the storyworld functions as an all-encompassing veil that fuels the player's imaginings and where his actions create fictional truths).[135] Consequently, it is only through a negotiation of both layers on part of the player that the newly formed 'reality' of the gameworld appears and is understood—as Nitsche formulates: the "new 'reality' gameplay constitutes."[136] This reality "is part perceived ... and part performed"[137] and requires the player to construct "mental images of both the game state and ... the game's storyworld."[138]

The reason why I am stressing this aspect is that both layers of the gameworld (or *virtualised storyworld*) are fundamentally *incomplete* and require *filling* in particular ways. While the virtual gamespace represents a "setting for the gameplay"[139] that necessitates the player's ergodic motions to dynamically change as a result of this interference, the fictional storyworld prompts the player to imagine a larger space than actually exists.[140] As a consequence, storyworlds can be considered wide "expanses of time" and "space"[141] that make references to the past, foreshadow the future, and evoke imaginings about the present state of affairs. In fact, a storyworld includes everything there is to know about the diegesis.[142] As a result of these interactions (on an ergodic, physiological, imaginative, and eman-

133 Egenfeldt-Nielsen et.al., *Understanding*, 232.

134 Nitsche, *Game Spaces*, 35.

135 In this respect I disagree with Tavinor who, similar to my observations, claims that "games are typically *both fictional and virtual*," but he deems "virtuality ... [as] a distinctive mode of depiction, whereas fictionality regards the ontological statues of what is thus depicted." (Tavinor, "Fiction," 438).

136 Nitsche, *Game Spaces*, 127.

137 Domsch, *Storyplaying*, 30.

138 Ibid., 29.

139 Egenfeldt-Nielsen et. al., *Understanding*, 206.

140 Nitsche, *Game Spaces*, 28, 33, 51.

141 Domsch, *Storyplaying*, 28.

142 Ibid., 28-29.

cipated level), the resulting reality of the gameworld cannot *only* resemble a cognitive image the player has helped to construct—as Domsch claims[143]—but goes further in constituting something graspable and real, which comes to life through play and in the oscillation between the fictive, the imaginary, the virtual, and the real (but differently to how Aarseth has described it).

Now, in order to apply this to the genre of the VGD, one has to regard the nature of the estranged gameworlds the player is confronted with. This will illuminate how the player makes sense of these worlds and acts upon his deliberations by filling in their indeterminacies and, by doing so, elaborates on the preconditions of aesthetic response. Paving the way to Part III (where my findings find application), the rest of this chapter will discuss theories of estrangement combined with a focus on the player's multifaceted involvement in a fictional gameworld. Finally, I will complement these through Iser's deliberations on the reader's process of ideation and the creation of the aesthetic object—a theoretical manoeuvre that adds to the foregoing discussion and which, to my knowledge, has never been conducted carefully in video game studies.

5.2 DEFAMILIARISED WORLDS AND PLAYER ACTIONS

"Where in the world am I? What in the world is going on? What am I going to do?"[144] Fictional worlds are always *estranged* to a certain degree. They confront the appreciator with things unknown to him and require effort to decipher. Within the genres of SF, utopia, and dystopia, this estrangement is magnified and sometimes taken to an extreme, confronting the appreciator with "a realist mode that is defamiliarized yet dynamic."[145] For Moylan, therefore, the questions stated above are essential when describing the reader's venture into an estranged world and an uncanny Other—and nowhere could they be more aptly asked than for the video game medium.

To approach the issue, I wish to begin with a detour into the *strangeness of gameworlds*, for they are often described as semi-detached from the empirical world.[146] Gameworlds "are totally constructed environments. Everything there was put on the screen for some purpose -- shaping the game play or contributing

143 Ibid., 30.

144 Moylan, *Scraps*, 3.

145 Ibid., 27.

146 Domsch, *Storyplaying*, 99.

to the mood and atmosphere or encouraging performance, playfulness, competition, or collaboration."[147] Imagine the open world of GTA V, which virtualises a fictional version of Los Angeles and its surroundings but alters the gamespace in such a way that it caters to challenge/gameplay[148] (by reducing the size of the map and rearranging the content). Gamespaces are often "reductive: they reproduce some features of the real world, but create their own rules in order to facilitate gameplay."[149] They are, in other words, "allegories of space," which "pretend to portray space in ever more realistic ways, but rely on their deviation from reality in order to make the illusion playable."[150]

However, it is not only space that creates distance from the empirical world but also a game's rules, which determine aspects of the gameworld. These create "an 'as if' situation"[151] similar to that of non-ergodic fiction and have the gameworld diverge from the empirical world in the same manner as "a fictional world is set apart by its fictional existents."[152] The "realist paradigm" is thus "never absolute,"[153] and encountering the newly created reality of fiction, players have to become familiar with these rules, as much as with the peculiarities of space. Consequently, to decipher these "disorienting 'otherworlds',"[154] a fair amount of "cognitive work"[155] is required from the player—which is especially so in spaces that seem "surrealist," "disfigured and distorted through" the "dream-like images"[156] they project.

Similar in scope and magnitude to the surreal spaces of horror games are dystopian gameworlds. While every gameworld in a sense differs from what one may

---

147 Jenkins, "Contested."

148 Gamespaces are often "contested spaces" that confront the player with a fierce struggle over space. Ibid. This is the case when the player tries to overcome environmental blockades (like the pipes in SUPER MARIO) or hazards such as large mountains, mazes, or snow storms, or enemies that roam the environment. (Ibid.; James Newman, *Videogames* [London: Routledge, 2004], 110-112).

149 Egenfeldt-Nielsen et.al., *Understanding*, 206.

150 Aarseth, "Allegories," 169; cf. Newman, *Videogames*, 118-119.

151 Domsch, *Storyplaying*, 14.

152 Ibid., 19; cf. 14, 19.

153 Ibid., 17.

154 Nitsche, *Game Spaces*, 47.

155 Ibid., 48.

156 Jenkins, "Contested."

call 'reality', the genres of SF, utopia, and dystopia push the degree of estrangement further to create a more complex relation to the empirical world.[157] As such, dystopian gameworlds are often considered as *other spaces of estranged nature*, [158] but it is rarely disclosed what exactly this means and entails. To answer this, let me start by thinking of the player of METRO 2033 (4A Games, 2010) who will experience first-hand the "disorientation"[159] the SF genre offers when he encounters a strange new world that was devastated by nuclear war and where the surface of the Earth is now contaminated by radiation, only traversable for those wearing gas masks, if they dare confront the mutants that roam it. The few who managed to survive escaped to the Moscow metro system, where several factions have formed—including bandits, fascist, and communists. Alliances were established to fend off the mutants and a trading network operates between the various stations. Most of the populace, however, has not learned from the mistakes of their predecessors. War continues to wage in the Metro, and an uncanny force called the Dark Ones causes fear and hallucinations. METRO 2033 is a critical dystopia of variant II in which the player undertakes a mission to an ominous tower where a decisive choice to determine the future awaits. But in order to accomplish the task to the full extent (to execute it and to relate such an action to the empirical world), the new reality of fiction has to be understood.

METRO 2033 and other VGDs stand in the tradition of many SF narratives that find "ways to explore and to go where others will not, might not, dare not go."[160] This is because the estranged gameworld the player encounters leaves him both awestruck and pondering the unthinkable. According to Suvin, such "a curiosity about the unknown"[161] is a basic feature of human nature and inherent to the SF genre. It is "where the thrill of knowledge join[s] the thrill of adventure"[162] and happens when the reader, spectator, or player encounter new environments and

157 On a basic level, dystopian fiction is a genre with tropes that the reader/player recognises: the destroyed/polluted world, cities in ruins, lack of people, gangs, etc. This does not mean, however, that more complex forms of estrangement may emerge—that the games of estrangement captivate the reader/player into creating connections to the empirical world that were previously unknown to him.

158 Frelik, "Video Games," 230; Jagoda, "Digital Games," 150; Newman, *Videogames*, 107-109; Nyman and Teten, "Lost and Found," 6-7; Schulzke, "Virtual," 1; Schmeink, "Dystopia."

159 Moylan, *Scraps*, 4.

160 Ibid., 4.

161 Suvin, *Metamorphoses*, 5.

162 Ibid.

their inhabitants. These range from hidden islands in unknown waters to isolated valleys and faraway stars and societies, inhabited by creatures and peoples unknown to the appreciator—from "aliens" to "strangers," from "utopians" to "monsters."[163] Such trial actions in estranged "elsewhere[s]" are highly pleasurable for the participant and reinforced in game fiction through the spatial representation of the world, which not only allows the player to navigate these spaces—thus increasing the "sense of alterity" of the scenario[164]—but also to experience their "awe, wonder, and estrangement" through "majestic vistas of alien worlds, futuristic cities, gigantic spaceships, and outer space."[165] It is as Jenkins puts it: "Games … may more fully realize the spatiality of these [SF] stories, giving a much more immersive and compelling representation of their narrative worlds."[166]

Figure 15: In METRO 2033, *the player encounters an inhospitable surface after the nuclear bombs have fallen, where people manage to survive in the Metro.*

METRO 2033, (4A Games, 2010), ch. Dead City 1.

163 Moylan, *Scraps*, 5; cf. 5.

164 Frelik, "Video Games," 230.

165 Ibid., 233; cf. 230, 233.

166 Jenkins, "Architecture," 122. In this sense, the SF genre and video games share a fundamental similarity: the pleasures of stepping into unknown words, exploring their boundaries and discovering their secrets through imaginative and ergodic action. As a result, they grant the player a sense of what Calleja calls *inhabiting* these worlds, which, in this magnitude, is a sentiment unknown to the reader, viewer, or appreciator. (Calleja, *In-Game*, 73-76).

However, although awe and pleasure are important emotions, the technique of estrangement goes further in that it offers the player a glimpse into the true nature of his reality. By showing him the familiar in an unfamiliar fashion, SF possesses a "most powerfully subversive potential, for if a reader [similar to the player] can manage to see the [empirical] world differently (in that Brechtian sense of overcoming alienation by becoming critically estranged and engaged), she or he might just … do something to alter it."[167]

Consequently, playing SF is by no means detached from the empirical world, although it aims to transcend it. The player is thereby confronted with a new world, a new reality that fiction has composed and which is constructed around the logic of a "cognitive innovation,"[168] "*a fictional 'novum'*."[169] This novum is of "totalizing" effect, because it "entails a change of the whole universe of the tale"[170] and creates distance to the author's empirical world. Thereby, various "degrees of magnitude can be discerned:" from simple inventions such as gadgetry, phenomena, or techniques, to entire settings, agents, or relations that are unknown to the empirical world.[171] As such, the fiction-building device of the novum can start as small as time working differently in BRAID (Number None, 2009) or the ability to alter time in Remedy Entertainment's QUANTUM BREAK (2016). It may involve the invention of a cloning tool in SWAPPER (Facepalm Games, 2013) or a portal gun in Valve's PORTAL (2007) that creates bizarre passages through space—a game that irritates the player "with uncommon, perhaps unpredictable, spatial rules."[172] Such agential powers change the old locus in a considerable fashion—but there are also entire worlds that are constructed around the novum's logic.[173] Take, for instance, the world of FEZ (Polytron, 2012) in which people are not accustomed to a third visual dimension or the many post-apocalyptic worlds in which after a certain incident the world as once known has vanished—due to a virus, environmental catastrophe, nuclear war, and so on.

Given these repercussions on the gameworld, there are two aspects to foreground: first, the novum assumes narrative hegemony in determining the logic of the SF storyworld (or gameworld) and sets it apart from the empirical world

167 Moylan, *Scraps*, 5.

168 Suvin, *Metamorphoses*, 64.

169 Ibid., 63.

170 Ibid., 64.

171 Ibid.; cf. 64.

172 Newman, *Videogames*, 108; cf. Jagoda, "Digital Games," 141, 146, 149.

173 Suvin, *Metamorphoses*, 79.

through estrangement;[174] second, and as a result of the foregoing aspect, the novum represents "a creative, and especially aesthetic category."[175] This is because the novelty induces the reader/player to engage in a complex *feedback oscillation* between empirical world and fictional world—not only to comprehend the latter but also to gain insight into the former. Suvin describes this important aspect as follows:

> Though I have argued that SF is not—by definition cannot be—an orthodox allegory with any one-to-one correspondence of its elements to elements in the author's reality, its specific modality of existence is a feedback oscillation that moves now from the author's and implied reader's norm of reality to the narratively actualized novum in order to understand the plot-events, and now back from these novelties to the author's reality, in order to see it afresh from the new perspective gained. This oscillation, called estrangement by Shklovsky and Brecht, is no doubt the consequence of every poetic, dramatic, scientific, in brief *semantic* novum.[176]

SF is thus not so much a genre about the future (this lies in the domain of futurology and is of secondary importance to SF) but rather reflects "on the author's own historical period and the possibilities inherent in it."[177]

In this sense, the world-building of SF, utopia, and dystopia is closely linked to the empirical world and dependent on the technique of *extrapolation*: "a scientific procedure [that] … is predicated upon a strict (or, if you wish, crude) analogy between the points from and to which the extrapolating is carried out."[178] Consider again METRO 2033, which takes as a starting point the ever-raging conflict between ideologies and extrapolates it into the future, into a world that now shows the ramifications of these conflicts, after the nuclear bombs have fallen. Instead of being an instrument of futurology, however, the game illustrates that these conflicts continue after the cataclysm and not only paints a bleak picture of what is to come but, further, of what already is, and what can still be averted by the player (in the virtual as well as empirical world). Such a conclusion underlines the effectiveness of many VGDs and points to the fact that the player's imaginings and

174 Ibid., 70.

175 Ibid., 80.

176 Ibid., 71.

177 Ibid., 76; cf. 28, 76.

178 Ibid., 76.

ergodic actions are never insulated from his empirical surroundings but are informed by them—as he both steps "away from a known world and yet [is] always in creative connection with it."[179]

As one can discern by my use of overlapping quotes, I do not distinguish between the reader's and player's experience in this respect. Both are confronted with an estranged world that shows their empirical world in a "grotesque distorting mirror"[180] which requires deciphering—a mirror that, as Suvin puts it, "is not only a reflecting one, it is also a transforming one, [a] virgin womb and alchemical dynamo … a crucible."[181] What is different, though, from the reader's experience is the player's extended involvement in the worlds of the video game. Similarly, Pawel Frelik warns that one cannot simply utilise "the set of generic protocols formulated from an analysis of literary SF" and apply them to the video game medium without having in mind the "consequences for the protocols themselves."[182] Consequently, when analysing VG SF, one shall consider the similarities and differences between media but focus on "a new set of approaches that recalibrate critical attention to those aspects that are inherent in the medium's unique deployment of the SF repertoire," which centre on "its species of visuality as well as its interactivity and performativity."[183]

I have formulated my own phenomenological approach to VGNs in chapter IV and will now integrate the specifics of the SF repertoire into my deliberations. For this purpose, it is necessary to expand on the notions of *estrangement* and *extrapolation* in games, and Patrick Jagoda's conclusions are helpful here. To him,

> [d]igital games alter the nature of cognitive estrangement, speculative thought, and world building that have been central qualities of science fiction literature. They do so through the introduction of gameplay mechanics, procedural capacities, navigable worlds, and multimedia interactions.[184]

Jagoda's conclusions are both generic and inspiring, for it is clear that with a new medium, new horizons await the researcher. However, they point to the fact that in the world-building of SF games, elements such as norms and conventions taken

179 Moylan, *Scraps*, 24.

180 Ferns, *Narrating Utopia*, 109.

181 Suvin, *Metamorphoses* 5.

182 Frelik, "Video Games," 227.

183 Ibid., 228; cf. 227-228.

184 Jagoda, "Digital Games," 150.

from the empirical world are redistributed horizontally in the game and in its perspectives: in its world, rules, processes, and the player's actions. As such, they are magnified in scope, distorted, and disrupt the empirical world's vertical hierarchy of these elements.

It is as Doležel claims: "The author creating a literary fictional world draws on the actual world in many ways: adopting its elements, categories, and macrostructural models."[185] This "material coming from the actual world has to undergo a substantial transformation at the world boundary," and "actual-world entities have to be converted into nonactual possibilities."[186] Such a transformation of the known into the unknown (and the potential) fabricates a defamiliarised world for the reader/player to decipher, and to illustrate this process is to approach the *aesthetic function of the SF genre*. This is an important step, because SF is often described as a form of *escapism* from the frustrations of the empirical world that enables the appreciator an imaginative/ergodic vacation from the real (which is only partially true). Contrary to this claim, Suvin remarks that the escapism SF grants

> is an optical illusion and epistemological trick. The escape is, in all such significant SF, *one to a better vantage point* from which to comprehend the human relations around the author. It is an *escape from constrictive old norms* into a different and alternative timestream, a device for historical estrangement, and an at least *initial readiness for new norms of reality*, for the novum of dealienating human history.[187]

Consequently, SF, utopias, and dystopias offer the appreciator pleasurable as well as subversive ventures into estranged elsewheres and are *firmly anchored in the empirical world*.[188] They vividly "participate in reality in an active and productive way"[189] and "foreground both the connection between utopia and reality … and the essential conflict between them."[190] As such, the connection between worlds is similar to "the *heyiya*, or hinge—most commonly depicted as a pair of interlocking spirals"[191]—and this relation must not "be regarded simply as 'metaphorical' (worse yet, 'mythic') retelling of the present moment" but instead "pursues

185 Doležel, *Heterocosmica*, 20-21.

186 Ibid., 21.

187 Suvin, *Metamorphoses*, 84; emphasis added.

188 Ferreira, "Biodystopias;" 52; Booker, *Impulse*, 19.

189 Ibid., 14.

190 Ferns, *Narrating Utopia*, 209.

191 Ibid., 216-217.

a more complex engagement that enters a *dialectical negotiation* of the historical tension between what was, what is, and what is coming to be."[192] It follows that to uncover the mysteries of SF depends on the appreciator's acts of ideation through which he will establish the connection between realities.[193]

To create the estranged world, then, there are primarily two ways of setting things up. The "trace," as Jameson argues,[194] between past, present, and "the not-yet-being of the future"[195] is created through the scientific strategy of *extrapolation*. This technique is the driving force behind the SF genre and occurs in *direct* or *analogic* form. In its first form, a "direct, temporal extrapolation"[196] from the present into a potential future creates worlds that "expose current problems and warn of the dire consequences."[197]

> Extrapolative SF begins with the current state of the empirical world, in particular the state of scientific knowledge, and proceeds, in logical and linear fashion, to construct a world which might be a future extension or consequence of the current state of affairs.[198]

This is also why extrapolation is suitable for dystopia. Consider Eidos Montreal's DEUS EX: HUMAN REVOLUTION—a 2011 critical dystopia of variant II—where in 2027 scientific advances in augmentation technology have enabled humankind to augment their physical and cognitive abilities through prostheses and implants. The game builds its premise on real-world augmentation technology (prosthetic limbs) and imagines the consequences of such technology and its use in a dark, cyberpunk gameworld. Corporate struggle for power, the hubris of pushing

192 Moylan, *Scraps*, 25; emphasis added.

193 Given this observation, it is not enough to describe video game fiction—and SF games in particular—in *a metaphorical relation* to the empirical world. Such an enterprise (pertaining to games in general) was conducted by Sebastian Möring, who claims that "games and play are always already metaphoric" (Möring, "Games and Metaphor," 57)—and "that there is not only a reciprocal relationship between play and metaphor but instead a triadic relationship between the concepts of play, metaphor, and representation." (Ibid., 106).

194 Jameson, *Archaeologies*, xv.

195 Ibid., xvi.

196 Suvin, *Metamorphoses*, 27.

197 Moylan, *Scraps*, 44.

198 Brian McHale, "Towards a Poetics of Cyberpunk," in *Beyond Cyberpunk: New Critical Perspectives*, ed. Graham J. Murphy and Sherryl Vint (London: Routledge, 2010), 4.

boundaries, and a two-tier society are the result, in which augmented people are driven into isolation and forced into addiction through the consumption of a drug that prevents the human body from rejecting the prosthetics. DEUS EX: HR is thus a good example of the novum's totalising hegemony. It virtualises a world that was meticulously built around a cognitive novelty and which is experienced by the player through both the world he perceives and the PC he manoeuvres—for Adam Jensen is highly augmented as well, and the player may experience the advantages and disadvantages of this technology through play.

Direct extrapolation thus creates a connection between empirical and SF world that is neither simple nor "flat"[199] but that rather stands "in a metonymic relation to the current empirical world."[200] Scientific accuracy are its hallmarks, and the technique is often employed in so-called *hard SF*, which requires meticulous deciphering by the appreciator.[201] Whether this results in the creation of an alternative history (BIOSHOCK INFINITE), a faraway future (DEUS EX: INVISIBLE WAR), or a near-future intensification[202] (WATCH_DOGS 1 and 2) is of no great importance, since cognitive estrangement is present in all of these, exerting its influence on the player.

There is, however, a second form of extrapolation that postulates a "more totalizing and more indirect"[203] relationship between realities, and this variant can be discerned in its analogic offshoot. This "richer, more complex model"[204] is sometimes connected to *soft SF* and "[s]peculative world-building."[205] It "involves an imaginative leap, positing one or more disjunctions with the empirical world which cannot be linearly extrapolated from the current state of affairs."[206] For Moylan, the analogic model does not stand in contrast to direct extrapolation but further, and in creative manner, defamiliarises the fictional world. He complements Suvin's deliberations[207] on the matter as follows:

> The simplest form of analogic modeling is close to the extrapolative in that it posits an alternative world that stands in almost a one-to-one relationship to the author's, even though

199 Suvin, *Metamorphoses*, 76.

200 McHale, "Poetics of Cyberpunk," 4.

201 Ibid.

202 What Seeber refers to as "Zuspitzung." (Seeber, *Selbstkritik der Utopie*, 38).

203 Moylan, *Scraps*, 44.

204 Ibid.

205 McHale, "Poetics of Cyberpunk," 4.

206 Ibid.

207 Suvin, *Metamorphoses*, 27-29.

the alternative space is as far back as the prehistoric or as far away as another galaxy or the end of time. … In the 'highest' form of analogic sf, the alternative world does not come across in a direct correspondence to the empirical world but rather is ontologically and epistemologically modulated and distanced so that rigorous work must be done to articulate the critical connections.[208]

Gleaning from this description, the analogic model is partially intertwined with that of direct extrapolation but requires intricate acts of ideation from the appreciator. This is so because the worlds he encounters are "quite fantastic (in the sense of empirically unverifiable) as long as they are logically, philosophically, and mutually consistent."[209] Thus, even gameworlds such as BEYOND GOOD AND EVIL's and far-off worlds like in HORIZON ZERO DAWN (Guerrilla Games, 2017) can be integrated into the realms of SF. Whereas the former is filled with anthropomorphic animals, the latter creates an intricate inversion in which humankind has been pushed back to tribal societies, whereas robotic dinosaurs roam the wilderness. Another example of analogy is TALES FROM THE BORDERLANDS. Here, corporations rule the planet Pandora from their towering skyscrapers (exemplified through the space station of Hyperion that harbours a variety of bureaucrats), while the populace suffers in the realms below, plagued by miserable conditions, violence, and greed: traits which are necessary to survive in such a world.

In sum, both direct and analogic extrapolation create gameworlds that depict not only *what could be* (or what could exist somewhere far away) but, in a sense, *what already is* (in the empirical world). They plunge the player into worlds unknown, creating "a sense of wonder"[210] established in diverse ways, and have him negotiate the connection between realities. However, it is sometimes difficult to discern whether the connection is purely direct or intermingled with analogic techniques, and many games employ a mixture of both forms.

What is of greater importance, though, is that the VGD involves the player in *games of estrangement* by placing him within an unfamiliar scenario he has to make sense of. These games do not exhaust themselves in spatial references (the layout, signs, or peculiarities of the gameworld) and its inhabitants (who seem strange to the player) but extend to the world's underlying processes and the player's actions. The aesthetic response of dystopia (and of all significant SF) can thus only be experienced through the player's acts of ideation and by closing the blanks between the perspectives he encounters and helps create. Consequently,

208 Moylan, *Scraps*, 44.

209 Suvin, *Metamorphoses*, 29.

210 Moylan, *Scraps*, 7.

and in a last theoretical manoeuvre, I now wish to turn to said discussion and to uncover the mysteries leading to the player's aesthetic response to dystopia.

5.2.1 Emancipated Play as the Player's Efforts of Ideation (Previous Research)

Throughout video game studies there has been a tendency to describe the player's combinatorial efforts while playing a game. What is surprising, though, is that while some of these theories mention Iser's contributions (and draw analogies to his work), his insights are employed only sparsely, although they promise insightful results.[211] Consequently, I will begin this section by giving an overview of previous research on the player's efforts of synthesis and ideation, to then develop my own model by integrating Iser's deliberations in a precise manner. These insights will be helpful to describe not only the forms of indeterminacy inherent in a gameworld and how the player fills them in through ergodic and imaginative action—what I will describe in terms of a creative dialectic with dystopia's implied player. In addition, they will illustrate how this process results in the creation/comprehension of *plot* and the *aesthetic object*, leading the emancipated player into the realms of significance.

I therefore wish to return to Calleja's discussion of *alterbiography*, an aspect I briefly mentioned in a footnote in Chapter IV. Calleja's observations are loosely related to Iser's phenomenology, and he describes the player's *process of synthesis* as the interaction "with both arbitrary and iconic signs (i.e., verbal text, images, and audio) as well as with the rules of the game."[212] This negotiation with what Calleja has deemed the *scripted narrative* (any aspect inscribed into the implied player) and the "stringing together [of] a series of causally related segments"[213] leads to the formation of alterbiography—which is to say, "the active construction of an ongoing story that develops through interaction with the game world's topography, inhabitants, objects, and game rules and simulated properties."[214] While touching on this important aspect, however, Calleja's observations fail to encompass the relations between the various elements—how the friction between them

211 For example: Engelns rejects Iser's theory on the basis that it seems too restricted to describe the dynamic medium of the video game. (Engelns, *Spielen und Erzählen*, 68-70).

212 Calleja, *In-Game*, 128-129.

213 Ibid., 129.

214 Ibid., 127; cf. 120-124, 127-129, 132.

create indeterminacy and compel the player to certain imaginings, actions, and the creation of the aesthetic object.

Nitsche proceeds likewise, with observations that are similarly based on Iser's process of ideation and describes how a game's structure can "foreshadow this ideation through the design of … *evocative narrative elements* [ENE], but it comes to life only through the work of the player" who weaves in the connections.[215] ENE include "anything and any situation encountered in a game world that is structured to support and possibly guide the player's comprehension."[216] Imagine the player of DEUS EX: HR arriving at his apartment after having suffered severe injuries and consequent modifications to his body. Entering the room, an artificial voice welcomes the player, while the electronic curtains of three large windows are raised. They give way to rays of light illuminating the room and a marvellous view of Detroit. No doubt the scenario triggers recollections of films like BLADE RUNNER and evokes a noir, cyberpunk image. The apartment is located in a skyscraper high above the grounds of the city and is filled with props that give the player insight into the PC Adam Jensen: the apartment's style, stacks of books on the floor, to a large television which hides a stash of weaponry, ammunition, and supplies. As such, the room evokes images of detective fiction, and in the adjacent room (with a bed, get-well cards, a box of cereal called Augmentchoo, and a book on prosthetic limbs) further insight into Jensen's psyche await: a smashed mirror in the bathroom suggests frustration after having seen his reflection after the modifications.

These props, ENEs, or perspective segments (whatever one wishes to call them) fuel the player's comprehension of this world and aid its *cognitive mapping*.[217] Every aspect in the room (and in the gameworld) was placed for a particular purpose and in a specific arrangement. What is lacking, though—what remains indeterminate—are the connections between the aspects, which have to be inferred by the player. Such a task is quite "demanding," since the gameworld often confronts the player with a "multiplicity of perspectives"[218] he has to negotiate and describes an activity that comes close to the reader's involvement in postmodern fiction, when he has to make sense of a highly fragmentary arrangement of events that each offer "but a glimpse of the overall story."[219] As such, it would

215 Nitsche, *Game Spaces*, 43; emphasis added; cf. 43.

216 Ibid., 37.

217 Newman, *Videogames*, 114-117.

218 Maziarczyk, "Playable Dystopia," 243.

219 Ibid., 243; cf. 243.

not be unfair to compare Nitsche's ENEs to what I have termed perspective segments, though his exclude ludic and procedural elements.[220] They can therefore be regarded as "the foundational building blocks"[221] of a game narrative, its "suggestive markings"[222] the player reads and combines into a coherent plot, but which he also "cocreates … and changes."[223]

ENEs thus exert a strong influence on player actions, and Nitsche comments on this by describing the end of DEUS EX: HR, where after a lengthy experience the collected impressions lead the player to action and the choice between three different endings—"each following a specific philosophy with its ethical qualities."[224] This decision was long in the making and stands as an example of how the player's imaginings influence his ergodic actions in the critical dystopia of variant II.

Consequently, although Calleja's and Nitsche's deliberations have gone further than others (which centre on how narrative is conveyed through gamespace), they only touch on the complex dialectic between fictional and empirical world a VGN affords. Instead, they remain caught up in the diegetic level of plot comprehension/creation and neglect to address how a game triggers an aesthetic response in the player. To move the investigation further into this direction—and into the realms of *significance*—it is beneficial to resort to theories of *environmental storytelling*, a world-building technique borrowed from theme park design that is concerned with the artful integration of story elements in the environment to create causation between them.[225] Such a technique is beneficial to the developers and the player alike, for the former may direct the plot and guide the player—as sort of "narrative architects" who "design worlds and sculpt spaces"—while the latter rejoices in the imaginings that are fuelled.[226] Although environmental storytelling theories focus on the player's comprehension of the diegesis, they nonetheless point in the right direction by suggesting that the gameworld formulates a *malleable space* which negotiates the past, present, and future of this world and creates

220 Nitsche, *Game Spaces*, 43.

221 Ibid., 37.

222 Ibid., 44.

223 Ibid., 45; cf. 37, 44-45, 52.

224 Ibid., 44; cf. 44-45.

225 Don Carson, "Environmental Storytelling: Creating Immersive 3D Worlds Using Lessons Learned from the Theme Park Industry," *Gamastura: The Art and Business of Making Games* (2000), https://www.gamasutra.com/view/feature/131594/environmental_storytelling_.php

226 Jenkins, "Architecture," 121; cf. 124-126.

ties to realms beyond it—in the form of allusions to other fictional worlds and to what players know from the empirical world.

One important theory in this respect is Clara Fernández-Vara's discussion of *indexical storytelling*. Becoming involved in a VGN, players enter "a game of *story-building*, since the player has to piece together the story, or construct a story of her own intention in the world by leaving a trace."[227] This game is neither confined nor exclusive to the gameworld, and an important aspect of Vara's discussion is the use of so-called *indices*, "a type of sign" that "can help the player situate herself in the world ... and relate her experience in the game with her previous knowledge."[228] Indices, as such, set the player in the role of a "detective"[229] and function as both "markers pointing the player towards what she has to do ... but [also] encourage her to interpret and reconstruct the events that have taken place in the space."[230] Vara explains this aspect using the example of BIOSHOCK, whose gameworld is filled with signs that point the player in the right direction or give her information: for example, signs naming different areas of the game and directional signs, or Plasmid advertisements and propaganda posters that grant players further insight into the world's ideology.[231] In addition, Vara mentions indices that point to the gameworld's history—and to Rapture's fall from a utopian into a dystopian society[232]—by leaving a trace of bygone events for the player to decipher. Such traces can also be left through the player's actions, who in retrospect will interpret them as being part of this world's recent history.[233]

It is undeniable that spaces which trigger imaginings about the gameworld's past are important to the critical dystopia, which lays emphasis on how dystopia came into being or has developed from a previous Utopia. This process is often fuelled by what Jenkins has called *embedded narratives*. These transform the gameworld into "a kind of information space, a memory place"[234] in that they prestructure player participation by embedding micronarratives into the environment

227 Clara Fernández-Vara, "Game Spaces Speak Volumes: Indexical Storytelling," *Proceedings of the 2011 DiGRA International conference: Think Design Play* 6 (2011): 1, http://www.digra.org/digital-library/publications/game-spaces-speak-volumes-indexical-storytelling/

228 Ibid., 5.

229 Ibid., 6.

230 Ibid., 5.

231 Ibid., 8.

232 Schmeink, "Dystopia."

233 Fernández-Vara, "Indexical Storytelling," 5-6.

234 Jenkins, "Architecture," 126.

which prompt his imaginings (see my discussion of JOURNEY in chapter IV).[235] Embedded narratives often stand in close relation to the environment they were placed in and may include environmental signs, specific architecture like ruins, text notes, audio logs, pictures, hallucinations, video recordings, retrospective narration, or playable flashbacks that show the present environment in a different light, and so on.[236] Such references refuse to lock the dystopian game "into an eternal present"[237] and lay the focus on the extrapolative trace between *what was* and *what is*. They thus confront the player with at least two societies he has to negotiate: the fictional world's past/present and the empirical world from which these stem.

For all the benefits of environmental storytelling theories, which lay the focus on the gameworld's past, one must not forget gamespaces' concern for the future and how players make predictions based on their previous experience within the game. As Tosca claims: "Experiences are always a combination of past, present, circumstances and future expectations" and have "aesthetic potential."[238] The participant makes sense of things through "operations" such as "anticipating, connecting, interpreting, reflecting, appropriating and recounting."[239] Consequently, regarding the gameworld as a space in constant flux between *what was, what is, and what will be* is essential to playing dystopia. This can be explained in that the gameworld opens up a space for "performative simulations, conveying a sense of the malleability of the future"[240] to the player, who aims to decipher the moment utopian enclave mays transform dystopia. Gameworlds, in other words, exhibit a plethora of *pregnant moments* (known from pictorial representations) that open up a "temporal window" into the past but also "let the spectator imagine what immediately … followed the represented scene."[241] This occurs when the player draws conclusions from past/present moments and anticipates the near or far-off future, which may include predictions about his own actions (depending on the current situation), the blood on the wall that might announce further confrontation, a lit

235 Ibid.

236 Ibid., 127.

237 Ibid.

238 Susana P. Tosca, "Transmedial Museum Experiences: the Case of Moesgaard," *NODO: Transmedial Narratives* 18 (2016): 58, https://www.researchgate.net/publication/312024183_nodo_18_3067-13176-1-PB

239 Ibid. Tosca bases these claims on a theory developed by (John McCarthy and Peter Wright, *Technology as Experience* [Cambridge, Mass.: The MIT Press, 2004], 6, 19).

240 Frelik, "Video Games," 229.

241 Ryan, *Avatars*, 15; cf. 15.

up barn that evokes imaginings of inhabitants, or dark clouds that herald a thunderstorm.

This dialectic between past, present, and future is an integral quality of dystopian games and helps the player decipher utopian enclaves hidden within the gameworld. Yet the connections he weaves are not confined to the gameworld and are inspired by his world knowledge, the cultures he is familiar with, and other fictional worlds or ludic conventions he has come to know. The prompters, thereby, may include characters, processes, or spaces that remind the player of other (fictional) personae, mechanisms, or architecture/styles he knows from the empirical world or previously visited fictions. Jenkins has called such spaces *evocative spaces*[242]—but these can easily be extended to include any element of the gameworld (characters, processes, or certain types of actions) that creates ties to other games, narratives, or things the player knows from the empirical world. This is the case, for example, in BIOSHOCK INFINITE, which through its damsel-in-distress structure and the tower in which Elizabeth is locked up reminds the player of the fairy tale Rapunzel, as well as other analogies to real-world entities like the Statue of Liberty or the World's Columbian Exposition are imaginable. Consequently, by involving the player in a world that seems strange yet familiar, *evocative elements* (to expand on Jenkins' term) either "enhance our sense of immersion with a familiar world or communicate a fresh perspective on that story through the altering of established details"[243]—which formulates an integral quality of aesthetically complex games and is an important incentive for emancipated play.

5.2.2 Utopian Enclaves as Vacancies Within the Gameworld

These deliberations pave the way for a discussion of utopian enclaves hidden within the gameworld that may lead the dystopian reality into a utopian direction. These may be found in a variety of aspects and references in the game, and they need not occur in a comprehensive form—one may recall in this respect *Nineteen Eighty-Four* and Winston Smith's resistance in the form of a diary and his sexual relationship with Julia in the woods.

Consequently, utopian enclaves may be found in embedded narratives and allusions to the past, when the world was a better place (such as Artyom's hallucinations in METRO 2033, which for brief moments transform the derelict world of the game into peaceful scenery by showing glimpses of how it was before, when

242 Jenkins, "Architecture," 123-125, 129.

243 Ibid., 129.

playgrounds were filled with life and the cheering of children). They may be found in diegetic music tunes (in the songs of the band 'Die Käfer' in WOLFENSTEIN: THE NEW ORDER, which satirically comment on the prevailing system) and references to other fictional worlds or places in the real world that promise hope (for example, when THE LAST OF US creates connections to ecological fictions like the optimistic works of the French writer Jean Giono or triggers imaginings about a life in balance with nature, untouched by the viral nature of modern capitalism).

In addition, utopian enclaves can be found in certain types of spaces that evoke specific emotions in the player, for instance, those influenced by "expressionism (which maps emotions onto physical space) [[244]] and romanticism (which endows landscapes with moral qualities)."[245] Both find their way into many VGDs, such as the posthuman fantasies of NIER: AUTOMATA (Platinum Games, 2017) that juxtaposes natural spaces with the ruins of human civilisation and consumption, and reflects on the emotions of androids (who the player controls) as they manoeuvre through these spaces. By embedding such spaces within the gameworld in an artful manner, utopian microcosms often shine with marvellous scenery, architecture, and labyrinthine structures. They create near-perfect design and behaviour that result in delightful though thought-provoking possibilities for play.[246] One could imagine here FLOWER (Thatgamecompany, 2009), which juxtaposes dreamlike nature sections that have a calming effect on the player with its bleak menu, depicting a rainy city. Whereas such spaces can only transform into dystopian ones through inappropriate player behaviour—Walz mentions SHADOW OF THE COLOSSUS (Team ICO, 2005), where the player disrupts the tranquil scenery through her actions by killing the behemoths—there are others that are inherently dystopian.[247] These still "provide pleasure in setting up entirely unenjoyable, i.e. frustrating places that must be playfully escaped, saved, destroyed, or equilibrated"[248] and can be found in game sections like arenas, multicursal mazes, or similar structures that afford pleasurable frustration.

All in all, these conclusions illustrate the range of hopeful possibilities that pervade the VGD and, at times, push the supposedly pessimistic genre towards a fully-fledged Utopia. The allusions and references to better times and places

244 Domsch calls such spaces "Seelenlandschaft," which depict "landscapes that reflect the mood of a protagonist, a scene, or a whole narrative (e.g. the fact that it is raining at a funeral)." (Domsch, *Storyplaying*, 103).

245 Jenkins, "Contested."

246 Walz, *Toward a Ludic Architecture*, 137-138.

247 Ibid., 140.

248 Ibid., 146.

thereby play an important role in the transformation of dystopia. But in order to effectuate successful change (that is, to escape the boundaries of the classical dystopia), more than pure imagination is needed, and the player's ergodic efforts of actualising the utopian horizon and setting the dystopian gameworld on a liberating route towards Utopia become of importance. Hence it follows that the enclaves to be found in the *gameworld's system* and the *player's agency* are the most promising.

To explain this issue, let me refer back to the systemic nature of a (dystopian) game and its system of rules that channels the player's movements and degree of agency through aspects such as labyrinthine structures, invisible barriers, cutscenes, obstacles, possibilities to act, and so on.[249] These constraints

> imposed by the games should be seen as having a central role in constructing dystopian narratives. The underlying message of the constraints on player freedom is that some situations may be so deeply flawed that they are beyond redemption. In such situations, even the most heroic and courageous character is powerless to fix the devastated world or to move it in a radically different direction.[250]

I have deemed this type of dystopia the classical dystopia and illustrated its confines with the example of THE STANLEY PARABLE and EVERY DAY THE SAME DREAM. Yet, with the critical dystopia, loss is not an option, and rather than succumbing to the confines of the game system, the player aims at discovering its glitches[251] and "the flaw of the system" that "invariably … underestimate[s] the player's agency" and eventually leads to its downfall.[252] This is the case in both variants of the critical dystopia, although in its second form, the player will be held responsible for finding it.

---

249 Domsch, "Dystopian," 398, 400, 402; Schulzke, "Virtual," 316, 327, 329, 330.

250 Ibid., 330.

251 Dystopian game systems *enclose the player within their confines* (to recall the player's failed escape in *The Stanley Parable*, chapter II). They do so by illuminating the system through "authoring arguments through processes" and making "claims about *how things work*" (Bogost, *Persuasive*, 29) but go further by prohibiting players "to raise *procedural* objections—that is, the player of a videogame is usually not allowed to change the rules of play." (Ibid., 37). In the VGD as critical dystopia, this prison house of procedural objection is flawed, however, and inscribed into the game for the player to find any glitches.

252 Domsch, "Dystopian," 402; cf. 400-402.

It follows that utopian enclaves with the potential to shatter the dystopian system are hidden within the processes of a game, its system of rules that affords them, and the player's agency to actualise these potentialities. They are *vacancies* in the game system and require the player's combinatorial efforts in deciphering them and his ergodic actualisation; otherwise the enclaves will remain hidden within the system (which is not to say that the player may not stumble upon them by pure chance). It is this "system-shattering agency" that makes the critical dystopia in video game fiction "utopian at heart"[253] and that requires a form of agency that has the player negotiate his possibilities for action within the context of a fictional storyworld and the narrative that fuels it (as described in chapter IV).

However, what begins in a process of realisation and the consequent measures to change the gameworld does not remain within the virtual space of the game but extends into the real world. In this sense, the actualisation of the utopian enclaves in the gameworld shows the quality of an *event* as described by Alain Badiou.[254] It creates a "rupture"[255] with the current state of affairs and in a risky but pleasurable act of transgression evokes an experience of "happiness" through "exceptional moments." One of these is when the player breaks with the repressive order in the gameworld and becomes sensitised to similar action in the real world. Such an action requires the activation of the player through *emancipated play*, which creatively negotiates what dystopia's implied player has offered. It occurs, for example, when the player of BIOSHOCK delays his need for immediate gratification and refrains from harvesting the Little Sisters to actualise a utopian horizon at the game's end (see chapter VI). This event powerfully extends into the real world (when the player realises that participation in useless consumption comes close to harvesting little girls or third-world workers) and illustrates the seductive power of fiction, when virtual actions create happiness and lead the individual to emancipation and to taking action in the real world.

Emancipated play, in other words, often entails precarious play, which is the result of an event that is risky but exceptional. It presupposes *a gap, a place of*

---

253 Ibid., 403.

254 Sicart similarly, though for a different purpose, employs Badiou's "événement" as "an experience of delimited boundaries with a series of imperatives that have to be assumed in order to become a subject" and considers it "operationally similar to" a game itself. (Miguel Sicart, *The Ethics of Computer Games* [Cambridge, Mass.: MIT Press, 2009], 71).

255 Alain Badiou, interviewed by Mira Davidson, "Happiness is a Risk that we Must be Ready to Take," *Verso*, https://www.versobooks.com/blogs/2032-alain-badiou-happi ness-is-a-risk-that-we-must-be-ready-to-take

indeterminacy that represents the player's primary desire for Utopia (what Jameson has called the unconscious utopian impulse), whose filling brings about the rupture needed to transform the current capitalist order. In order to sustain the event's effect on political thought and action, however, a "fidelity" to the event is necessary that manifests itself in "the continuous creation of the rupture itself"—that is to say, in its constant renegotiation "by way of a collective discussion."[256]

According to this argumentation, *aesthetic response* (created in and through the act of play) may transform into political action. In order to sustain the response's effect, however, a continuous negotiation between all parties involved becomes necessary and a willingness to change that requires a *self-sacrificing stance* from the player (in both the virtual and real world). It is the first step towards Utopia (as conducted by the ethical player of BIOSHOCK) that illustrates an endeavour which is, nonetheless, often aggravated through finding suitable actions to break with the status quo.

Fiction can thereby evoke additional confusion, for the distorting realities it creates may have players ponder the true meaning of the gameworld, its processes, and the goals they pursue. Think about the processes and player actions that occur in the BORDERLANDS series, and that the constant renegotiation of images these evoke transforms an initially fun experience of pursuing goals, conducting strategies, killing bad guys, or hunting down the treasures of Vaults into one of alienation. As a consequence, players will realise that what they are participating in is a defamiliarised form of capitalism's most brutal excesses. This state of affairs can be partially explained with the form of estrangement that marks VG SF—when a game "raises speculative questions through the immediate player experience of agential actions and interactive processes rather than the exposition of a science fiction novel" and combines those with "often-experimental mechanics."[257]

In other words, the experience of gameworld processes and certain player actions is distorted in the VGD, since they often do not mean what they denote. Let me briefly resort to animal play to explain this phenomenon, which, as Iser claims, is similar to a reader's involvement in the literary text in that the "signifier is fictionalized … because denotation is laid to rest and what is said is not meant."[258] Iser bases this claim on Gregory Bateson's observations that in animal play actions are generally not what they seem:

256 Ibid.

257 Jagoda, "Digital Games," 141.

258 Iser, *Imaginary*, 248.

> Paradox is doubly present in the signals which are exchanged within the context of play, fantasy, threat, etc. Not only does the playful nip not denote what would be denoted by the bite for which it stands, but, in addition, the bite itself is fictional. Not only do the playing animals not quite mean what they are saying but, also, they are usually communicating about something which does not exist.[259]

For Iser, the result of this process is a "split signifier [which] signals that this is play" and suspends it "denotative function."[260] This vacancy gives way to something new. "By removing something, it makes something else possible" and thereby liberates the original signifier's "implications."[261]

Transferring these insights to playing dystopia is fruitful, since it clarifies why processes and player actions (and those of NPCs) often do not mean what they denote. It will help explain why the lock picking in BIOSHOCK INFINITE does not simply stand for unlocking the next area of the game but, on a further level, illustrates the unlocking of Booker DeWitt's unconsciousness. It will help illustrate why harvesting a Little Sister no longer means to simply kill a little girl but to succumb to the internal mechanisms of a rapturous capitalism. Finally, it will help explain why Elizabeth's tower not only resembles a prison under the guise of freedom for her but, on a grander scale, for any of those in the empirical world who relish their imprisonment in the iron cage of Anti-Utopia. What follows from these observations is that player actions are akin to *shifting images*, where signifiers lose their denotation and give rise to "free-floating implications."[262] These places of indeterminacy can be filled and deciphered, however, bringing to the fore unexpected implications—and it is through the player's acts of ideation and by closing the blanks between the various perspectives that this becomes possible.

5.2.3 The Repertoire of the Game and the Oscillation Between Completion and Combination

Before coming to the image creation in the act of play, it is necessary to point out that playing dystopia demands considerable work from the player and that the games of estrangement not only test his combinatorial and ludic skills but also his

259 Gregory Bateson, *Steps to an Ecology of Mind: Collected Essays in Anthropology, Psychiatry, Evolution, and Epistemology* (Lanham, Maryland: Jason Arson, 1987), 188.

260 Iser, *Imaginary*, 248.

261 Ibid.; cf. 247-248.

262 Ibid., 249.

world knowledge. This aspect is readily accepted by the player, for to understand the gameworld, he orients himself in the *familiar aspects of the game*: what he knows from the real world or from other fictional (game)worlds (a prototypical chair or landscape, certain norms and processes, or character behaviours). These aspects are called the *repertoire of the game*. They facilitate the player's comprehension of the game and enable him to establish links and associations with what he already knows.[263] "Understanding the repertoire is a matter of competence,"[264] and it is needless to say that the emancipated player's state of knowledge will ultimately determine the complexity of his interactions with the implied player.

(Dystopian) games thus demand a few competences from the player, and the first necessary requirement is the ability to play a game. Calleja has termed this form of involvement *kinesthetic involvement*, since it is only when players grasp the controls of a game that they are able to move around its environment and act out their fantasies.[265] A basic level of skill thus formulates the premise for (emancipated) play—and of further benefit is the player's knowledge of ludic or narrative genres (which fosters his understanding of the gameworld, its rules and conventions; raises expectations and creates ties to other games/narratives) or philosophical aspects (that have him ponder about the meaning of the experience).[266]

The repertoire of the game thus informs the player's playthroughs and narrows certain actions/imaginings/interpretations.[267] It helps him fill in the gameworld's indeterminacies and create closure through ergodic and imaginative interaction. This desire for the resolution of unanswered questions/non-actualised possibilities (or the frustration of such) is a basic inducement to the player's involvement. Every player will thereby communicate with a game differently by using his private stock of knowledge and expressing who is through play (to recall the different player types).This dialectic with the implied player is due to the structural openness of work worlds and oscillates between the desire for *completion* and *combination*.

Iser touches upon this important difference in the reader's involvement in the literary text and distinguishes between two categories of indeterminacy: 1) *gaps in the determinacy of the storyworld* and 2) *blanks that arise between the textual perspectives*. Such a distinction comes suspiciously close to what the player encounters in video game fiction, and, therefore, I wish to quote Iser at length:

---

263 Engelns, *Spielen und Erzählen*, 87-88; Egenfeldt-Nielsen et al., *Understanding*, 204.

264 Ibid., 206.

265 Calleja, *In-Game*, 71.

266 Egenfeldt-Nielsen et al., *Understanding*, 204-206, 216-217, 232-233.

267 Ibid., 206.

> What we have called the blank arises out of the indeterminacy of the text, and although it appears to be akin to Ingarden's 'places of indeterminacy,' it is *different in kind and function*. The latter term is used to designate a *gap in the determinacy* of the intentional object or in the sequence of the 'schematized aspects'; the *blank*, however, designates a vacancy in the overall system of the text, the filling of which brings about an *interaction of textual patterns*. In other words, the need for *completion* is replaced here by the need for *combination*.[268]

Iser comes to this conclusion from the premise that "effect and response arise from a dialectical relationship between showing and concealing,"[269] and that the literary text indulges the reader in this dialectic through that "which is not formulated"[270] and has "the reader … work it out for himself."[271] I have already touched on what this means for the VGN, which, similarly though not equal to the literary text, evokes the player's participation through *what remains indeterminate/virtual* in a game (unspoken utterances and hidden aspects that remain potential for the player to actualise).

The difference between *gaps* and *blanks* as indeterminacies of the gameworld becomes apparent once one regards the incompleteness of the gamespace, which awaits the player's imagination and ergodic actions for completion. Gamespaces are the virtualised parts of the greater storyworld and resemble a "stage" that, as Nitsche argues, "is too big and too active to be fully comprehensible."[272] For this reason, game environments are often smaller than they appear.[273] They leave many gaps for the player to fill in and imagine the gameworld. Consider BIOSHOCK INFINITE's Columbia, where the player can only traverse small parts of the city and enter a handful of buildings, while in the background further regions supposedly await. In this respect, gamespaces are akin to the decorative function of film sets, and complete (contiguous) spaces are rare, if non-existent.[274] What the player instead encounters are "discrete areas physically unconnected with those already

268 Iser, *Act*, 182; emphasis added.

269 Ibid., 45.

270 Ibid., 47.

271 Ibid., 46.

272 Nitsche, *Game Spaces*, 106-107.

273 Calleja, *In-Game*, 78.

274 An exception is Hello Games' NO MAN'S SKY (2016), which has the player explore a procedurally generated universe that simulates completeness, where only the interior of planets is not visible.

present"[275]—and this includes open world spaces, which are often separated into individual parts, be it through mountains, rivers, or other barriers.

Such *environmental gaps* speak to the player's innate desire for completion and evoke in him the urge to cognitively map the gameworld.[276] But often, these gaps are not to be filled only by imagination, and there are a plethora of game areas that remain indeterminate only so long as the player has not discovered them—for example, the player can imagine the inside of a house only so long as he has not entered it. Consequently, besides closing the gameworld gaps through imagination, the player's ergodic efforts of exploration and task fulfilment are vital to finding closure in VGNs. They speak to his desire to discover hidden parts of the gameworld and the completion of its tasks and, thereby, create important perspectives in themselves. Imagine an area the player discovers which gives him further information about the gameworld, a character who tells him a personal story, or an entire city or society which awaits exploration. As such, to attain closure in a VGN, the player is either handed information in the ongoing plot (through cutscenes, scripted events, dialogue, text, or entire playable sections[277]) or is encouraged to do the work himself, by completing quests or other chores that will reward him with bits and pieces of the total picture.

Filling in the gaps in VGNs thus differs from what appreciators have come to know in non-ergodic fictions and oscillates between what Egenfeldt-Nielsen et al. describe as the "story" and "action level."[278] Play is "informed by: our cognitive, often-unconscious filling-in-the-gaps, our sense of what we think we have to do within the game, and the hand–eye coordination that gives us our playing ability."[279] To illustrate their claims, they describe a section of RESIDENT EVIL: CODE VERONICA (Capcom, 2000) in which the player asks himself questions that will be answered by subsequent events: "Who is the guy? (open)—Umbrella guard who

275 Newman, *Videogames*, 105.

276 Calleja, *In-Game*, 78-79.

277 In this respect, Hensel mentions a gap in the plotline of THE LAST OF US that is filled in by the DLC of the game: THE LAST OF US: LEFT BEHIND (Naughty Dog, 2014). Not only does the DLC illuminate Ellie's back story, but it also closes an ellipsis the original game left unanswered, which describes (and has the player enact) how Ellie managed to treat the injured Joel. (Thomas Hensel, "Zwischen *Ludus* und *Paidia*: *The Last of Us* als Reflexion des Computerspiels," in *New Game Plus: Perspektiven der Game Studies. Genres – Künste – Diskurse*, e Benjamin Beil, Gundolf S. Freyermuth, and Lisa Gotto [Bielefeld: Transcript, 2015], 160-164).

278 Egenfeldt-Nielsen et al., *Understanding*, 217.

279 Ibid., 219.

captured me (close); Why is he opening the door? Maybe he will kill me, maybe he is a spy (open)—he opens it because the base is under attack and wants to give me a chance (close)."[280] Because it is a game, however, such gap filling on a story level is complemented by the player's urge for completion on a ludic level—what I have termed the creation of perspectives through ergodic action.

> We are in a cellblock. Start off by going to your item screen. Go to the LIGHTER you see and use "Select" with the action button … grab the HANDGUN BULLETS. On the desk near the unconscious jailer is a COMBAT KNIFE. Once you have all the items, you may exit the room.[281]

Although the distinction between story and action level is viable for analysis, I will not distinguish between them, for I consider them part of a coherent whole. It nonetheless illustrates that the gap filling in VGNs differs from what the appreciator is used to in non-ergodic art. By activating the player's cognition and having him intervene in the plot, he attains closure through a combination of imaginative and ergodic action.

This completionist activity is however complemented by the *drive to combination*. Negotiating the perspectives offered by the game and co-created by the player formulates a vital aspect of the player's participation and will ultimately lead to the creation of the aesthetic object. Sicart comes closest to my deliberations in this respect as he describes *frictions* that may occur between the different aspects of a game: between "the narrative or the characters, the gameworld, the rules and systems, or the play settings and presence of other players."[282] However, in order to explore this aspect any further, Iser's deliberations cannot be neglected and shall inform the phenomenology of play I am proposing for the VGD (and VGN).

280 Ibid.

281 Ibid.

282 Sicart, *Beyond*, 110.

Table 8: Forms of indeterminacy/incompleteness in VGNs

Gaps:	**Drive for completion**
Environmental	• Inaccessible parts of the gameworld (parts that were not constructed and function as decorative props). • 2D gameworlds, which hide certain regions. • Evocative spaces.
Abstract regions	• Reductions, condensations, or magnifications of space. • Physically unrealistic structures.
Plot progression	• Questions about the plot (what happened, how it will proceed). • Ellipses (distortions and jumps in time). • Parts of the gameworld that refer to its past or allow predictions about its future (through environmental storytelling). • Parts of the gameworld that refer to other narratives (embedded narratives).
Character personalities	• Character traits and personalities that are not explicitly elaborated and have to be inferred.
Possibilities for action (actualisation)	• Virtual possibilities for action (affordances in the gameworld that allow player participation).
Blanks:	**Drive for combination**
Between the perspectives	• Help the player understand the gameworld and the ongoing plot. • Allow him to compose images on the level of significance.
Between the gameworld and empirical world	• The level of estrangement (forms of referentiality) between gameworld and empirical world.

5.3 THE FLUX OF IMAGES AND THE PLAYER'S CREATION OF THE AESTHETIC OBJECT IN Metro 2033

It has been clarified so far that representational art—and the genres of SF, utopia, and dystopia in particular (through their games of estrangement)—denies direct access to the empirical world. Instead, representations involve the appreciator in creative games of make-believe. They urge him into imaginings of a certain kind and the player to action and grant access to their worlds through acts of ideation. This feedback oscillation between fictional and empirical world is largely encouraged by the creation of *images* and their negotiation as the player closes the blanks between the perspectives he encounters and co-creates. I have previously outlined the player's creation of images with the example of JOURNEY, and I now wish to go into further detail by resorting to Iser:

> The imagistic vision of the imagination is … not the impression objects make upon what Hume still called 'sensation'; nor is it optical vision … it is, in fact, the attempt to ideate … [*vorstellen*] that which one can never see as such. The true character of these images consists in the fact that they bring to light aspects which could not have emerged through direct perception of the object.[283]

The discussion of image creation in the act of play becomes of benefit to understand the player's venture into dystopia. It will not only illuminate how he is able to decipher the distorted dreamworlds but, in addition, will clarify the mechanisms behind the emancipatory window the genre offers—when the player catches a glimpse of the truth behind the experience and of the opaque nature of his empirical surroundings.

METRO 2033 does not deviate from this fact, and it would be hard to deny the naturalness of its aesthetic effect—sending the player on a cathartic journey to enlightenment and having him ideate the images that compose and arise from his experience. These are in constant flux and their ever-changing signifieds urge the player to reconsider his previous imaginings and actions. In this respect, the *tower* at the game's beginning and the image of the *Dark Ones* are of considerable importance. For, as Iser would say, they urge the player to imagine something that their iconic signs have not yet denoted but what is nonetheless guided through that denotation. Denotations transform into connotations, guided by the structural finesse of the game's strategies, which allows the player to see the gameworld in a

283 Iser, *Act*, 137.

different manner and to create unexpected connections to his empirical surroundings.[284]

Figure 16: The image of the tower will ideate in the player's mind and grant new insights into both the gameworld and the empirical world.

METRO 2033 (4A Games, 2010), ch. Prologue.

Perception and ideation are thus "two different [yet not mutually exclusive] means to access the world:" the former requiring an object's "presence," the latter its "absence or nonexistence."[285] Although video games differ in this respect from literary works, as the player perceives large parts of the world (similar to the spectator or the viewer, whereas the world of a book can only be imagined), it is still necessary for him to ideate the truth behind these impressions and interactions. Such an enterprise will ultimately lead to the creation of the *aesthetic object*, whose initial "insubstantiality ... spurs on the reader's [/player's] imagination" and induces him to partially complete "its shape."[286] The creation of images, thereby, is by no means an arbitrary process but is guided by the textual positions and strategies that demarcate the reader's journey and which formulate the "lines along which the imagination is to run."[287] Following this train of thought, the *strategies of a game* become of fundamental importance to the participation process,

284 Ibid., 65-66, 137.

285 Ibid., 137; cf. 137.

286 Ibid., 92.

287 Ibid.; cf. 92.

since they guide the player on the lines of the ergodic and the imaginative, and set him in a dialectic with the work's implied player.

5.3.1 The Blank and Its Ideation-Inducing Function as Positive Hint or Negation

To better understand the player's acts of ideation, let me again resort to METRO 2033 in which the player takes on the role of twenty-year-old Artyom and embarks on a dreamlike journey of disclosure and disguise towards a mysterious tower. The tower is where the plot begins, at the story's end, and eight days into the future. Artyom and Miller are about to finish their mission of dealing with the Dark Ones, a race of hostile creatures that through their psychic abilities induce nightmares in human beings and attack their stations. The first perspective segments contribute to this image. The player is about to reach the surface of what used to be Moscow and puts on his gas mask when a shadow, resembling a werewolf, bids him welcome to a frozen world. The threat is palpable, and once the player reaches the tower, his military convoy is attacked by vicious creatures. These first moments of the prologue (after which the game jumps back in time) introduce the player to the gameworld of METRO and anticipate future events. They are pregnant moments that create a space of uncertainty (indeterminacy) and function as "existential presuppositions" that, similar to the opening of a novel, indicate what *can be found* in this world and what is *still to come*.[288]

What the game achieves by doing so is not simply enabling the player's understanding of this world but setting him on journey towards the truth. This journey is structured by the form of indeterminacy I have termed the *blank*. Blanks, as Iser argues, "are the *unseen joints* of the text, and as they mark of schemata and textual perspectives from one another, they simultaneously trigger acts of ideation on the reader's part."[289] In doing so, the blank controls communication with a text/game and sets into motion the perspectives (segments)[290] encountered and created by the player. The drive for completion is here intertwined with that for

288 Doležel, *Heterocosmica*, 175.

289 Iser, *Act*, 183, emphasis added.

290 The blank not only conditions the links between the perspectives of a text/game but also the indeterminate space between perspectives segments themselves. In combination, these compose a greater perspective: for example, if the reader/player is given only segments of a character's personality and composes a coherent picture of that character through combinatorial acts. (Ibid., 182-183).

combination, and this process of "passive synthesis"[291] induces the reader/player to slowly build up a *gestalt*, to create consistency by closing the blanks between the signs/perspectives.[292] The blank, as such, not only prompts the reader/player into imagining something that is not—the absent image upon which the player can act, not the reader—but it also structures this process in a decisive manner.[293]

Devising certain strategies, choosing how to proceed in a game, or imagining the gameworld's particulars and plot details designate processes that are fuelled by the blank's structure. In this form it is akin to what Doležel calls "positive (hints)"[294]—that is to say, it helps the player comprehend the plot and the game's ludic structure and enables him to make informed decisions from these deliberations. In METRO 2033 this means understanding that the Dark Ones pose a threat and that for the sake of the gameworld and to complete the game, it is important to tackle the goal of defeating them, while comprehending how it could be done. The strategies of the game inform this process, and various perspectives aid the player's comprehension, while virtualised potentialities enable him to actualise certain imaginings he deems possible and fruitful to enact.

This first function of the blank is complemented by what Doležel calls "negative (lacunae)"[295]—and herein lies the blank's "aesthetic relevance."[296] It is when play is at its most exciting that any attempts at "*good continuation*"[297] in the comprehension process and the act of play are shattered. This is not to say that in order to create aesthetic complexity a game should be unplayable. But what it should do is have the player ponder problems, his tactics, and not present him with premature solutions to both ergodic and imaginative issues. In such cases, blanks "break up the connectability of the schemata, and thus they marshal selected norms and perspective segments into a fragmented, counterfactual, contrastive or telescoped sequence."[298] The result is a confusing array of perspectives that often contradict each other and stand in opposition in an intricate game of *agôn*. As such, they

291 Ibid., 135.

292 Ibid., 118-120, 169.

293 Ibid., 135-136.

294 Doležel, *Heterocosmica*, 174.

295 Ibid.

296 Iser, *Act*, 186.

297 Ibid.

298 Ibid.

withstand convergence/synthesis—and this runs against the player's habitual dispositions in that his expectations of the game are shattered, or at least refuted, while he tries to solve the conflicts he is presented with.[299]

Such hindrances to play may occur in basic and more complex forms. For now, however, it suffices to say that art (or at least complex art), "impedes the acts of ideation which form the basis for the constitution of meaning"[300] and spurs the appreciator into creating a sequence of images that move on one another. In constant flux, they negotiate what is presented and in conflict and what is created by the player, and devise an unprecedented newness. This occurs through "various types of negation," which "invoke familiar or determinate elements only to cancel them out"[301] and coerce the reader/player into discarding previously composed images. First degree images turn into those of a "second degree,"[302] and it is the latter to which readers/players respond most intimately. For the potential of a perennial response lies in one's own creations, through acts of negotiation and revision and by imagining the unthinkable.[303] The player's increased involvement in a game intensifies these games of proximity and distance, there is no doubt, by testing and validating the created images through ergodic efforts and acting upon them.

METRO 2033 is a well-suited example to illustrate these issues and to address the guiding function of the blank as well as the ideation-inducing hindrance of negation. When the plot moves back eight days in the past, the player has already glimpsed the hostile world that awaits him. First impressions have formed, and these continue to be informed by perspective segments once we set out to discover Artyom's home station Exhibition. The section begins in Artyom's room, where postcards of the old world are pinned to a wall. He is informed that a man called Hunter is on his way to the station and sets out to find him. On the way there, the player is made familiar with life in the Metro. There is chatter about disease and mutant attacks, while a child cries in the background. People in Exhibition blame the Dark Ones for their current situation, mentioning how they damage their prey's minds through hallucinations. The existential fear of the Other is palpable in every respect, and the player passes a hospital area with men wounded by the attacks. What can be the solution to these issues, the player may ponder. For the situation is desperate and the station will no longer endure it.

299 Ibid.

300 Ibid., 187-188.

301 Ibid., 169.

302 Ibid., 186.

303 Ibid., 186, 189.

It is obvious that these perspectives contribute to the negative image of the Dark Ones formed in the prologue, and further insights intensify these impressions. Finally, the player meets Hunter, who seems to know Artyom, as he brings him a postcard of the Statue of Liberty to complement his collection. Hunter is a high-ranking Polis Ranger and constitutes the first important character perspective, with a clear motto: "If it's hostile, you kill it."[304] The conversation is interrupted by a sequence in which the mutants attack. The player quickly gathers a weapon and ammunition, tensely awaiting the upcoming action. Exhibition is saved for now—in a brutal skirmish—and an additional perspective that complements the horizon of past perspectives is created through the player's actions. For now, the situation seems clear: The Dark Ones pose a threat that needs to be dealt with, and the perspectives the player has gathered and co-created strengthen this insight. Various blanks were closed in the process, which guided the player's involvement in the game, and led to the premature solving of the conflict. These deliberations and actions are further propelled by another perspective created through the informational distance between Artyom and the player. Although these share a similar point of view (first person) and most of the action Artyom conducts (except in cutscenes), the player does not know Artyom. He lives in a world that is unfamiliar to the player, which is emphasised by Artyom's retrospective narration of the events leading up to the tower. This arrangement of affairs leads to the creation of the most interesting blank between the two, which comes to the fore at the game's end.

Blanks in VGNs thus arise between the perspectives the player encounters in the gameworld and co-creates through his actions. They are closed by the player's acts of ideation (and in imaginative games of *alea*), which are informed by his world knowledge and contribute to his understanding of the gameworld, and also to his ability to perform in it. The strategies of the game organise this involvement, as they structure "both the material of the text [the repertoire of the game, the familiar context it draws on] and the conditions under which the material is to be communicated."[305]

The conditions thereby refer to the aesthetic arrangements of the perspectives, which are cancelled once the experience of the game is narrated afterwards. What this means for the strategies is important, since they help the player understand the gameworld through employing his real-world knowledge but, additionally, expose it to meticulous scrutiny. They thus constitute the juncture with the empirical world—and so familiar norms, conventions, or references to other fictional worlds

304 METRO 2033 (4A Games, 2010), ch. Hunter.

305 Iser, *Act*, 86.

are reorganised horizontally in the game's perspectives and by virtualising potential actions and processes. Consequently, it is only through the player's acts of ideation that these blanks can be closed.[306]

5.3.2 The Player as Wanderer Between Sensorial Impressions and Playful Actions, Themes and Horizons

The process of understanding the fictional gameworld and creating connections to the empirical world is thus informed by "a *background-foreground* relationship, with the chosen element in the foreground and its original context in the background."[307] This is to say, familiar norms and conventions "establish a frame of reference" and background for gameplay. As they are liberated from their original surroundings, however, they are depragmatised in the fictional gameworld and allow for "hitherto unsuspected meanings."[308] Such a relationship is similar to "that of figure and ground in Gestalt psychology"[309] and helps the player not only to comprehend the plot on a basic level, but also the levels of concept or significance.

To do so, the reader/player creates a "primary gestalt [that] emerges out of the interacting characters and plot developments,"[310] the gameworld occurrences, and the player's actions within it. This primary *gestalt* is more diverse[311] than it could be in a novel or film, because different players create a great variety of plots arising from the same story (if the game allows them to). However, at times the creation of the primary *gestalt* runs into hindrances, as the games of *agôn* juxtapose seemingly incommensurable perspectives.

As a consequence, the creative function of *negation* begins to exercise its effect, which is complemented by the vivid games of *mimicry* METRO 2033 plays with the player. These doublings and distortions contribute to the primordial force of the fictive and begin to affect the player's involvement on a basic plot level. I have clarified before that the perspectives help the player get his bearings in the gameworld, but they may also stand in conflict to one another and negate them-

---

306 Ibid., 86-87.

307 Ibid., 93; emphasis added.

308 Ibid., 93; cf. 93.

309 Ibid.

310 Ibid., 123.

311 At the same time, it can, however, be described as less diverse. For when experiencing a novel, different readers may imagine the storyworld in different ways.

selves. This is because, similar to reality, watching a film, or reading a book, playing a game confronts the player with a panoply of vistas or perspectives, out of which only a fragmentary number can be discerned at any given moment. These are nonetheless "interwoven in the text [/game] and offer a constantly shifting constellation of views"[312] that bewilder the player. The "*theme*" thereby designates "the view" or action the player is "involved with at any particular moment."[313] It is substituted by additional themes that emerge in the course of play and moves into the background and the "*horizon*"[314] of previously encountered/enacted perspectives. The horizon of past perspectives thus includes both the inner perspectives of the text/game and the outer perspectives, which link the game to the empirical world, and conditions the player's subsequent actions and imaginings based on the information he has gathered before (from the fictional and empirical world).[315]

METRO 2033 aggravates these games of *agôn* with a distorted dreamworld in which the true nature (or meaning) of the perspectives remains oblique. This is the case with the created image of the Dark Ones and the current gameworld situation that induces the player to handle the supposed threat. Such is the inevitable conclusion at this moment of play as the player draws from the horizon of past perspectives to compose it. Yet this image is fragile and will change, since with new perspectives, new impressions inform the player's acts of ideation.

> The continual interaction of perspectives throws new light on all positions linguistically manifested in the text, for each position is set in a fresh context, with the result that the reader's attention is drawn to aspect hitherto not apparent.[316]

This statement gives a viable explanation as to why player actions are prone to assume different meanings in the aftermath of their execution—for example, when the player of THE WALKING DEAD feels guilty about actions he previously deemed noble but which, through the encounter with new perspectives, turn out to be quite the contrary, or at least ambiguous. Consequently, with each new perspective—whether it is a player action, a character telling her news, a sign or gameworld

312 Ibid., 96; cf. 96, 116.

313 Ibid., 97; emphasis added.

314 Ibid.; emphasis added.

315 Ibid., 96-97.

316 Ibid., 97.

process, "a retroactive effect on what has already been read [/played]"[317] is provoked, which results in potential "enrichment, as attitudes are at one and the same time refined and broadened."[318]

Iser has called this process of continuous revision the reader's "wandering viewpoint,"[319] an insightful and romantic term that is, nonetheless, not sufficient for the player. Because what he experiences rather resembles the venture of *a wanderer between sensorial impressions and actions*, between the floating of the spectator's imagination and the ergodic participant's navigation of and action within the gameworld. Such a feeling is unknown to a reader/viewer—think of how the player moves the virtual camera and catches a glimpse of an extraordinary event or discovers parts of the world that require his intervention. The perspectives he thereby encounters and co-creates are mapped to an entire panorama of sensorial impressions and actions that compose the horizon of perspectives that inform his subsequent actions and their potentiality (for players gain a feeling for what they can do in a game based on their previous knowledge/experience). This process, as Iser has remarked for the reader, is driven by a constant alternation between "retention and protension," between what *was* and what *is about to come* (once the player acts).[320]

What follows from these observations is that the reader/player's expectations are either confirmed by the newly encountered/co-created perspectives (this narrows down the semantic potential of the text/game; and to a degree the ergodic one) or they are frustrated and renegotiated in the flux of further perspectives in games of *alea*. The second option predominates in the literary text (in aesthetically complex literature) and forces the reader to constant reshaping of memory and the restructuring of the aesthetic object, when *alea* breaks open the semantic veil of the text and evokes the games of *ilinx* in the reader.[321] Such an initial frustration and semantic ambiguity occurs as well in METRO 2033 in the constant renegotiation of the image of the Dark Ones, whose formation is influenced by perspectives that stand in opposition. These hinder the player's comprehension of the plot (and choice making in this respect) as well as his acts of ideation on the level of significance. They nonetheless drive the player to a constant renegotiation of the aesthetic object.

317 Ibid., 111.

318 Ibid., 99.

319 Ibid., 135.

320 Ibid., 111; cf. 110-111.

321 Ibid., 111-112.

5.3.3 The Emergence of the Aesthetic Object Through the Creation of a Secondary Gestalt

Various hints and possibilities to action point at a different conclusion concerning the Dark Ones, and these begin with the hallucinations they evoke in Artyom. These are unclear to him but imply that the intentions of the supposedly hostile race may be peaceful after all. This manifests itself in that the Dark Ones try to convince Artyom of their pacifism through pre-war images of a better world or sections in which they help him overcome the paranoia the Metro tunnels induce.[322] However, the hallucinations are not a reliable vantage point for now—for neither Artyom nor the player—and they could be a trick to stop Artyom from destroying the Dark Ones.

Figure 17: The image of the Darks Ones remains ambiguous. Are they attacking or surrendering?

METRO 2033 (4A Games, 2010), ch. Chase.

322 Indeed, the conflict between the Dark Ones and the humans was caused by a misunderstanding, when the former approached the humans with peaceful intentions, but the humans were arrogant in their concept of humanity. Too afraid of change and driven by the anxiety of evolutionary defeat, the humans declined to accept the posthuman solution the Dark Ones promised. As *homo novice,* the Dark Ones are better adapted to the new world, but they are a life form humankind does not understand (or does not wish to understand)—all of which the player is unaware of for now.

Indeed, it seems that the games of estrangement and *mimicry* METRO 2033 involves the player in are the main reason for his confusion. To blame here is the technique of "inversion,"[323] which creates "discrepancies" in the player's acts of ideation that "make him dispute his own gestalten" and induce him to constantly revise them.[324] This is because none of the game's points of view are entirely reliable (including the prologue and the section at Exhibition). Consequently, the *gestalts* the player is initially composing (of the plot, gameworld, and character relations) are flawed, which affects the creation of a *secondary gestalt* that complements the first and extends it.[325]

> [T]he plot is not an end in itself—it always serves a meaning, for stories are not told for their own sake but for the demonstration of something that extends beyond themselves. And so a gestalt that represents the plot development is still not completely closed. The closing can only come about when the *significance* of the action can be represented by a further gestalt.[326]

The potential for confusion here is considerable, because both *gestalts* are necessarily intertwined, and one should not underestimate the importance of any one of them for the player's acts of ideation. However, it is true that the secondary *gestalt* (which moves the player into the realms of significance and helps him understand the fictional world's relation to the empirical world) is more flexible than the first and allows for a variety of fillings. This can be discerned in the fact that people seem to discuss more vividly the interpretation of a narrative (and what it means to them) rather than its plot details.[327]

The creation of the aesthetic object is thus a complex process, although one that works naturally if one shows the necessary willingness to do so. This fact and the necessity for an emancipated player should not be dismissed so easily—specifically for the genres of SF and dystopia. For as Moylan argues, the SF story involves the reader in a complex world he must take seriously and whose logic he needs to understand in order not to misinterpret it. Otherwise, and in "[a] refusal of an engaged, cognitive reading process," he may "only find ... [his] own position and prejudices bounced back at" him.[328] Playing dystopia proceeds similarly, and

323 Iser, *Act*, 131.

324 Ibid.; cf.131.

325 Ibid., 123.

326 Ibid.

327 Ibid.

328 Moylan, *Scraps*, 25; cf. 24-25.

there are a plethora of instances in which the naive player can misplay these games and resist their aesthetic function (see the discussion of player types in chapter IV). In METRO 2033, this entails the loss of the revelatory effect at the game's end, which I will come to shortly. The emancipated player resists such premature playing and, to experience dystopia's aesthetic effect, he composes two intertwined *gestalts* during play.

This interplay between images on the plot level and their significance can be discerned relatively early in the game with the psychoanalytic connotation of the Metro tunnels. These hide a dark secret and set the player within the deepest regions of the human unconscious, which is characterised by the innate fear of Otherness. Such an image is fostered once the player connects the narrow stretches of the Metro tunnel to Freud's deliberations on the night-time dream, while he passes through them with the phallic symbol of his weapon extended. In order to see this image more clearly, the acts of ideation need to be fed with additional perspectives. I have clarified that people in the stations suffer from deep trauma—not only because of the nuclear annihilation that caused the current state of affairs but more so in their mistrust of anything that seems different. Their lives underground are marked by isolation within the stations, and although there are trading networks, conflicts between the factions (ideologies) are ongoing.

An example of this is the never-ending battle at Cursed Station in which monsters (the Nosalises, a common enemy in the game) continuously attack the humans. They stand for those parts of the self that humankind cannot get rid of and that fundamentally revolve around the instinct of survival and the mistrust of the Other (an image the player will steadily compose). After a barbaric slaughter, the section ends with the player entering a shrine at the tunnel's end. It is guarded by shadows of fallen men, and access to it is only granted to the virtuous and pure at heart. Khan, who accompanies the player through this part of the game, leads him in. The experience thus serves as a reminder of the player's deeds, by reconfiguring their image and having him ponder their ethical justification.

The second image I wish to stress is when the player passes a bridge that is contested by both the Red faction and the Nazi faction. Not even the apocalypse could stop them from bloodshed, and the bridge creates a terrifying but, at the same time, beautiful image of the futility of these conflicts. The player may choose to either sneak below the bridge or participate in the frenzy of combat—but he will certainly ask himself whether mankind is doomed to fight forever.[329] Consequently, through the experience of the factions in the game and how the player deals with them, he not only further comprehends this world but has already begun to compose a secondary *gestalt* by linking the enacted events to facts about his empirical world. This process is supported in that the factions and the rules by which they work are inspired by those of the empirical world—but they are crammed into the microcosm of the Metro stations and distorted in perception to have the player decipher this connection.

Important stations in the game thus include Hansa, the wealthy and capitalist trading centre in the metro, where a free market has been established and which is heavily guarded from the outside world. In addition, there is Polis, where the player encounters the head of the Rangers Miller. Polis is the centre of science and knowledge in the Metro. It is situated beneath the former Moscow State Library (a revelatory juxtaposition), and many scouting missions to extract its treasures are undertaken. However, although the best and brightest reside in Polis, they refrain from intervening in the issue of the Dark Ones, and, thus, the supposedly reasonable turn a blind eye to ethical issues. Miller, on the contrary, has a clear opinion. He exhibits militaristic characteristics of the empirical world and will accompany Artyom to the tower, after the latter has secured the D6 documents from the library—a place of knowledge and virtue, which the player scavenges for missile documents in the attempt to accomplish the game's goal.

It is revelatory that METRO 2033 outlines various images for the player to compose, which include the negotiation of what he encounters and what he participates in. The stations are thereby important points of orientation. By understanding their ideologies, the player gains vital perspectives that will inform both the decision that awaits him and the analogic connection of the enacted events to the real world. In playing dystopia, he may thus get a glimpse behind the opaque nature of empirical reality by testing its norms, conventions, and processes in the condensed and defamiliarised version of the fictional gameworld. In this context the player's actions also stand, which imply various signifieds: from a phallic symbol of power to the related one of waging a blind war against the Other, which has caused disaster before.

These possibilities notwithstanding, the player does not need to fall into this trap of naiveté, of following orders (or game goals) without pondering their ramifications. In this regard, there is an important perspective that may help him see

329 Here again, I wish to stress the difference between the (uncritical) playthrough of the gamist or achiever player and that of the emancipated player, for the former will probably not come to such a conclusion.

things differently. During the journey, he eventually meets *the tempter of this story*—the enigmatic figure *Khan*—who will become Artyom's and the player's mentor. Khan understands the psychic phenomena of the Metro and warns Artyom on various occasions that force is not the answer and that to "break this vicious circle one must do more than just act without any thought or doubt."[330] The Khan chapters thus lead the player into the darkest parts of the Metro and into the deepest spheres of the human unconsciousness. They illustrate that even beneath humankind's ugliest parts lies hope in the search for compassion and in a cathartic cleansing of the aggressions towards the Other. Shortly before the game's final moments and the player's ascension of the tower, Khan reminds him of these truths. He therefore exhibits those characteristics of empirical society that aim at prudence and dialogue, not war and destruction.

What enhances this image is that the conversation takes place in an old church underneath Moscow, which creates a beautiful inversion in that hell has extended to the surface of the planet, while only a few parts beneath it remain untouched. Khan's perspective is thus to be seen in terms of an altruistic world view that promotes benevolence towards the Other and which stands in strong contrast to that of Miller, who has established a base of operations within the church. As such, the player is presented with two opposing perspectives that have plastered his way before and is set between the seemingly incommensurable fronts of war and peace and in an intricate game of *agôn* to which he can ergodically react. However, METRO 2033 does not make it easy for the player to have a say in this choice after all, for only the virtuous and pure-hearted may take it and instigate a successful *counter-narrative* to this dystopia.

Of critical importance in this respect is how the player behaves in certain situations in the game, and METRO 2033 pays close attention through a subtle morality system. It is the big choices but even more so the little choices that matter in a world of despair—and, consequently, decisions such as passing through certain areas without engaging in conflict, being generous to those in need by handing them ammunition or rejecting rewards for helping them, or listening to people's problems (even to those of supposedly evil factions like the Nazis) will reward the player with morality points. A most precious example of such is when Artyom walks past a derelict playground and a vision of the Dark Ones fills it with life and the playing of children. If the player chooses to relish this moment, and even takes off his gasmask, he will receive a morality point, or even two. If he rushes through it, or shoots a bullet for any reason, the vision will end with a negative entry in the morality system. These are only a few instances in which the morality system takes

330 METRO 2033 (4A Games, 2010), ch. Sparta.

grip of the player, but it suffices to say that only if he has a positive balance in the system is he able to choose the future of the Dark Ones at the game's end.

The choice between killing the Dark Ones by the use of nuclear weapons or saving their species and averting a potential genocide thus soon awaits the ethical player. It is guided through various perspectives that negate each other and also (and probably more importantly) through virtualised potentialities inscribed in the gameworld. These are *utopian enclaves* the player can actualise or not—and it is only when he realises the stakes early enough that a utopian horizon to this dystopia can emerge. The image of the Dark Ones thereby remains ambiguous until the very end, and, consequently, the player is confronted with "incomplete information" and a choice that is based on "conflicting arguments for and against … that might have probabilities, but no certainties attached to them."[331]

The decision, in other words, creates a space of "uncertainty"[332] (indeterminacy) about its outcome (*alea*). While the game's primary goal is clear (defeat the Dark Ones), it also ascribes a major value to it, or what Domsch has called a "theological attitude" and "a clear hierarchical valorisation to the options offered."[333] The emancipated player, however, knows no such fetters of the "ludic"[334] and refrains from the relentless pursuit of the primary goal. Instead, he embarks on an "explorative" and "paidic"[335] route and enables the imaginary (the games of *alea* and *ilinx*) to permeate his body by closing the blanks presented by the fictive in a creative and playful manner. This attempt to attain a "transcendental viewpoint" over the "positions" of a text/game (which nonetheless can never be reached entirely) is tantamount to approaching the aesthetic object.[336]

METRO 2033 thus has a persuasive effect on the player. It is evoked by what Sicart has called "ethical cognitive frictions" that "might encourage a thoughtful kind of play"[337] and have the player close various blanks in the process. The thereby composed image of the Dark Ones remains incomplete, but it may tempt the ethical player to save them from annihilation and to pursue an open-minded route towards the posthuman. For he understands that in order to overcome this world (to not let history repeat itself), humankind has to evolve and to free itself from the suspicion and fear of the Other. The Dark Ones have paved the way for

331 Domsch, *Storyplaying*, 115.

332 Ibid.

333 Ibid., 116.

334 Ibid.

335 Ibid.

336 Iser, *Act*, 98; cf. 98.

337 Sicart, *Beyond*, 96.

it, but, in the end, it is up to the player to comprehend their message and to act accordingly.

Even in the game's final moments such emancipatory thoughts are put to trial in games of *alea*, and the player's abilities to think reasonably and to suppress irrational emotions or premature accusations are scrutinised. This can be discerned in his ascension of the tower, which evokes the psychoanalytical image of the phallus and humankind's greed for power in the survival of the fittest. Having almost reached the primary game goal, Miller and Artyom are attacked by a demon, a flying monster resembling a bat. Miller is nearly killed in the encounter, but Artyom succeeds in defeating it. The player may now be filled with fictional anger for having almost lost a dear friend, while Artyom's victory seemingly gets to his head. He now believes he is the strongest predator on Earth, and this frenzy of emotions is easily transferred to the player and furthered by the rapid ascent of the elevator, which like a seminal fluid shoots Artyom to its peak. Consequently, through this game of proximity and distance between the player and his PC, a blank emerges that can be closed in primarily two ways: succumbing to irrational instincts and the frenzy of combat (such as Artyom in this instance) or remaining calm by having in mind the greater picture (what Khan told the player).

Even the tranquil race of the Dark Ones are becoming nervous in the face of potential extinction. Although their reaction depends on whether the player shows a positive or negative morality balance, they utter doubt in both instances—going as far as trying to stop Artyom by inducing hallucinations in him. Playing METRO 2033 is thus precarious. It is driven by uncertainty and the loss of a potential Utopia, and yet it appears inescapable that the player has not become aware of these facts. Stepping into a creative dialectic with the game's implied player, he was exposed to a system of perspectives that he has helped create. Various images were evoked through this involvement and by closing the blanks between the perspectives while at the same time negotiating the plot and its significance. These images of a secondary *gestalt* differ from player to player, but they are nonetheless outlined by the game's strategies.

All in all, the experience of dystopia leads to the climactic moment where the player's acts of ideation are tested—even when he is declined the choice due to previous failures. The magic, then, lies in a revelatory moment at the game's end, which has the player ponder humankind's true nature, which is deeply flawed. The game has outlined several images to come to such a conclusion and provide the insight that only in overcoming these primitive urges, will it be possible to create a sustainable future for generations to come.

5.4 THE CREATED REALITY OF FICTION AND ITS AESTHETIC EFFECT

This chapter has shown that for the player to communicate with a fictional world and to create the connection to empirical reality requires the construction of images on both the level of the plot and significance (for the latter requires the former). Although this creation of experiential *gestalts* is similar to how humans perceive and process the real world, it differs from them in that these initial *gestalts* remain incomplete, for the fictional world is hard to grasp through its doublings and distortion (*mimicry*). Consequently, the closing can only come about by the creation of secondary *gestalts* that further negotiate the composed images. Thereby, the blank, and its function as negation, assumes utmost importance. It sets the perspectives of the game in conflict (*agôn*), which initially frustrates the player to then break open its semantic veil (*alea*). The result is an experience of subversiveness and vertigo (*ilinx*) that eventually results in the formation of the aesthetic object and the insight into the true nature of both the fictional and empirical world. The blank, as such, is not only responsible for affording and outlining the player's involvement in a game (on all levels) but spurs his imagination to experience art's aesthetic effect. It is here that the *fictive* permeates the implied player and outlines an estranged gameworld by drawing on the *real*, and the *imaginary* negotiates what the *fictive* has presented by manifesting itself in the player's psyche, imaginings, and ergodic actions, to then re-incorporate its results into the *real*. The following table illustrates these interrelations:

Table 9: Playing dystopia as the interplay between the fictive, the imaginary, and the real.

The Real	Inspires the *fiction-making* process that feeds back into the empirical world through *aesthetic response.*
The Fictive	***Agôn*****:** sets in conflict the perspectives (norms, processes, characters, actions, conventions, etc.). ***Mimicry*****:** aggravates the player's perception/understanding of the gameworld through games of estrangement (doublings and distortions).
Important Perspectives of METRO 2033	**Rangers of the Order (Hunter, Miller)**: militaristic, suspicion of Other. **Dark Ones:** ambiguous, posthuman, dangerous, misunder-stood. **Station perspectives:** Hansa (capitalism), Exhibition (des-pair), Cursed station (never-ending strife), Black station (Nazi occupied), Polis (power and wealth). **Khan:** ethical, altruistic, tempter, reflective. **Artyom:** partially in flux, depending on player actions. **Player choices** with negative (unethical, selfish) or positive (ethical) impact on the moral system.
Blanks between the perspectives	**Closing** through imagination/ ergodic action
The Imaginary	***Alea:*** breaks open the interrelations between the perspectives and allows for blank closing. ***Ilinx:*** games of subversiveness, precariousness that result in the player's aesthetic response by re-incorporating their re-sults into *the real* (empirical world).

Part III:
Playing Dystopia

Preface to Part III

Part I and II served to create a viable framework for describing the player's venture to dystopia from a phenomenological standpoint and constituted the premise for a close playing of dystopian games.

Part I thereby laid the groundwork for the theory of the player's aesthetic response to dystopia in that it distinguished between four subgenres of the VGD: the anti-utopia, the classical dystopia, and the critical dystopia in variants I and II. Going into detail concerning their respective plot structures and differing approaches to the negotiation of hope, a viable initial framework for the VGD was established, which revolved around the opposing structure of official narrative and counter-narrative (the player's experience of the dystopian regime and eventual revulsion against it). Finally, Part I came to a close by alluding to the VGD's aesthetic potential and describing the genre's real-world targets.

Building on these insights, Part II intensified the discussion of the VGD's aesthetic complexity by scrutinising the preconditions leading to the player's aesthetic response. It incorporated the conclusions on dystopia's plot structure into the comprehensive framework of the implied player. Chapter IV thus combined a multi-faceted discussion of representational art, fictionality, structuralist narratology, and reader-response theories with related insights from video game studies and concluded by describing the perspectival network of a game. The chapter illuminated the existence of the VGN as a genre and pointed at strategies of player involvement (or guidance) at a game's disposal. These were described as guiding the player on an ergodic and imaginative level through a system of perspectives and contributing to his understanding of the gameworld and plot. In addition, they offer the player a gateway to a more complex dialectic that extends into the realms of significance and creates ties to his empirical surroundings. Through various images created in acts of ideation, the player steadily builds a secondary *gestalt* that complements the first, which arose out of plot developments and gameworld events. This secondary *gestalt* was illustrated as more diverse and as tantamount to an approximation of the aesthetic object.

The gained insights from chapter IV were extended in chapter V, where the general reflections on the VGN were combined with specifics of the SF/dystopian genre. Consequently, strategies of estrangement and extrapolation were given priority and the methods by which these involve the player in a distorted dream-world—which is organised by the fictive and made palpable by the imaginary permeating the player's mind and actions. By involving the player in confusing games of estrangement and fictionality, the VGD was described as creating an aggravated referentiality to the empirical world—and to close the blanks between both worlds, the player's activation in the act of play (as an emancipated being) became necessary, which is governed by a feedback oscillation between fictional and empirical reality.

All in all, the conclusions here are twofold: first, there is a naturalness of aesthetic response to playing dystopia, once the player engages with a game with an open-minded attitude. Second, the insights into the act of play helped construct a framework for analysis that outlines/schematises play's underlying processes, the player's reactions to them, and makes them tangible for the critical observant. The structure of the blank is thereby of utmost importance as is how various perspectival arrangements give rise to it. Blanks constitute the basis for all sorts of communication, and in this respect, Iser's distinction of four types of perspective arrangements in narrative fiction become of interest: "counterbalance, opposition, echelon, and serial."[1] In order to prevent premature conclusions, however, the discussion of perspectival arrangements was postponed to Part III.

Consequently, the following and last part of this study elucidates my theoretical deliberations by applying them to two thematically diverse VGDs. These close playings complement the many examples I have dispersed throughout this study. Hence, I will describe two critical dystopias of variant I: BIOSHOCK INFINITE and THE LAST OF US: REMASTERED. These games address issues such as religious fanaticism, oppressive regimes, ecological catastrophes, and capitalism and convey their warnings by juxtaposing several perspectives and placing them in dialectical opposition and within echelon structures. Through the player's continual renegotiation of previously composed images, the experienced meanings diverge from player to player but are nonetheless outlined by the perspectival arrangements of these games.

1 Iser, *Act*, 100.

6 Night-Time Dreams and Wish-Fulfilment: The Struggle for Utopia in BIOSHOCK INFINITE

> **DeWitt:** What are all these lighthouses? Where are we ... who are … ?
> **Elizabeth:** They're a million, million worlds. All different, all similar. Constants and variables. There's always a lighthouse. There's always a man, there's always a city …
> **DeWitt:** How do you know this?
> **Elizabeth:** I can see them through the doors. You, me, Columbia, Songbird … But sometimes, something is different … yet the same
> *BIOSHOCK INFINITE (Irrational Games, 2013), ch. Sea of Doors.*

BIOSHOCK INFINITE (Irrational Games, 2013) builds its gameworld around the premise and fictional novum of the *multiverse*, in particular Lewis Everett's notion of the *many-worlds interpretation*. According to this understanding of quantum physics, any potential choices and outcomes (taken or not taken) exist in a state of superposition, with each of them creating an alternative universe that follows their logic. In other words, the many-worlds interpretation suggests that the universe is composed of limitless possibility, an infinite array of potential futures, presents, or pasts that have become actualised worlds.

In BIOSHOCK INFINITE, this revelation is reserved for the game's end, where the player encounters a *sea of doors* composed of an infinite array of lighthouses and docks that connect them. These represent entrances to the Utopias behind them, all of which are similar but somehow different. It is a key motif permeating the BIOSHOCK series, which deals with the question of *human agency* (or the lack thereof) in the face of ideological confinement. This struggle for self-expression was experienced by the player in the first two games—BIOSHOCK (2K Boston, 2007) and BIOSHOCK 2 (2K Marin, 2010)—which had the player explore the capitalist underwater Utopia of Rapture, an alternative history of the early 1960s. The events of BIOSHOCK INFINITE predate these and lead back to 1912 where the player lays foot in a city in the sky by the name of Columbia. It is a Utopia based on the principles of religious piety and American exceptionalism, in which the player meets a woman called Elizabeth who will help her overcome the dystopia(s) she is facing.

Figure 18: The BIOSHOCK *multiverse is composed of an infinite array of worlds, each of which is represented by a lighthouse. These formulate entrances to the Utopias behind them, which were built around a noble dream.*

BIOSHOCK INFINITE (Irrational Games, 2013), ch. Sea of Doors

What unites the BIOSHOCK games, then, is a *vicious circle* of choices that led to the dystopias the player is now facing. As such, BIOSHOCK INFINITE gives context to the previous games as it reveals their inner connections, but it also sends the player on a journey to aesthetic response and self-sacrifice that brings an end to the circle of ideological delusions. This intimate bond between the BIOSHOCK games makes it difficult to discuss INFINITE in isolation, and I will integrate the first BIOSHOCK and the two episodes of BURIAL AT SEA (Irrational Games, 2013, 2014) into my argument later on. For only then can the characters' and player's *search for forgiveness for having lost Utopia and the struggle to regain it* be understood in a nuanced way.

In the following, I will thus describe BIOSHOCK INFINITE as a critical dystopia of variant I that targets the dangers of *ideological delusions* and *oppressive regimes*. Its warning thereby revolves around the perils of theocracy, American exceptionalism, racism, and free market capitalism, which occur in an intertwined fashion. This selfish desire for power, supremacy, and the greed for capital paints a bleak picture of humankind and gives rise to the game's *official narrative*. To counteract this dilemma, the player takes on the role of private detective Booker DeWitt who is sent to Columbia with the task of saving a mysterious woman named Elizabeth. DeWitt is initially unaware of the circumstances surrounding his journey, which is explained by the side effects of trans-dimensional travel. Yet there is a more precise explanation of the events.

In this interpretation, which is influenced by Freud's deliberations on the night-time dream,[1] DeWitt suffers from a *neurosis* stemming from his involvement in the Wounded Knee massacre. He drifted off into alcoholism and gambling, and supposedly gave away his daughter to settle his debts. When the game begins, DeWitt suffers from nightmares, and the player becomes involved in an anxiety dream at whose core lies the *wish for forgiveness*. As such, BIOSHOCK INFINITE's counter-narrative follows a creative path where the player has to decipher an estranged dreamworld and where the struggle for Elizabeth/Anna is symbolically intertwined with that for Utopia. This twofold role is adopted by the player, who has to become aware of these facets, and revolves around a *father's guilt due to alcoholism* and the *failure of humankind in losing Utopia because of racist atrocities*.

In this regard, Elizabeth assumes a pivotal role, for she will lead DeWitt and the player on a journey of gradual realisation. This route will have them discern DeWitt's true agenda of coming to Columbia (his wish for forgiveness) and lead to the cathartic insight that the world as they know it is plagued by the delusive powers of ideologies and the supposed Utopias these create in their image. Such a wake-up call prompts the player to change her empirical surroundings in the same manner as it startles DeWitt at the game's end, when he wakes from his nightmare to hear Anna in the adjacent room. It is an ambiguous ending, which requires a self-sacrificing attitude from DeWitt in his struggle for Elizabeth, and evokes in the player the insight that it is not too late to affect change in the empirical world through selfless action.

To outline such an aesthetic response (which in this case is also ethical), BIOSHOCK INFINITE employs the strategies of *inversion/negation* to conjure up *a flux of images* in the player's mind. Thereby, utopian images pervade the scenery

1 Freud, *Dreams*.

initially, but the game gives subtle hints at another truth. This truth will be revealed more forcefully by further perspectives the player encounters and co-creates, only to negate it one more time when she reaches the game's end, to have the constructed images shown in an ambiguous light.

In the following, I will describe the perspectival arrangements that afford these images and contribute to their renegotiation. Thereby, important images include: the lighthouse and shining allure of Columbia, Monument Tower (a statue built to imprison Elizabeth), and the character relations between DeWitt/Comstock, Elizabeth/Anna, and her guardian, Songbird. In addition, there are two factions that struggle for supremacy in Columbia: the Founders and the revolutionary group of the Vox Populi. These gameworld and character perspectives are complemented by gameplay processes and interactions that fuel the player's acts of ideation. The strategies of inversion/negation affect gameplay in that the game begins with a leisure tour around Utopia to then expose this society's true nature by reappropriating tranquil spaces into a venue for combat and atrocities. All of these images are informed by the player's experience of DeWitt's anxiety dream, which buries the wish to save Elizabeth within all its distortions and doublings and the desire to re-attain Utopia.

To give the chapter a clear structure, it will be subdivided into five areas of investigation:

1. The first part will investigate the player's venture to and tour around the utopian city of Columbia and describe basic perspectival arrangements including the city's architecture, events, and potential player actions.
2. Second, I will focus on how the strategies of inversion and negation expose the dystopian nature of Columbia and lay bare its official narrative as DeWitt's struggle against a dark part of his self.
3. Third, I will explicate Elizabeth's process of realisation and the resultant counter-narrative. The young woman adopts the role of the temptress and will lead DeWitt and the player to seeing beyond the delusions of dominant ideologies.
4. Fourth, this process of realisation is guided by the factions of Columbia and how the characters and the player experience these. In addition, DeWitt's role in the game and his nightmarish delusions will be given further attention.
5. Finally, I will discuss BIOSHOCK INFINITE's ambiguous ending in the context of the BIOSHOCK multiverse and its individual games (thereby addressing the first BIOSHOCK and INFINITE's DLCs BURIAL AT SEA: EPISODE 1 and 2).

6.1 THE UTOPIA OF COLUMBIA AND BASIC BLANK STRUCTURES TO LURE IN THE PLAYER

When the player is introduced to the world of BIOSHOCK INFINITE, she finds herself on a rowing boat towards an unknown destination. The sea is in turmoil, and the boat is steered by a lady and a gentlemen who are engaged in an argument about thought experiments. They hand DeWitt a box with his name on it and the regiment he served in: "7th Cavalry, Wounded Knee."[2] In it he finds information about the job at hand—a gun, postcard, three symbols, a key, and a picture of a woman—while in the background the glaring of a lighthouse can be discerned. It is a mission briefing players have come to know from other games, yet its symbolism points to the dreamlike character of the events: to the repression of memories in the darkest regions of the protagonist's unconscious but also to the key to redeem these, which is the struggle for Elizabeth and Utopia.

These initial perspective segments foreshadow the events that follow. They include the visual environment, character dialogue, in-game artefacts (such as signs and writings), the labyrinthine structure of the level, and player interactions. Between them several blanks emerge on the level of the *plot* and *gameworld* and evoke questions in the player such as: where am I and for what purpose am I here? Who is the character I am playing and who is the woman in the picture?[3] Consequently, already from the beginning, BIOSHOCK INFINITE's perspectival structure is meticulously designed and guides the player's involvement in the game. Thereby, images on the level of *significance* are already given contours, and once the player reaches the lighthouse, the process continues when she knocks on the door and encounters a note that issues an unmistakable warning: "DeWitt – BRING US THE GIRL AND WIPE AWAY THE DEBT THIS IS YOUR LAST CHANCE!"[4] Blood is smeared on the note, and, stepping through the door, religious symbols and writings complement the image.

The first sign in this respect is located in the player's field of view upon entrance. It is situated above a basin and holds: "Of Thy Sins I Shall Wash Thee."[5] The player is able to glimpse DeWitt's reflection in the water, which is followed by an upward camera movement towards the sign and the latter's utterance: "Good luck with that, pal."[6] These perspective segments create an initial blank whose

2 BIOSHOCK INFINITE, (Irrational Games, 2013), ch. Lighthouse.

3 Maziarczyk, "Playable Dystopia," 252.

4 BIOSHOCK INFINITE, (Irrational Games, 2013), ch. Lighthouse.

5 Ibid.

6 Ibid.

magnitude is opaque to the player and will be closed late in the game. They are self-referential, but the player will only understand them once she has composed an image of the plot and DeWitt's struggle for *forgiveness*. For now, the player's route continues upwards on a spiral stairway, where signs in the environment are highlighted by in-game lighting. These include further religious writings, such as "From Sodom Shall I Lead Thee" and "In New Eden Soil Shall I Plant Thee,"[7] a map of Columbia's itinerary, and a corpse leaking blood. The initial image of the plot thus remains blurry, but it points to a dangerous yet exhilarating world.

Complementing this sense of wonder is the fact that the journey to Utopia begins with a *lighthouse*, which triggers imaginings about distant locations and a safe haven from the raging sea.[8] In BIOSHOCK INFINITE, the lighthouse grants the player entrance to wondrous city in the sky and comes close in function to More's tower in Utopia in that it works "as a nexus, … an entry point to all possible parallel worlds"[9] in the storyworld of the game. For the player this insight remains to be uncovered in a *process of realisation* that will lead her from initial unawareness to seeing beyond the facades of the false Utopia of Columbia. Such a response is anchored in the game's structure, which guides the player through the strategies of *negation/inversion* of composed images. Thereby, the lighthouse assumes a pivotal role, since its positive image will fade and allow the player to discern the gameworld and empirical reality differently.

Before coming to this issue at the chapter's end, I wish to exemplify the route towards it, which continues when the player breaches the clouds into Columbia and is greeted with a magnificent view of the floating city. A gigantic angel's statue appears at the centre of the screen. It is surrounded by city districts illuminated by fireworks and the glaring sun. A mellow piano version of *Will the Circle Be Unbroken* (Charles H. Gabriel and Ada R. Habershon) is playing, underlining the serene panorama. It is as if the player has reached heaven, but one that was achieved by the marvels of quantum technology. An airship passes the player's way now as the capsule descends into the shadows of a tower. It is adorned with pieces of cloth designed in the style of the American flag, and one can discern a portrait of Father Comstock on it. Like a traveller in a stagecoach, who witnesses her surroundings from an observing point of view, the player is introduced to the

---

7 Ibid.

8 Nyman and Teten, "Lost and Found," 7-8; Andra Ivănescu, "The Music of Tomorrow, Yesterday! Music, Time and Technology in *BioShock Infinite*," *Time and Technology in Popular Culture, Media and Communication* 7, no. 2 (July 2014): 53, http://ojs.meccsa.org.uk/index.php/netknow/issue/view/34

9 Ibid.

Columbian society. Thereby, important landmarks are brought into focus, since the player may only move the camera in a limited way while in the capsule. This linear route leads into the undergrounds of the Welcome Center, where the religious undertone is brought into the foreground. It is a first *inversion* from *brightness into darkness* that vaguely implies Columbia's true nature.

Upon entry, the pilgrim disembarks from the stagecoach to investigate Columbia's past and the story behind the prophet, and is sent on a linear journey to *baptism*. This religious tradition is a prerequisite to enter Columbia and represents a first "forced choice"[10] that foreshadows the player's *lack of agency* in the game and the confinements of theocracy alike. What contributes to this sense of *aporia* is the game's linear structure, which largely follows the trajectory of a unicursal labyrinth. As such, the player is led past glorified graven images of Father Comstock and his wife Lady Annabel Comstock to the aforementioned baptism. The place is flooded with water, and candles are placed besides the streams, reminding one of a shrine in which the Utopians kneel before their prophet, while a gospel version of *Will the Circle Be Unbroken* resounds through the halls.

Will the circle be unbroken
By and by, by and by?
Is a better home awaiting
In the sky, in the sky?[11]

The song will reappear in the game in different versions. But whereas in the beginning it evokes imaginings about Utopia, during the game its image and the blank it creates will change, and the *circle to be unbroken* will assume a different addressee.

To enter the city, then, DeWitt undergoes *baptism*, and the preacher almost drowns him in the process. Here again, the dreamlike character of the events comes to the foreground, for it continues with a scene in DeWitt's office outside of Columbia. It is untidy place, littered with alcohol bottles, while the game is now depicted in black and white. A man is knocking on his door, with a reminder for DeWitt to fulfil their arrangements. When the player opens it, the scenery changes to the rooftops of New York City, where Columbia wages war against the Sodom below. DeWitt is hit by a rocket and awakes in the gardens of the flying city.

---

10 Robert Jackson, *BioShock: Decision, Forced Choice and Propaganda* (Winchester: Zero Books, 2014), 193.

11 BIOSHOCK INFINITE, (Irrational Games, 2013), ch. Welcome Center.

Given this introductory sequence (and what is to follow), *religion* (in the form of Christianity) and the dangers of *theocratic regimes* have often been deemed one of BIOSHOCK INFINITE's prime targets.[12] Frank Bosman, for instance, links Columbia to "a decadent and violent" version of John "Winthrop's idealistic city upon the hill',"[13] which shows this concept in a perverted manner in direct extrapolation from our times.[14] Still, one has to be careful in ascribing the role of the culprit to religion here—or "Christianity"[15] in particular—and a precise investigation suggests a specific target. In this respect, Jan Wysocki argues that although religion in BIOSHOCK INFINITE is depicted in a negative manner, it is rather its *use* that is blamed. Religion, in this sense, is used to marginalise the Other and as a justification for misanthropic behaviour. The aim of the Columbians is to build a pure form of Christianity exempt from the sinners of the Sodom below, and, therefore, the game's point of attack can rather be found in a protestant, evangelical-fundamentalist branch, which creates a sect-like microcosm in Columbia. This society combines characteristics of millennial groups, ultra-nationalism, and racism in the twisted world view of Father Comstock and his followers.[16] For now, the player has only glimpsed at this truth, which will be extended by further analogies to the empirical world.

This implicit comparison to the player's empirical surroundings continues in a telling image after DeWitt awakes from the baptism. Three statues of the U.S. Founding Fathers— Benjamin Franklin, George Washington, and Thomas Jefferson—are worshipped by the Utopians. They hold the symbols of the *key, sword,* and *scroll*, which are symbolically passed over to the Utopians as a reminder of their great nation's values: *wisdom*, *courage*, and *righteousness*. The place is

12 Jackson, *Forced Choice*, 174; Nyman and Teten, "Lost and Found," 8; Maziarczyk, "Playable Dystopia," 241; Buinicki, "Nostalgia," 722; Lizardi, "*BioShock*;" Frank G. Bosman, "'The Lamb of Comstock'. Dystopia and Religion in Video Games," *Online Heidelberg Journal of Religions on the Internet* 5 (2014): 177, http://heiup.uni-heidelberg.de/journals/index.php/religions/article/view/12163; Ryan F. Peters, "The Global Dystopian: Twenty-First Century Globalization, Terrorism, and Urban Destruction," (PhD diss., Loyola University Chicago, 2015), 124, http://ecommons.luc.edu/luc_diss/1964/; Jan Wysocki, "Die Präsentation amerikanischer Religion im Spiel *BioShock Infinite* am Beispiel des Charakters des Propheten," (master's thesis, University of Heidelberg, 2014), 49, 60-61.

13 Bosman, "Lamb of Comstock," 177.

14 Maziarczyk, "Playable Dystopia," 241.

15 Bosman, "Lamb of Comstock," 177.

16 Wysocki, "Religion im Spiel *BioShock Infinite*," 26, 80-84.

called The Garden of New Eden and resembles "a refuge for believers"[17] in which the Utopians immerse in prayers. It is a sterile paradise, linked to the glories of U.S. American history, whose linear route leads to a large door into the city. Upon entry, this bewildering marriage of *religious piety* and the *exceptionalism of a nation* (the United States' unique position and supposed superiority in comparison to other nations) is fortified.[18] For only slowly does the door open, granting the player a majestic vista on the district. It is a kernel event that follows the typical utopian plot "of an outsider coming to no-place."[19]

As a stranger to an unfamiliar world, the player thus assumes the role of a tourist and is guided by the level's labyrinthine structure in a linear fashion through a city district called New Eden Square. It is the annual celebration of Columbia's and the Prophet's magnificence, and the events remind the player of a "vision of America" that, as Ken Levine remarks, "a lot of politicians think existed. This perfectly idealised summer day"[20] and dream of "a simpler, more perfect society and time."[21] To outline such an image, the game organises its perspectives in a precise manner: from the docking of the city parts and cheerful Utopians, to a gigantic statue of Father Comstock and the dialogues that praise him. Meanwhile, a parade glorifies the city's creation and presents the Comstock family in a religiously idealised way. It depicts the moment when the Archangel Columbia descended to the peasant Comstock with the task of building a city in the sky.[22]

What follows from these observations is that the *stereotypical guide of the utopian narrative* is replaced by BIOSHOCK INFINITE's discourse and perspectival

---

17 Nyman and Teten, "Lost and Found," 10.

18 Wysocki, "Religion im Spiel *BioShock Infinite*," 23, 47; Bosman, "Lamb of Comstock," 175-176.

19 Maziarczyk, "Playable Dystopia," 239.

20 Colin Campbell, "The Big Ideas of *BioShock Infinite*: An in-depth look at Major Themes with Ken Levine and Shawn Robertson," *IGN*, January 30, 2013, accessed March 7, 2014, http://www.ign.com/articles/2013/01/30/the-big-ideas-of-bioshock-in finite

21 Buinicki, "Nostalgia," 726.

22 See Wysocki's *classification of Father Comstock's roles* in the game: 1) as Moses, who leads the Israelites to the promised land; 2) as saint, for he is depicted in posters with an aureole; 3) as patriotic hero in a glorification of his deeds at the Boxer Rebellion; 4) as sect leader and omniscient father figure who enjoys the absolute loyalty of his followers; and 5) as a simple peasant to whom the Archangel of Columbia descended (see above). (Wysocki, "Religion im Spiel *BioShock Infinite*," 64-70).

arrangements. In this way, the game's strategies create *a positive image* of Columbia that is reinforced by the city's pompous architecture and how the game channels the player through it. Thereby, it is no coincidence that the experience of Columbia evokes reminders of the great World Fairs of the late 19th and early 20th century, and specifically of the *World's Columbian Exposition* in 1893.[23] The Chicago World's Fair, as it is also known, was initially built to celebrate the anniversary of Columbus' discovery of the 'new world' in 1492, and became a symbol of American extravagance. Of specific interest in this regard is the visitor's tour through the White City, which begins with a trip to the island on a steamer and progresses further in a linear fashion through it. Right away, visitors are greeted with striking vistas that conjure up a sense of the exceptional, which is intensified in that their vantage points are low compared to the enormous buildings and statues that surround them.[24]

Such a strategy to guide the visitors in an exhilarating manner is also common to theme parks, and BIOSHOCK INFINITE takes both examples as inspiration. In contrast to the real-world fairs, however, the steampunk world of Columbia is a floating one, designed as an ecology in itself: a Utopia "of spatial exceptionalism from the discursive norms below."[25] This extravaganza is now experienced by the player, whose imagination[26] is guided around the city and whose ergodic actions are evoked by the gameworld affordances. The *range of potential interactions*, thereby, includes: the movement within a unicursal labyrinth, which at times opens up to a multicursal one; talking to the Utopians (NPCs), investigating shops, buying, stealing, or picking up objects such as food or Silver Eagles (Columbia's currency). Moreover, the player may watch kinetoscope films about Columbia's past, listen to Voxophones (which are recorded by Columbia's citizens and include personal stories or information about the city), or participate in the activities at the ongoing Raffle and Fair.

---

23 Jackson, *Forced Choice*, 171-173; Peters, "Global Dystopian," 123-125.

24 Norman Bolotin and Christine Laing, *The World's Columbian Exposition: The Chicago World's Fair of 1893* (Urbana: University of Illinois Press, 2002), 29, 34-36.

25 Jackson, *Forced Choice*, 167; cf. 167.

26 Lange underlines the imaginative evoking force of the BIOSHOCK games, which requires an aesthetic-symbolic interpretation from the player. (Carsten Lange, ">>Der Freie hat die Wahl, der Sklave gehorcht.<< Entscheidungsfreiheit und Determinismus in *BioShock* und *BioShock Infinite*," in *>>I'll remeber this<< Funktion, Inszenierung und Wandel von Entscheidung im Computerspiel* (Boizenburg: Werner Hülsbusch, 2016), 107).

Figure 19: Columbia takes inspiration from the World's Columbian Exposition in Chicago, 1893. Especially Monument Tower, its design, and the perspectives on it, is inspired by the Statue of the Republic.

Agricultural Building at the World's Columbian Exposition.

Given the pleasures of these interactions and the extravagance of the Columbian architecture, the image of Columbia's magnificence is easily evoked. These perspective segments are complemented by the level's labyrinthine structure that leads the player through a narrow gamespace but, nonetheless, evokes the sensation of large open spaces. Such a strategy is common to the creation of gameworlds and links them to utopian architecture, which usually contrasts "small, restricted spaces of work and living"[27] to "massive public spaces dominated by the regimes ideology."[28] BIOSHOCK INFINITE makes use of this method and accentuates the seemingly open spaces through vibrant colours and intense lighting from the sun. According to Levine, this follows the purpose of depicting "a world that's ... bold and ... idealized,"[29] which makes it reminiscent of the great public spaces

27 Gregory Claeys, *Searching for Utopia: The History of an Idea* (London: Thames & Hudson, 2011): 113.

28 Ibid., 114.

29 Campbell, "Big Ideas of *BioShock Infinite*."

in ancient Rome that "were designed to impress if not overwhelm, to wed religious, political and military symbols and, above all, to inspire patriotism."[30] The result is an experience of extravagance that in BIOSHOCK INFINITE is magnified through gigantic buildings situated on miniature islands, oversized statues, or balloons of the Founding Fathers. The player's POV is thereby similar to that of the visitors to the World's Columbian Exposition, which creates a feeling of *insignificance* within the confines of the unicursal labyrinth.

Of particular interest in terms of architecture is *Monument Tower*, a colossal statue that was built in the image of the Archangel Columbia. It can be seen from nearly any part of the city and functions as a symbol of hope for the Utopians, for it harbours *The Miracle Child* who shall lead Columbia to greatness. Especially in the game's early hours, the tower is foregrounded to the player in stylised ways: when she turns a corner or moves through a narrow alley, Monument Tower bombastically re-enters her field of view. Such a presentation of important buildings is known to the tourist in cities like London or the theme park visitor, who are equally surprised when they catch an unexpected view of the Palace of Westminster or Cinderella's Castle in Disney World—the latter which is visible to the visitor from many vantage points and functions as a landmark for orientation. Monument Tower exhibits a similar function and facilitates the player's orientation in that it is situated higher than any other buildings in the game and appears larger than it actually is.[31]

As such, the guiding function of Monument Tower helps the player to reach the macro goal of finding Elizabeth—yet its most prominent function is yet to be revealed. Even at this point in the game, it is inevitable that the player connects Monument Tower to similar architecture in the empirical world: whether these include the Statue of the Republic—the main attraction at the World's Columbian Exposition[32]—or better known symbols such as the Statue of Liberty. Most of these inspire hope and the promise of freedom, and it would not be unfair to claim that the libertarian symbolism in Columbia prevails at this moment of play and plunges the city behind the veil of a Utopia. This insight has been contoured by various images in the player's mind: from the lighthouse that granted her entrance to a safe haven and taking pleasures in the extravagance of the city, to a noble mission to save a damsel in distress, and the enthralling vistas of Monument Tower. Still, despite all the utopian beauty of these initial sections, the images

30 Claeys, *Searching for Utopia*, 115.

31 Campbell, "Big Ideas of *BioShock Infinite*."

32 Bolotin and Laing, *World's Columbian Exposition*, 62.

created are fragile. Subtle clues point at another truth, and the player may suspect that there is something wrong in this candy wonderland.

6.2 THE SHADOWS OF UTOPIA: STRATEGIES OF DISNEYFICATION AND THE CAPITALIST-RACIST DYSTOPIA OF COLUMBIA

There is a general consensus among scholars[33] that it does not take long for the player to recognise that in BIOSHOCK INFINITE "[t]he dystopian lurks in the shadows and fault lines of the utopian."[34] I have already described certain clues in the first levels that point in this direction, and the *image of dystopia* continues to foster itself in a part of the city called New Eden Square. Here, the player may eavesdrop on conversations that stain the city's immaculate veil. Two examples of such are a couple staring at Comstock's statue with the woman exclaiming: "Hmm. Oh that statue … I just think it fails to capture Father Comstock's absolute … you know … divinity", while a mother on a picnic with her son teaches him a lesson in discipline: "'Like' does not matter to a Liberty Scout. There's no room for preference, only 'duty'."[35] Even though these perspective segments are satellite events, they give the player hints about Columbia's true nature, in which strict *order* and a religiously inspired *racism*[36] dominate the scenery and where the ideological infatuation can be discerned in the *paralysis of the NPCs* after they speak their lines.

This discovery continues when the player reaches the *fairground* and in a playful manner glances behind its shimmering façade. The fair is all a child can hope for, and the adult Utopians are also savouring the ambiance. A multicursal labyrinth now awaits the player, in which she may move around freely, marvel at the attractions, and give some a try. The fairground thus first of all functions as a *tutorial* for the game, where the player is introduced to the world of Columbia and

33 Peters, "Global Dystopian," 115, 124, 126; Buinicki, "Nostalgia," 725, 728; Maziarczyk, "Playable Dystopia," 241-242.

34 Peters, "Global Dystopian," 126.

35 BIOSHOCK INFINITE (Irrational Games, 2013), ch. Raffle Square.

36 Racism is another target of BIOSHOCK INFINITE that has been widely acknowledged by scholars. For example: Lange, "Der Freie hat die Wahl," 107; Jackson, *Forced Choice*, 197; Buinicki, "Nostalgia," 726; Peters, "Global Dystopian," 124; Wysocki, "Religion im Spiel *BioShock Infinite*," 49-50.

key game mechanics. There are, for example, automatons to illuminate the technical sophistications of this steampunk world such as mechanical horses or a Handyman—a physically enhanced human being. In addition, the functionality of Voxophones is explained and fairground games invite the player to participate—such as *Cast out the Devil*, where the *Vigor* Bucking Bronco can be tested in a game to save a mother and her child from a devil. To do so, the player uses the ability to airburst the devil from his hiding place and, if successful, is awarded a prize. Vigors are a key game mechanic and a cornerstone of Columbia's scientific marvels. They are advertised by posters[37] and endow the PC with biotic abilities such as freezing parts of the gameworld, spitting fire on enemies, or intruding into their minds to have them turn on each other.

Another game worth mentioning is *Hunt Down the Vox*. At this moment of play, the player has heard of a revolutionary group called the *Vox Populi,* who cause unrest in the 'peaceful' society.[38] The Vox Populi, as the player will learn, constitute the working class of Columbia and is comprised of immigrants from Ireland, Asia, Africa, as well as Native Americans. They are disregarded by the general, white population—*The Founders*—and this mistrust of the Other reaches the Utopians as children. Consequently, by participating in the fairground shooting gallery (Hunt-Down-The-Vox), the player creates a perspective on the scenario that, when combined with the remaining perspectives of the fairground, grants her insight into the gameworld's mechanisms. The game resembles an eerie fairy tale where the Vox are hiding in a dark forest. For every hit the player receives points, and now and then the leader of the Vox and 'anarchist' Daisy Fitzroy emerges. Shooting her down grants the player bonus points, which will raise the amount of Silver Eagles she receives as a prize.

As a result, the experience at the fairground gives rise to the contours of an important image and involves the player in a game of estrangement. This image not only familiarises her with the shooting mechanics of the game, but it also fosters an *enemy image* of the Vox through the challenging yet pleasurable encounter. The leaders of Columbian society, so it seems, are teaching their citizens who are the 'good' and 'bad' guys from an early age through playful interactions. These

37 The advertising is a first hint at the *Vigor industry* the player will come to see later. (Buinicki, "Nostalgia," 725).

38 Two hints and perspective segments of the Vox can be found by the player if she overhears two conversations at New Eden Square. In one, there is a woman enquiring about the meaning of the term Vox Populi. She quickly loses interest when her husband claims that it is a Latin term. Meanwhile, another couple describes the Vox as brutal savages who attack the peaceful Columbians and with whom one cannot reason.

run the risk of remaining unquestioned by the Utopians and the player alike, because of their pleasurable nature. As such, the game puts such an attitude in perspective and targets the *gamist and achiever player type*, who is interested in achieving the game's goals and gathering points in an efficient manner. This playing style is afforded by the game through its possibilities for action and induces the player to participate in pleasurable combat and scavenging for items. Yet it also creates a perspective on the gameworld that is important for the creation of those images that expose the game's major themes such as humankind's complicity in capitalism and racist atrocities. This form of estrangement through play may go unnoticed by the player if she does not question the agenda behind such processes—but if deciphered in acts of ideation, it makes a lasting impression on her.

Consequently, the player may come to the conclusion that the *scapegoating of societal problems to the Vox* develops between the glamour foregrounded by the game. It is hidden within the satirical nature of the Raffle and Fair section and also comes to the fore in propaganda posters, which alternate between religious, racist, and capitalist symbolism. In addition, the aforementioned statues are reminiscent of Orwell's Oceania in *Nineteen Eighty-Four*, and the Voxophone and kinetoscope messages/films incrementally reveal Columbia's past.[39] Hence, the player steadily gains insight into the dystopian nature of this world by interacting with both the *gameworld's present* and *past* that uncovers a sinister ideology.[40]

In this manner, BIOSHOCK INFINITE fulfils the criteria of the critical dystopia as it illuminates the origins of dystopia and exposes the gameworld for what it is. It then combines this layer with present affairs and induces the player to weave connections to her empirical surroundings through acts of ideation. Thereby, the above-mentioned technique of estrangement is employed in a creative fashion. It aggravates the referentiality of the fictional to the empirical world and coaxes the player through enticing gameplay in "a vast capitalist dream machine dedicated to colonizing the imagination in the interest of the maintenance of the status quo."[41] This claim rests on the observation of the fairground processes and the capitalist-authoritarian agenda behind them, which might go unnoticed by players, for they are generally accustomed to certain gameplay conventions that they conduct automatically. For example, having arrived in Columbia, the player is tempted by the shimmering sight of Silver Eagles or other flickering objects such as food supplies and ammunition. She picks them up to satisfy her need for scavenging and

39 Jackson, *Forced Choice*, 174, 197, 199-201; Maziarczyk, "Playable Dystopia," 242, 250-251.

40 Ibid., 250-251.

41 Ferns, *Narrating Utopia*, 231.

uses the currency to participate in the attractions. She *playfully consumes* without questioning the dehumanising industry behind these façades—buying Vigors, health, ammunition, or other utensils from vending machines—and has become victim of what Suvin calls the strategy of *Disneyfication*: this "case of a dystopian misuse of eutopian images," to be found in "the edulcorated fables and fairy tales of Disneyland."[42] These resemble "a privileged *pars pro toto* of the capitalist and especially U.S. admass brainwash"[43] and can be regarded as "a shaping of *affective investment into commodifying which reduces the mind to infantilism*."[44]

BIOSHOCK INFINITE employs this strategy to the fullest and involves the player in interactions that mitigate the critical nature of the experience but which, at the same time, create a vital perspective on the gameworld. To do so, the game aligns the mechanisms of gameplay to those of people's daily routines in a *capitalist/consumerist*[45] world such as buying food or goods without questioning their origins. It does so in a distorted manner, however, by hiding these empirical world processes in gameplay activities like scavenging or combat interactions. The result is a form of *estrangement through play* that is critical to bringing out the VGD's warning and the novelty a work of art promises.

6.2.1 From a Peaceful Visit to Utopia to a Brutal Escape from Dystopia

Whereas initially, BIOSHOCK INFINITE involved the player in subtle games of estrangement, it will soon become more obvious in the *negation of utopian images*. At the Raffle and Fair only a few perspectives conjured up the *image of dystopia* in the player's mind, and the game did so by involving the player in a defamiliarised theme park of capitalist excess. This truth will become more obvious now and is fortified through an *inversion in terms of gameplay* that is about to occur at the raffle. The route there prepares the player for it and continues to foster an image

42 Suvin, "Theses," 194.

43 Ibid.

44 Ibid., 195.

45 Other scholars who have implicitly or explicitly seen *capitalism* and *consumerism* as a target in BIOSHOCK INFINITE include: Lizardi, "*BioShock*," 9; Jackson, *Forced Choice*, 171-172; Peters, "Global Dystopian," 126, 129; Maziarczyk, "Playable Dystopia," 250; Buinicki, "Nostalgia," 725.

that has been subject to heated debate in the BIOSHOCK games: the player's *lack of agency* and the issue of *determinism.*[46]

After the baptism, the player is shrouded in a veil of agency while exploring the fairground and interacting with the attractions. This sense of freedom is about to be negated more forcefully, when the player again encounters the lady and the gentleman. These will later be unmasked as the Lutece twins (Rosalind and Robert), and they invite DeWitt to participate in a game of *heads or tails* whose outcome will always be the same. To proceed in the level, the player's sole choice is to press the square button on the PlayStation 4 gamepad to have DeWitt choose tails. The result of the coin flip will be heads, however, no matter how often the player replays this section. In addition, the many strokes on Robert's chalkboard suggest that this was and will always be the sole result, implying that DeWitt has taken this journey many times before.[47] Combining these perspectives and closing the blanks between them, the player may view her previous interactions (and those that are to come) differently, and in terms of *aporia*. What complements this image is that the scene is presented hilariously, and it is fair to say that the Lutece twins are the game's most powerful satirical element. Whenever the player meets them, the music changes to a stand-up comedy tune, and their dialogue suggests they are playing a game of their own with DeWitt.

Having completed this event, the player's route continues towards the raffle and is guided by a chanting crowd. A poster is placed in the middle of an alley. It shows the devil's hand and warns of a False Shepherd by the mark *AD*: an abbreviation of the Latin term Agnus Dei, the Lamb of God.[48] When approaching it, DeWitt raises his right hand, and the player can discern the same mark on it. As such, the event not only forebodes the plot twist later in the game, where Elizabeth is revealed to be DeWitt's daughter, Anna, but it also foregrounds the dreamlike nature of the game. This *surreal impression* will intensify once the player reaches the raffle, where she is surrounded by a crowd whose eyes centre on her. A woman winks De Witt over now. She distributes the lottery baseballs to have him pick the winning number 77. It is a divine number, sometimes connected to Jesus Christ or to the 77 generations that, according to the Lucan genealogy of the New Testament

46 Jackson, *Forced Choice*, 191-193; Lizardi, "*BioShock*," 10-11; Peters, "Global Dystopian," 119; Maziarczyk, "Playable Dystopia," 251; Lange, "Der Freie hat die Wahl," 120-121.

47 Lange, "Der Freie hat die Wahl," 120.

48 Bosman, "Lamb of Comstock," 177.

(the Books of Enoch), were necessary to pass from Adam to Jesus.[49] Shortly after, the announcer (Jeremiah Fink) rewards DeWitt with the first throw at an interracial couple. They are strapped to poles, and the stage scenery includes monkeys hanging from a tree and an ape version of Abraham Lincoln decorating the background.

Evidently, the event exposes the *racist undertone* of the Columbian society, its "xenophobia and elitism"[50] and paves the way to a more *fundamental inversion in terms of gameplay*. To do so, it confronts the player with the choice to throw the baseball at the interracial couple or Fink himself. Such a choice is more obvious than whether to participate in the Hunt-Down-the-Vox game, and has long-term consequences, however minor they may seem. For if the player throws the baseball at Fink, she will meet the couple later in the game and they will express their gratitude. In any case, the player is able to express herself through ergodic action, and this potentially develops the *emergence of a counter-narrative* to an ever-clearer image of dystopia.

Thus far, the *official narrative* of the Columbian society has remained in the shadows, but with DeWitt's intervention at the raffle and his exposure as the False Shepherd, an outburst of violence marks the scene. This *inversion of the use of gamespace* turns the multicursal labyrinth into a *battlefield* and changes the range of actions the player may perform—which have come a long way from the leisure tour around Utopia and the playful interactions at the fairground. They now cater to a brutal skirmish and include the use of the Sky-Hook, a device that enables the player to melee attack opponents and use the Sky-Lines. In addition, firearms are now at the player's disposal, and the gamespace is littered with items such as health, ammunition, or Salt (a substance to recharge the Vigors, and a religiously inspired term). It is this standard equipment of first-person shooters that transforms the scenery and gameplay processes into a venue of intensity, scavenging, and combat. Diverse combat strategies are thereby available to the player, as the Sky-Lines, melee attacks, firearms, and Vigors can be used in various combinations. Moreover, vicious enemies, like the Firemen, now plunge the streets of Columbia into a burning chaos. They are reminiscent of Bradbury's eponymous characters in *Fahrenheit 451*, as their agenda is contrary to their job description.

What follows from this inversion of *gameplay and the use of space* is that the player becomes overtly involved in an *official narrative*. She is only about 40-60

49 Paul C. Schnieders, *The Books of Enoch: Complete Edition*, trans. Robert H. Charles (Las Vegas: International Alliance Pro Publishing, 2012), ch. 10.

50 Peters, "Global Dystopian," 125; cf. 125-126; Lange, "Der Freie hat die Wahl," 121; Buinicki, "Nostalgia," 726-727.

minutes into the game (with a total length of approximately 15 hours), yet her imaginings have already been subjected to a rollercoaster ride, as the horizon of past perspectives and creation of images are continually revised. The blanks closed thereby inspire the player's actions and result in a wandering between sensorial impressions and ergodic actions, themes and horizons.

6.2.2 Saving the Damsel in Distress: DeWitt's Unconscious Struggle for Forgiveness

I now wish to continue with the player's first encounter with Elizabeth and the events leading up to it. Meeting the young woman is a kernel event in the ongoing plot and on the conceptual level of DeWitt's unconscious struggle for forgiveness. This struggle primarily involves two levels: 1) a *personal* one in which DeWitt seeks forgiveness for the agonies he exposed Anna to and 2) a *universalised* one that revolves around his guilt about having lost Utopia by participating in the racial atrocities of Wounded Knee.

In this regard, it is no accident that the path leading towards Elizabeth is accentuated with marvellous vistas of Monument Tower, which, as the player comes closer, gains in pompousness. It is not easy to get there, since police forces are gathering to stop the False Shepherd. One microcosm the player has to pass here is *The Fraternal Order of the Raven*, Columbia's version of the Ku Klux Clan and its darkest underground operation.[51] The experience there makes explicit what the player's imaginings have implied before, and sets in context the worship of the Founding Fathers. While inside the Order, the player encounters a sombre location and organisation whose purpose is the protection of the white race. There are hints at a rejection of Native Americans and experiments in phrenology, while in one room, a portrait of a figure resembling the devil awaits discovery. On closer inspection, the player may discern Abraham Lincoln, because of the recognisable hat. This perspective segment is complemented by a statue of John Wilkes Booth, the known murderer of the former president, who is presented in a glorifying pose: outstretched forearm and a gun in his hands.

Closing the blanks between these perspectives, the player might come to the conclusion that the game creates a *reversal of the positive myth around Lincoln*, which explains the pompous image of the Founding Fathers. For from the Columbian view, it was Lincoln who accepted the risks of war in a heretic attempt to abolish slavery.[52] The attentive player comes to see this *distortion of history* and

51 Wysocki, "Religion im Spiel *BioShock Infinite*," 44, 84.

52 Ibid., 44.

the racist perspective on it, while a Voxophone message by Father Comstock gives further context.

> What exactly was the Great Emancipator emancipating the Negro from? From his daily bread? From the nobility of honest work? From wealthy patrons who sponsored them from the cradle to grave? From clothing and shelter? And what have they done with their freedom? Why, go to Finkton, and you shall find out. No animal is born free, except for the white man. And it is our burden to care for the rest of creation.[53]

Demonising Lincoln in this manner and having the Founding Fathers worshipped by the Utopians places the latter under a negative image and construes them as antagonists. It is as Lizardi and Buinicki hold: in BIOSHOCK INFINITE "accepted and beloved icons of US history are called into question because of their complex backgrounds, such as the game's transformation of the images of Washington and Lincoln into violent religious icons."[54] This inversion interrogates "the construction of history itself" and reveals "the dystopic quality of US history ..., its violence concealed beneath conventional imagery of patriotism and nationalist mythology."[55]

Guiding the player's imaginings in such a way and confronting her with a brutal regime fuels the urge to save Elizabeth even before she meets the young woman. To reach her, the player needs to take Sky-Lines to Monument Island. These ensure rapid transportation, and the police forces are using them to pursue DeWitt. As a result, the player becomes involved in hectic combat on a linear route upwards or in three-dimensional multicursal arenas, which at times disrupts the unicursal labyrinth. Moreover, barricades are placed in strategic positions to hinder her progress, and these perspectival arrangements allow a certain interpretation of the events, especially if taken up from a psychoanalytical point of view.

In Freud's deliberations on the night-time dream, the Austrian psychoanalyst describes the formation of dreams originating in the human *unconscious* to then chart their way to perceptibility. It is reached once the *dream-thoughts* (the original *wish*) arrive in the human *consciousness*, but the route there is plastered with hurdles and disruptions. This is because a filtering apparatus, the *preconscious*, ensures the dreamer's sanity from repressed memories—and so the original wish needs to undergo distortion in order to pass said barrier (see chapter I).[56] In

53 BIOSHOCK INFINITE (Irrational Games, 2013), ch. Comstock Center Rooftops.

54 Lizardi, "*BioShock*," 9.

55 Buinicki, "Nostalgia," 723.

56 Freud, *Dreams*, 382-383, 410, 415.

BIOSHOCK INFINITE it is known that DeWitt struggles from past deeds as a solider of the 7th Cavalry, where he took part in the massacre at Wounded Knee, and as a Pinkerton agent. He buried these memories through alcoholism, which is what supposedly led him to give away his daughter, because of accumulated gambling debts. During the game events, DeWitt is initially unaware of these wrongdoings, and only a few soliloquies hint at another conclusion, which points to a *repression* of past deeds.

Participating in the game events, the player will thus explore DeWitt's unconscious and uncover the *wish* together with her PC. The result is a creative use of the dystopian narrative's process of realisation, which can be linked to *the interpretation of DeWitt's nightmarish delusions* stemming from the previously described neurosis. The original wish behind them is opaque, and the *dream-work*—which distorts the original wish and turns it into the perceptible *dream-content*—hinders the player's progression *and* coming to awareness.[57] Consequently, the hurdles and combat she encountered before (and will continue to encounter) prevent the player and DeWitt from unmasking the latter's well-hidden secrets. Naturally, they also function as ludic obstacles, but in doing so, they conjure up images on the level of significance.

One of these is an exploration of DeWitt's past as a mercenary, whereby the player engages in violent combat processes herself. She thus assumes the role of mass-murderer DeWitt, who always puts his mission objective first. Such a perspective specifically comes to the fore when the game is approached with the attitude of a gamist or achiever, who strives to attain their game goals most efficiently. This player may fall prey to the pleasures of combat and scavenging, without having in mind the greater picture, and, therefore, runs the risk of an uncritical playthrough and succumbing to the trap of mass media entertainment.[58]

To evade such a pitfall, BIOSHOCK INFINITE offers the player several perspectives that lay a different light on the combat processes. This occurs, for example, when Comstock accuses DeWitt of his wrongdoings. It is an eerie scene in which the prophet orders his followers to stand down to then appear on a projector. To trigger the event and to uncover the truth behind the dream-content, the player has to pull a lever.

---

57 Ibid., 169, 351.

58 Equally, *narrative* or *wanderer* player types, who engage with the game to explore its gameworld and plot, may downplay the importance of the combat processes and scavenging perspectives.

I know why you've come, False Shepherd. I see every sin that blackens your soul. Wounded Knee. The Pinkertons. The drinking. The gambling. And, of course, Anna. And now, to repay your debt, you've come for my lamb. But not all debts can be repaid, Booker …

What brought you to Columbia, Booker? 'Bring us the girl and wipe away the debt?' This will end in blood, DeWitt. But then again, it always does with you, doesn't it?[59]

Comstock, so it seems, knows about DeWitt's sins. Intuitively, this knowledge could be explained with the player believing Comstock to be a prophet, but paying attention to the gameworld events enables a more appropriate conclusion. In this scenario, DeWitt and Comstock are one and the same person existing in different universes, and DeWitt is addressed by a *repressed part of his own self.* Although the game implicitly plays with such a conclusion, DeWitt fails to acknowledge this truth for now—and so, potentially, does the player.[60]

Step by step, a more comprehensive image of DeWitt's struggle for *forgiveness* emerges and paves the way to the encounter with Elizabeth. The young woman will later turn out to be DeWitt's daughter, but neither of the two (nor the player) can know about this at this moment of play. As such, the encounter forebodes an image that deals with *the culpability of the father towards his daughter*.[61] This fact can be explained in that DeWitt gave Anna away to wipe away his gambling debts, while Comstock imprisoned Elizabeth in a tower. Such a constellation leads back to Freud's deliberations on the night-time dream, who describes the original wish, "existing in repression," to be "of infantile origin."[62] The child's relation to its parents thus becomes the focus of investigation, in particular the

59 BIOSHOCK INFINITE (Irrational Games, 2013), ch. Comstock Center Rooftops.

60 Another example are the events surrounding Comstock's speech, where the player may create an additional perspective on the described scenario. Once Comstock's voice appeared through the loudspeakers, the enemies in the area kneeled down before their prophet and started praying to the Founding Fathers. Being astounded by the sudden lack of challenge, the player now has the possibility to walk past them or execute them most brutally. The Utopians will not resist in the latter case, showing the player how far these people might go for their beliefs. Moreover, this action creates an important perspective on DeWitt and the player, because they have become complicit in the slaughter of defenceless people, similar to how DeWitt solved the issues at Wounded Knee and as a Pinkerton agent.

61 Lange, "Der Freie hat die Wahl," 119.

62 Freud, *Dreams*, 392.

son's relation to his mother and the daughter's to her father.[63] However, when extended with Ferns and Jameson's deliberations (see chapter I), there is a *more fundamental repression* at work here. It can be vindicated by the game's upcoming events and revolves around *humankind's yearning for Utopia*. Consequently, to decipher the wish's true nature and to uncover DeWitt's neurosis, the player embarks into the depths of her PC's unconscious—not only to playfully enact this struggle for Utopia but also to awaken from the false dreams of ideology herself.

This gradual journey to aesthetic response continues once the player arrives at Monument Tower and encounters a sealed-off area resembling a version of Dr. Frankenstein's laboratory. As before, barricades hinder the player's progression, and signs warn about the interaction with a specimen. It is as if a dark secret lies hidden within the tower, and its inside confirms the worst imaginings. The place is littered with medical equipment and suggests that experiments on test subjects, particularly Elizabeth, have been conducted. In fact, the player is given a comprehensive introduction to the young woman before she meets her. There are projector films that documented Elizabeth's steps—while she was dancing or trying to lock pick her way out of her imprisonment—and a large blackboard illustrates Elizabeth's morphology from ages one to seventeen.

The player may ponder the reasons for Elizabeth's imprisonment as she penetrates further into the tower. It fortifies the negative, mysterious image of the Columbian society and points to Elizabeth's importance (or danger) to the ruling ideology. It is thereby striking that the place is deserted, as if DeWitt would have disabled the dream-work's defence mechanisms that bury the secret behind them. Given this *semantic charging*, it is little wonder that in a second playthrough, the closing of the blanks works differently here. For the player has now accumulated enough information about DeWitt and Elizabeth to see the tower's true function, which refers to the cruelties an alcohol-addicted father may inflict on his newborn child. In fact, it would seem that DeWitt here visualises his own *manifestation of guilt* and the agonies he exposed Anna to. These come under the guise of an angel's statue and eventually pave the way to the charming encounter with the young woman and the inception of her complex coming-of-age story.

The moment the player meets Elizabeth is perfectly staged. Once DeWitt arrives in the Specimen Observation Room and pulls the lever, the iron curtain is raised and rewards the player with a glance at the young woman. Elizabeth is in

63 In this sense, the events of BIOSHOCK INFINITE could be interpreted as Elizabeth's unconscious wish for paternal security. This is expressed by her desire to go to Paris and see the Eiffel Tower (a phallic symbol *par excellence*), while it hides a struggle for emancipation from such wishes.

the Dressing Room, looking at herself in the mirror and immersed in reveries about Paris. She runs off, and the player follows her to a place where a child's drawings of a bird and a further painting of the Eiffel Tower decorate the walls. Elizabeth opens a *tear* now, a dimensional window into another period and time, and finds herself in the Paris of the future. As the player will soon find out, the young woman resembles what Peters calls a "cyborg,"[64] equipped with *posthuman abilities* that will be revealed as *wish-fulfilments*. For the moment, however, the player meets a literate young woman who is vulnerable and innocent yet enthusiastic about discovering the world outside the tower. She thus resembles one of Disney's princesses, such as Rapunzel or Cinderella, who were also imprisoned by a family member.

The scene culminates in Elizabeth's and DeWitt's escape from Movement Island, which marks the beginning of their getaway from dystopia. It is a turbulent ride in which the two are pursued by Elizabeth's guardian *Songbird*, a large mechanical automaton. Elizabeth has a special relation to Songbird, who is both her friend and jailer—and this constellation gives rise to the inception of an image that, according to Lange, revolves around the thought experiment of uniting two contrary aspects or persons.[65] In this respect, the imagery of *the canary and the cage* and its inversion in BIOSHOCK INFINITE becomes of importance. For the player has experienced the twisted relation between Elizabeth and Songbird—with the woman caged up in the angel's tower, while the canary makes his rounds in the Columbian skies to guard the ruling ideology's secret.[66]

As such, the created image is of semantic ambiguity, and Elizabeth's tower and her relation to Songbird assume the role of *floating signifiers*. From the symbol of hope in the beginning of the game to a place of imprisonment, and to a manifestation of another father figure who lovingly encages his daughter in the ramifications of parental failures, narcissistic hubris, or alcohol abuse. Once the player closes the blanks here, a strong connection between Songbird, DeWitt, and Comstock is created. These imaginings notwithstanding, the tower's semantic range is far from exhausted, and its most prominent point of reference has yet to be established. This image is closely linked to Elizabeth's attempts to lock pick

64 Peters, "Global Dystopian," 130.

65 Lange, "Der Freie hat die Wahl," 105-124.

66 Peters, "Global Dystopian," 121-122; Toh Weimin, "The Limits of the Evolution of Female Characters in the *BioShock* Franchise," *Proceedings of DiGRA 2015: Diversity of Play: Games – Cultures – Identities* 12 (2015): 12-13, http://www.digra.org/digital-library/publications/the-limits-of-the-evolution-of-female-characters-in-the-bioshock-franchise/

her way out of her imprisonment and manifests itself not only in the attempt to free herself from father figures—as some critics have suggested[67]—but also in the emancipation from the machinations of ruling ideologies and from the iron cage of Anti-Utopia.

Such a conclusion illustrates and underscores the diversity of the blank in its function. Various meanings emerged from the perspectival constellations on the level of the plot and significance, which differ in a second playthrough. This may come as a surprise to the player—because one is not always specifically aware of semantic ambiguity at first glimpse—and leads to the conclusion that the closing of the blanks offers fuel to the player's process of realisation. Consequently, through its complex structures, BIOSHOCK INFINITE bewilders the player yet guides her towards aesthetic response. For there are manifold ways one can construe the events so far, and the game will continue this semantic ride until its end.

6.3 ELIZABETH AS THE FIGURE OF THE TEMPTRESS AND HER PROCESS OF EMANCIPATION

Before jumping ahead to the game's point of convergence, I wish to further illustrate *Elizabeth's role*, who designates a fundamental perspective for the player's acts of ideation. Once outside the tower and at a safe distance from Songbird, Elizabeth is eager to explore the new and unfamiliar world.[68] This curiosity marks the inception of a *realisation process* that will affect not only the young woman but also leave its mark on the player, who is taking on the role of DeWitt. Scholars have described Elizabeth's role in this regard as one that shows "significant character development"[69] from *insecurity* to *capacity/tenacity* in a struggle for emancipation that will lead the young woman to see "with greater clarity the world around her; in particular … *the wires of the cages* working together to form a system of dominance."[70]

---

67 Ibid., 12; Catlyn Origitano, "'The Cage is Somber': A Feminist Understanding of Elizabeth," in *BioShock and Philosophy: Irrational Game, Rational Book*. Blackwell Philosophy and Pop Culture Series, edited by Luke Cuddy (New York: Wiley Blackwell, 2015), 40-41.

68 Ibid., 43.

69 Weimin, "Limits of the Evolution," 13.

70 Origitano. "Cage is Somber," 45; emphasis mine; cf. 41-45; Weimin, "Limits of the Evolution," 10-11; Franziska Ascher, "Es gibt immer einen Leuchtturm …" *Paidia: Zeitschrift für Computerspielforschung* (2013): 7, http://www.paidia.de/?p=2499

In the following, I will thus describe Elizabeth's diverse roles, which revolve around ludic, narrative, and emotional characteristics, and the emancipatory task of *freeing herself from the confinements of ideological structures and the sweetened images these employ to hide their agendas*. This relentless struggle for Utopia will lead Elizabeth from initial naiveté to enlightenment, to seeing beyond the falsifications and unspoken truths of the fictional reality. In doing so, she offers the player an *affordance and appeal structure* the latter may accept or decline, but which may affect her judgement of empirical reality. For the player has experienced a distorted version of it in the game events and has witnessed Elizabeth's reactions to the ideologies they encountered. As such, the charming young woman assumes *the role of the temptress*, who seduces the player to certain imaginings and ergodic actions within and outside the diegesis.

For Elizabeth, this process of realisation begins with the flight from Songbird, who is tearing down Monument Tower. It is a hectic scene characterised by a rapid descent on the Sky-Lines and Elizabeth and DeWitt falling into the depths of an ocean region. A brief interlude in DeWitt's office once more highlights the dream-like character of the events—since in dreams, one tends to awake after intense sections, such as falling or drowning. When the game returns to Columbia, DeWitt is swept to the shores of Battleship Bay, an artificial beach and place of leisure for the Utopians. Battleship Bay is part of a larger array of mini-islands that form the amusement district Soldier's Field. Upon entry to Battleship Bay, the player encounters a gateway to the beach in the form of a turtle head, while the Utopians are engaged in activities such as relaxation, sports, dance, or flirtation. Elizabeth is fond of these activities and invites DeWitt to dance. Again, the player witnesses a vibrant young lady who wishes to escape the Columbian society and go to Paris, but who is also vulnerable to the pleasures that await. This expresses itself if the player shows patience and allows Elizabeth to participate in activities such as flicking stones into the ocean or gymnastics. Even the player may participate here, by mimicking the movements of the Utopians (move the right analogue stick up, down, right, left, or jump, in various combinations).[71]

71 Meanwhile, the musical score of the level underlines Elizabeth's newfound freedom. Once DeWitt intervenes in Elizabeth's dance and pulls her from it, a cover version of Cyndi Lauper's *Girls Just Want to Have Fun* (1983) is played on a calliope—replacing the Irish folk music from before. This change of tone grants the player a further perspective on the events. Given the song's lyrics—which deal with women's emancipation from father figures and men—it underscores Elizabeth's wish to escape the confines of male imprisonment and to discover the wider world. (Ivănescu, "Music of Tomorrow," 58-59).

Notwithstanding these ludic pleasures, Battleship Bay's function within the Columbia society is quite the contrary. A slogan places the inception to this image, which holds "STRENGTH THROUGH LEISURE,"[72] while the gameplay processes of the activities are juxtaposed with it. To close this blank, the beach's garish presentation and the symbolism of its name (Battleship Bay) fuel the player's acts of ideation and have her connect the section to the previously encountered Disneyfication strategy. A further perspective is added to the conundrum, however, as it now includes Elizabeth. She was raised in this ideology but savours its candy pleasures for the first time—which places her in conflict between the desire to flee this society and its enticing, childlike pleasures.

This becomes clearer in the upcoming events, where the Lutece twins offer Elizabeth a choice between two necklaces: the one adorned with a *bird* motif, the other with a *cage*. Elizabeth seeks advice from DeWitt which one to choose, and this reminds the player of the symbols' inversion and the (seeming) artificiality of choice in this respect.[73] However, one can discern the game the Lutece twins are playing. While at first glance they seem indifferent to the ever-repeating events, Robert, at least, has not lost hope. This can be seen in how on several occasions he adopts a more optimistic stance than Rosalind, as he believes in the thought experiment he desired (bringing DeWitt to Columbia to make things right). For example, when Elizabeth picks the cage instead of the bird, he is frustrated. Such a satellite event is not easy for the player to catch, but it adds to her understanding of the gameworld and to the conceptual level of the struggle for Utopia.

This struggle is first and foremost reflected in Elizabeth, yet it begins with a naive young woman who is torn between conformism to the ideology she grew up with and the wish to question and later oppose it. The player can detect this conflict when Elizabeth witnesses the collapse of her angel's statue and in the events at the Arcade. Here, the player is introduced to the cartoon characters *Duke & Dimwit*, a pair of toy soldiers who remind the young Columbians to be valiant and determined instead of clumsy cowards. As one of the arcade machines narrates:

> Are you a Duke or a Dimwit? When Duke sees a suspicious-looking character, he reports him to his authorities straight away. When Dimwit sees a suspicious-looking character, he ignores him, and focuses on his new scooter. Remember, boys and girls, don't be a Dimwit.[74]

---

72 BIOSHOCK INFINITE (Irrational Games, 2013), ch. Battleship Bay.

73 Lange, "Der Freie hat die Wahl," 121-122.

74 BIOSHOCK INFINITE (Irrational Games, 2013), ch. Battleship Bay.

Elizabeth is enthralled by them, for she grew up with their stories. Full of excitement, she operates the arcades. It is a scene in which the young woman can be the child she was never allowed to be, but where she also succumbs to the pleasures of the Disneyfication machine. Duke & Dimwit are a fundamental method of coaxing young Utopians into joining the military and teaching them order and discipline, thus downplaying the atrocities of war.

Figure 20: Soldier's Field closely resembles the structure of Disney World, Florida, and involves the player in a vital inversion of gameplay and space.

BIOSHOCK INFINITE (Irrational Games, 2013), ch. Soldier's Field.

Still, Elizabeth does not succumb entirely to the candyfloss spinning around her imagination as she questions the existence of two separate bathrooms: one for white people, the other for Irish and coloured. The construed image is complemented by a Kinetoscope short film of Father Comstock that explains how to deal with the Irish problem. The Irish are depicted as drunkards, which comes as a contradiction to the player, who before entering the Arcade potentially met the friendly interracial couple from the raffle. Given these facts, the racist overtone of the Columbian society merges even more with its capitalist and cult-like structure. What remains to be shown in this respect is how the labyrinthine structure, together with the ludic encounters (interacting with the Utopians, participating in the attractions, scavenging, or combat) contributes to a reversal of the gameplay the player has experienced before. This reapportion of space took place at the Raffle and Fair, and a second inversion is about to occur when the player enters *Main Street*, the central section of Soldier's Field.

On Main Street, the player experiences a vast place that fortifies the negative image of the Columbian society. It is no coincidence, thereby, that Soldier's Field's map—and the names of its microcosms—resemble the structure of Disney World, Florida. Main Street itself is designed as a spacious multicursal labyrinth in which the player may move around freely, witness a Duke & Dimwit show, go on a carousel ride, or explore shops that sell ice cream and propagandistic war toys. Moreover, the place takes inspiration from modern malls in which, as Jackson notes, a central hub connects the different parts, allowing swarms of people to consume its goods and attractions.[75] Again, the young are brainwashed by leisurely participating in these Disneyfication processes, which mitigate the ramifications of war and encourage them to sign up for duty.

At this point, the strategies of *negation/inversion* once more grasp the player's attention. Similar to the game's beginning, these involve the player in a leisure tour around Main Street to then invert these processes by reinterpreting the multicursal labyrinth into a venue of atrocity. During the initial tour, hints at another truth are hidden within the environment such as propaganda posters, dialogues, and the eerie impression of the place. However, once the player wishes to leave the area to the First Lady's Aerodrome and calls a gondola by pulling a shining lever, a more explicit inversion awaits. That the gondola does not arrive as planed is a gameplay device to trigger combat, for one has to obtain the Vigor Shock Jockey to power it. DeWitt's way now leads to the Hall of Heroes and to the cathartic experience of revisiting his deeds at Wounded Knee. As such, the dream work intensifies its efforts to keep DeWitt from accessing this painful experience. It litters the place with enemies, ranging from police forces and soldiers using heavy weaponry to Zealots of the Lady (a dark figure with a coffin strapped to his back) or Motorized Patriots (automatons equipped with machine guns that resemble George Washington). Moreover, items are scattered throughout the environment and affordances help the player to succeed in combat such as puddles of oil she can incinerate. The result is a section of rapid combat that again exposes the ruthlessness of the Columbian society and lays the focus on DeWitt's inner struggle to uncover the secret behind the nightmare.

Elizabeth assumes a crucial role in this respect, as she helps DeWitt to access further areas of the game and his unconscious. She does this by lock picking doors DeWitt cannot open and by providing assistance in combat. Elizabeth therefore fulfils a "functional,"[76] ludic role. She never comes as a burden to the player but instead helps him obtain Silver Eagles or supplies such as ammunition or health.

---

75 Jackson, *Forced Choice*, 179-181.

76 Ibid., 182.

In addition, she will later be able to open tears in the environment, which grant the player access to supplies and combat strategies such as ammunition, blockades, a gun turret, or additional hooks in the environment the player can grab on to manoeuvre in strategic positions, and so on.[77]

Describing Elizabeth's function in such a way is appropriate in terms of gameplay rationale, yet it overlooks her role as temptress and how interacting with the young woman creates an emotional bond to DeWitt and, by extension, the player. Evidently, Elizabeth is vital to the game's plot, and her role invites the player to reflect upon aspects of the gameworld. Thereby, Elizabeth's experience of the ideologies in Columbia not only affects her perception of this world, leading to a pessimistic view of it, but also the player's, who witnesses the changes in the young woman and role-plays DeWitt's struggle against a dark part of his self. Combining these perspectives, a clearer image of the struggle for *forgiveness* emerges. Its origins can be traced back to the agonies DeWitt (Comstock) exposed Anna (Elizabeth) to, yet this also involves something greater. Especially when combined with the political perspectives of Columbia, *the struggle for Utopia* is foregrounded and *humankind's guilt* for succumbing to animal instincts and taking pleasure in the atrocities of war and racism.

6.4 THE ROUTE THROUGH THE MAZES OF IDEOLOGICAL PERSPECTIVES

BIOSHOCK INFINITE lets the player enact the story of an alcoholic, depressed neurotic who struggles for forgiveness for his past deeds. These involve not only his treatment of newborn Anna, who suffers from the repercussions of her father's alcoholism, but also the racist atrocities DeWitt committed as a soldier of the 7th Cavalry and while he was a Pinkerton agent. What it means to follow this path will be experienced by DeWitt and the player in the estranged and dreamlike gameworld of Columbia, where Father Comstock established a supposed Utopia that follows the principles of a religiously inspired nationalism. Similar to the player, Elizabeth experiences this world afresh and absorbs the newly found impressions, which gradually change her character. The route is both terrifying and enlightening as the player passes through the *microcosms* that compose the Columbian society. Thereby, like those of the World's Columbian Exposition—which was divided into regions and headquarter buildings, each of which reflected

77 Ibid., 181-186; Weimin, "Limits of the Evolution," 10-11; Origitano, "Cage is Somber," 41-43.

"the heritage and character of the state it represented"[78]—each area of the city fulfils its own function. I have already addressed the Welcome Center and the Raffle and Fair, which introduced the utopian traveller to a picturesque world, only to have her experience inverted by events and places that followed. These included the Order of the Raven and two areas at Soldier's Field—Battleship Bay and Main Street—and I now wish to address further, important ones.

The experience of moving through these islands and microcosms of Columbia stimulates the player's acts of ideation in a similar manner to how Elizabeth's perceptions of them are shaped. This route will now lead through a museum of DeWitt's sins in the Hall of Heroes, to the capitalist glamour of Finkton and its underground region Shantytown, where the working class of Columbia lives, to the encounter with Comstock, the story's high priest, in Emporia. Here, Elizabeth will come to see that the problem fundamentally lies within human nature itself, in the agenda for power and the ruthlessness necessary to obtain it.

6.4.1 The Hall of Heroes as DeWitt's Journey into the Horrors of his Past

The *Hall of Heroes* awaits the player with an enormous statue of Father Comstock (DeWitt) erected on top of its roof. In essence, the place can be described as a museum that sends the participant on a tour into the past and has her immersed in the narrative of DeWitt's deeds at Wounded Knee and Comstock's at the Boxer Rebellion. These are presented in a glorifying manner, and the location thus holds the function to remind the Utopians of the Chinese and Native American threat. Yet, as the player will see, "The Hall of Heroes is also a graphic reminder of how historical narratives are shaped and refashioned, often to cover up racial injustice."[79]

For this purpose, the environmental storytelling of the level proceeds in meticulous detail. There are at first two routes to follow, which deal with both of the above-mentioned historical venues, respectively, and have the player move past animated cardboard cutouts or stone statues that illustrate the events. The colour shade of the environment depicting the Native Americans and Chinese is comprised of a mixture of threatening red and dark tones, while statues of Comstock embody the Prophet in a heroic fashion. To complement the arising image, the player becomes involved in combat against the veterans of the 7th Cavalry under the leadership of Cornelius Slate, and in a *re-enactment of the brutalities* at

78 Bolotin and Laing, *World's Columbian Exposition*, 32.

79 Buinicki, "Nostalgia," 729; cf. 728-729.

Wounded Knee and the Boxer Rebellion. Of course, the enemies are different (Slate's men who wish to die an honourable death), but combining the perspectives the player witnesses (the historical narrative in form of the environmental storytelling) and those she enacts (the combat), the player comes to see the connections between her relentless mowing down of enemies and the historical massacres.

As a result, this form of "procedural rhetoric"[80]—which involves the player "in a situation where he has no choice but to respond with violence to the hostility of the environment"[81]—unfolds its subversive potential only if seen in the context of the remaining perspectives offered and afforded to the player. An additional perspective is a conversation between Elizabeth and DeWitt, in which the latter offers an interesting insight:

Elizabeth: Booker … I can tell what Slate said bothered you. You showed me – sometimes you have to do what's necessary to survive.
DeWitt: There is survival … and then there is *finding pleasure in the act*.[82]

In this scene, DeWitt not only speaks to himself (for having found pleasure in the acts at Wounded Knee) but implicitly addresses the player, who most probably enjoyed the frenetic and challenging combat action before. As such, BIOSHOCK INFINITE targets again the gamist player type in an estranged form and links her to a non-reflective being, unwilling to view her ludic actions within a greater ethical context. What is more, this facet plays with the *relative distance and proximity* between DeWitt and the player. One part of the truth is that the player, as a being outside of the diegesis, is free to criticise DeWitt for having participated in the atrocities at Wounded Knee.[83] The other part holds up a mirror to her and creates an intimate connection to the PC through gameplay. These conclusions are the result of a blank between DeWitt and the player that can be closed differently depending on the player's actions and imaginings, while having in mind the ethical context or not.

80 Maziarczyk, "Playable Dystopia," 253.

81 Ibid., 252-253.

82 BIOSHOCK INFINITE (Irrational Games, 2013), ch. Hall of Heroes Gift Shop; emphasis mine.

83 Buinicki, "Nostalgia," 729.

6.4.2 The Allures and Undergrounds of the Capitalist Dream Machine in Finkton

Throughout this chapter, I have described a faction of Columbia that goes by the name of the Founders. They are the ruling party in Columbia and are led by Comstock, whose conviction paved the way for an ideology built around the myths of American exceptionalism and a sect-like belief in the Founding Fathers. This supposed Utopia was uncovered by the game's perspectival arrangements and the player's acts of ideation to be a dystopia based on zealotry and xenophobia. Also, it is supported by a capitalist production machinery that commodifies aspects of the Columbian dream, which have an irresistible allure for the Utopians and coax them into participating in the system. Elizabeth and the player have potentially fallen for this trap, if the player participated in the candy wonderland of the earlier levels and gave Elizabeth the time to do so. Once the player arrives in Finkton, she will now be shown the industry behind the façade of the previous levels, which might come as a surprise to her or confirm previous expectations.

The microcosm of Finkton is named after "the city's most powerful industrialist and its wealthiest citizen," Jeremiah Fink. Fink is a "businessman and criminal, capitalist and baron"[84] who built a manufacturing district within Columbia. Finkton is thus a pompous location where factories and clock towers abound. They remind the Utopians of discipline and punctuality, and the constant ticking of clocks and the crunching of cogs creates a chilling scenario in which workers move according to their rhythms. In the meantime, golden statues embody the grand capitalist. It is, as Peters claims, "a self-indicting horror show of neoliberalism" in which "Big Business and the Church" join forces and "are presented as poisonous, violent institutions with monomaniacal leaders."[85] Such an observation can be justified in that people are participating in tombolas over jobs and spend the few Silver Eagles they earn on goods from Fink Manufacturing. This greed for power and the oppression of the Other is combined with a superstitious belief in higher beings and formulates a dark triangle of human nature. It fosters a pessimistic image of the Founder ideology and presents the Vox Populi as victims of white male supremacy.

84 Peters, "Global Dystopian," 127.

85 Ibid.

6.4.3 Descending into Abyss of Shantytown: The Route to Revolution

The next section initially supports this observation. When taking an elevator down to Shantytown, the player experiences the underground of the Columbian society. It is a dirty place and home to most of the Vox Populi. Thieves roam the streets by night, and there is sickness and a shortage of food. One scene especially illustrates the precariousness of the situation, in which a boy is scraping out the remainder of a food can, while opposite him a field of tomato cultivation is guarded by a steel fence. A sign on it explains that it was seized by the Columbian police to punish sympathisers of the 'terrorist' Vox Populi faction. Closing the blanks between these perspectives fortifies the image of the Founders as antagonists and potentially affects the player's ergodic actions.

For this purpose, the possibility space of the level affords the player several ways to proceed. The player's goal is to reach a gunsmith called Chen Lin to supply the Vox with weapons for a revolution against the Founders. Before doing so, she may however help these people in asking Elizabeth to open a tear that will provide people with food. Alongside this ethical possibility, there is also the choice to become a thief who exploits the workers in a manner not much better than Fink. Here again, the gamist attitude is linked to unethical behaviour, where the player's interest lies in personal gains rather than a critical observation of the greater picture.

The perspectives, so it seems, are straightforward and suggest the need for a liberation from the Founders. Yet there are hints that suggest a different truth. Right at the level's beginning, the player experiences a satellite event of utmost magnitude. A preacher is giving a speech to denounce the Founders for exploiting the people of Shantytown and letting them suffer from starvation and sickness. He promotes Fitzroy's cause, and various editions of a book called "The People's Voice"[86] are lying on the table he is standing on. As such, it would appear that his words are genuine, but if the player takes a closer look, she might discover another truth. During his speech, the preacher is holding a book whose cover he is cautiously hiding. Now and then he reveals it for less than a second, giving the player time to discern the cover through the scope of her sniper rifle—which grants a clearer look at it. The book turns out to be "The Word of the Prophet,"[87] the religious manifesto of the Founders and a subtle hint that the Vox ideology might not be much different than that of their oppressors.

---

86 BIOSHOCK INFINITE (Irrational Games, 2013), ch. Shanty Town.

87 Ibid.

Elizabeth still believes in the revolution, however, which is expressed in dialogue between her and DeWitt, while the latter is doubtful, for he has experienced such conflicts before. Unconsciously, Elizabeth opens a tear now to the wish she seeks and to another dimension in which the guns can be supplied to the Vox. The event is complemented by the following dialogue:

DeWitt: Come on, let's head back to the gunsmith and finish our deal. I aim to be on that airship, and well shut of this place for long [in a decisive voice].
Elizabeth: Booker, if the Vox get their weapons, there's going to be a revolution just like *Les Misérable* [Victor Hugo, 1862]! These people are gonna have better lives [in an enthusiastic, naive voice].
DeWitt: Yeah [in a doubtful, indifferent voice].[88]

Once the player steps through the tear, she experiences a world in flames where DeWitt has led to the Vox to greatness and has become a martyr. The gameworld is now covered in red cloth curtains—which reminds of the Red Army Faction[89]—while posters of DeWitt plaster the player's route. She might be flattered at the recognition her PC is given, but this revolution is by no means just. For the Vox Populi have slaughtered the Founders and are now posing for pictures on their dead corpses. Elizabeth (and maybe the player) are shocked at this sight, while DeWitt already suspected what was to come.

Besides these sensorial and participatory perspectives, *diegetic music* underlines the negation of the Vox perspective as liberators of Columbia. Shortly before Elizabeth and DeWitt stepped through the tear, the player can encounter a scene in a basement where DeWitt takes a guitar and plays *Will the Circle Be Unbroken*, while Elizabeth sings. The event creates a utopian enclave within the dystopian microcosm of Shantytown—and there are additional musical pieces that follow the same direction. In this respect, Ivănescu has observed that Elisabeth Cotton's *Shake Sugaree* (1967) is performed by three kids in the streets of Shantytown and comments on the misery of their lives, while the "anti-war anthem" Fortunate Son (1969) by John Fogerty and Creedence Clearwater Revival is sung by a woman after the revolution has taken place.[90] All of these songs and their lyrics contribute in different ways to the counter-narrative and to the formation of blanks that can be closed by suggesting that "the new rule is just as bad as the old one."[91] It is a

88 Ibid., ch. Bull House Impound.

89 Peters, "Global Dystopian," 128.

90 Ivănescu, "Music of Tomorrow," 59-50.

91 Ibid., 60.

first glimpse at the true meaning of *the circle to be unbroken*, which will become clearer by the game's end.

This insight, that "power corrupts strong leaders,"[92] leads Elizabeth to a *turning point*. Similar to Comstock, Fitzroy is not the leader she promised to be, and when she threatens to execute a child (Fink's son), Elizabeth intervenes and kills the Vox Populi leader. This brutal event is a vital step in Elizabeth's *coming-of-age*. As a sign of the young woman's loss of innocence (or virginity), Elizabeth cuts off her pony tail and changes into a more revealing dress.[93] To come to this point, Elizabeth was influenced by the gameworld perspectives, the same perspectives the player experienced and co-created. She thus assumes a role exemplary to the player, who may either accept or decline the ethical offer her character affords. In any case, Elizabeth has gone through a process of realisation that has come far from the enthusiastic, naive woman she was before. In contrast, DeWitt works on a more subtle note, and the player may easily dismiss his utterances as of disinterest or striving for the personal goal to wipe away his debt. This task, however, is more intricate and far-reaching than one might believe, and DeWitt's character steadily changes.

6.4.4 Entering the Darkest Regions of DeWitt's Unconscious in Emporia

Throughout this chapter, DeWitt's character was described as riven by guilt from the atrocities he committed at Wounded Knee and the agonies he exposed Anna to. He re-experienced these deeds in distorted form in the dreamworld of Columbia, where the journey will now lead into the darkest regions of his unconscious and to the encounter with a repressed part of his self: *dystopia's high priest*, Father Comstock. In these final moments, the player has reached *Emporia*, a district where the rich and powerful reside, and a word that implies both the terms *above* and *market place/bazar*. The macro goal is to reach Comstock House to obtain a means of stopping Songbird, who has fervently pursued DeWitt and Elizabeth ever since they escaped Monument Tower.

Once the player lays foot in Emporia, the noble residence of the upper class is covered in plumes of smoke. The presence of the Vox Populi is terrifying the Founder population, and ships with refugees are leaving the area. Emporia is a further manifestation of DeWitt's guilt, for in this reality another version of himself led the Vox revolution. What was left in its wake is the sight of a massacre,

---

92 Peters, "Global Dystopian," 128.

93 Ascher, "Leuchtturm," 5-6.

corpses and scalps. It is as if the game would like to remind the player one last time of the dangers of ideological delusions and how easily people turn into savages while following a supposedly noble cause. These themes of *guilt* and the search for *forgiveness* have accompanied the player throughout the game, and, to confirm my hypothesis, can be seen on two levels.

On the one hand, there is DeWitt's *personal guilt* for the way he treated his daughter Anna, but this individual level is inexorably connected to a universalised one. Consequently, combining the perspectives of DeWitt's deeds at Wounded Knee, Comstock's in Columbia, and the player's re-enactment of brutalities in the gameworld, the blank to be closed here evokes the image of Utopia's loss. It speaks to a *universal guilt* in humankind that concerns not only the characters of the fictional gameworld but also those of the empirical world. In this light, Elizabeth can be regarded *as a symbol for Utopia*, or for the struggle for it, which begins with an escape from the confines of her iron cage and leads her to see beyond the delusions of ideologies. Both DeWitt and Elizabeth are thereby in need of each other to attain the forgiveness they seek, for Elizabeth has also become involved in atrocities she cannot easily discard.

As such, the young woman facilitates DeWitt's journey into his unconscious in that she helps her father uncover his secrets by lock picking certain doors that open further regions of the game. In Emporia, this lock picking intensifies, and the doors the player encounters are fortified with golden bars and shining locks. In the meantime, combat encounters have worsened and dark clouds are lying over the region, underlining the dreamlike nature of the events.

Indeed, Emporia not only showcases humankind's violent nature but also their innate greed for power and money. When the player passes through the Bank of the Prophet, the gamist and achiever playing styles are again linked to mindless consumerism in an estranged manner—by gathering points and scavenging the environment for treasures. In fact, BIOSHOCK INFINITE offers so much unnecessary loot that the player will find herself hammering on the square button of the PS4 gamepad to gather supplies or Silver eagles. Because these abound, the player rarely runs out of them, but she may succumb to the addiction of collecting them—and this frenzy of scavenging reaches its climax in The Bank of the Prophet, where a myriad of safes and lockers conjure up the player's lust for money.

To strengthen this impression, the game utilises additional perspectives to help the player ideate the image of her own succumbing to the pleasures of capitalist exuberance. This is done via Elizabeth noticing that "[f]ifty percent of everything people earn here goes right to Comstock as tithe" and a sign on the wall that says "hoarder."[94] It was smeared on it by the Vox in a vibrant red and addresses both Comstock for imposing excessive taxes on the Columbian citizens as well as the gamist player, who has participated in this gluttony and in the accumulation of things she does not need.

6.5 THE GUILT OF HAVING LOST UTOPIA AND WAYS TO REGAIN IT

The foregoing descriptions of BIOSHOCK INFINITE have followed the trajectory of the game's plot and progression structure. This was conducted in such a manner as to simulate the player's experience of the game and explain her acts of ideation in a chronological manner. Various images have formed during this process on both the level of the plot and significance, some of which were negated or inverted by subsequent events. They will now lead the player to the game's plot twist and climax, and it is here that DeWitt and the player will face their inner demons and encounter a dark part of their selves against which they struggle. As a symbol of oppression and the loss of Utopia (for which all of humankind is to blame), the gameworld has darkened and black clouds surround Comstock House. They eventually pave the way to the encounter with dystopia's high priest, Father Comstock.

This last part of the chapter will lay focus on how the individual strings of the BIOSHOCK multiverse converge at INFINITE's ending, thereby considering the first BIOSHOCK and INFINITE's DLCs BURIAL AT SEA: EPISODE 1 and 2. In this context, I will continue to explain the previously mentioned *levels of guilt and forgiveness*, which involve a *personal level* and a *universalised* one. The first level revolves around DeWitt's guilt for having exposed a child to the cruelties of alcoholism and bad fatherhood. It is extended in BURIAL AT SEA to Elizabeth herself, where she seeks revenge on Comstock in an alternate universe. For this purpose, she uses a Little Sister, called Sally, to lure him out of his hiding place and subsequently suffers from the guilt of having exploited the child. In all the BIOSHOCK games, then, "self-sacrifice" and "the atonement of … past deeds"[95] assume the means *to challenge dystopia* in different ways. They create *utopian enclaves* within the possibility spaces of these games and conjure up the question whether ideological Utopias, however noble their initial dream was, are worth *the suffering of even one child*?[96]

---

94 BIOSHOCK INFINITE (Irrational Games, 2013), ch. Downtown Emporia.

95 Maziarczyk, "Playable Dystopia," 252.

96 Sargent, "Flawed," 226-227.

The characters' (and the player's) individual guilt is thus inexorably linked to a universalised level and to the struggle for Utopia of which these children (Elizabeth or the Little Sisters) are symbolic of. The question, now, is whether such deeds can be forgiven at all—be they private or affect the majority of the population, as with Comstock's establishment of a theocratic regime and Andrew Ryan's free market capitalism—and is shifted into the foreground. Therefore, and to answer the question of the utopian enclaves, it is necessary to regard the individual strings of the BIOSHOCK multiverse and how they converge at *Infinite*'s climax. I thus claim that to interpret BIOSHOCK INFINITE in a meaningful manner, both its predecessor (BIOSHOCK) and subsequent DLCs (BURIAL AT SEA: EPISODE 1 and 2) have to be taken into account. For only if seen in conjunction can the pessimistic though ambiguous ending of INFINITE and BURIAL AT SEA be explained in a nuanced way, with the characters DeWitt, Elizabeth, and, by extension, the player showing significant change and *the will to self-sacrifice in the struggle for Utopia.*

6.5.1 Self-Sacrifice and the Will to Agency as the Struggle for Utopia

The plot twist and climax in BIOSHOCK INFINITE takes time to build up and begins with Songbird recapturing Elizabeth. Once the young woman falls back into Comstock's hands, she slowly transforms into his heir and fulfils her role of drowning New York City in flames. Many years have passed, and the indoctrination in the asylum has consumed the woman. What really broke her, though, was time and the lack of a father. It is another example of DeWitt and Elizabeth's mutual dependence in their struggle for forgiveness. When DeWitt finally locates the now elderly Elizabeth, the player is back to the vision from the game's beginning, where Columbia bombards the Sodom below, under Elizabeth's control. She now hands DeWitt the melody to control Songbird, who always managed to stop them in any of the constellations, and sends him back in time to undo the events. The screen changes back to the year 1912, and the player sees a cage with a rose in it and the 'Word of the Prophet' beside it. This juxtaposition of perspectives once more shows the sophistication of BIOSHOCK INFINITE's perspectival arrangement. Certain combinations evoke concrete imaginings in the player and guide her acts of ideation—in this particular instance, aligning the destructiveness of war with a certain ideology and the cage within which it encloses its followers. This ever-repeating image is the most prominent and nuanced in the game and builds up to the climax where Elizabeth will free herself from the confinements of ideology.

Before turning to this universalised level, DeWitt's personal struggle for forgiveness continues in the encounter with a dark part of his self and with dystopia's high priest. The event is surrounded by the game's most intense combat sections, on board an airship, where De Witt and Elizabeth eventually find Comstock in a tranquil shrine.

Comstock: DeWitt, I'm a fool, I've sent mighty armies to stop you. I've rained fire on you from above. I did all that to keep you from her, when all I needed was to tell her the truth. Ask him child. Ask him what happened to your finger. Ask DeWitt. [Comstock grabs Elizabeth's arm violently]
Elizabeth: Let go of me …
Comstock: Ask him, ask the False Shepherd. Tell her, False Shepherd! Tell her the truth!
Elizabeth: My hand! Please let got!
[DeWitt/the player intervenes and smashes Comstock's occiput against the font and drowns him—this happens by the push of a button]
DeWitt [in anger]: She's your daughter, you son of a bitch! And you abandoned her! Was it worth it? Huh? Did you get what you wanted? Tell me! Tell me!
Elizabeth: Booker … Booker …
Comstock: It … is … finished [Comstock's last words].
DeWitt: Nothing is finished! You lock her up for her whole life. You cut off her finger, and you put it on me!
Elizabeth: Booker, stop it! You killed him.[97]

In this scene, DeWitt is not only confronted with a horrible secret he is unconsciously aware of (that he is Comstock), but it is also illustrates a first step towards redemption, in killing a part of himself that is culpable for abandoning his newborn child, Anna.

The difference between DeWitt and Comstock here is remorse. In contrast to DeWitt, who has changed over the course of the game's events, Comstock's life ends on the last words from Jesus Christ on the cross: *it is finished*.[98] It shows that the prophet believes in the Utopia of his creation and dies for this conviction. DeWitt, on the other hand, has taken steps towards the forgiveness he seeks, which will culminate in a self-sacrificial act for Elizabeth. To get there, he needs to destroy the *Siphon*, a device built inside Monument Tower that holds back Elizabeth's powers. Elizabeth compared the Siphon to a "leash" her father put on her and that begs the question: "What would happen if I [Elizabeth] took off the leash, and I found I was … as obedient as ever?"[99] The answer to this question is revealed

97 BIOSHOCK INFINITE (Irrational Games, 2013), ch. Command Deck.

98 Bosman, "Lamb of Comstock," 177.

99 BIOSHOCK INFINITE (Irrational Games, 2013), ch. The Atrium.

once DeWitt uses the notes to control the Songbird and orders him to tear down Monument Tower.

It is a metaphorical act to cut his daughter loose from the confines of daughterhood and ideological imprisonment alike, and completes Elizabeth's emancipatory journey. The counter-narrative has led the young woman to a specific insight. For with the destruction of the iron cage of Anti-Utopia, Elizabeth turns into "a force that can move in and out of these systems of control,"[100] a being that sees beyond the realities of the BIOSHOCK multiverse. She is thus a *utopian enclave* in herself, one who can look behind the façades and lies of dominant ideologies and has served the player as a role model.

Such a realisation is a first crucial step towards Utopia, and Elizabeth's role becomes more explicit once she leads the player to a moment that merges the individual threads of the BIOSHOCK games. With the Siphon destroyed, the young woman brings DeWitt to the underwater city Rapture, where the player witnesses an intimate scene between her and Songbird, whom she kills in an act of liberation by drowning the automaton in the ocean.[101] DeWitt and the player are astounded by the interconnectedness of Rapture and Columbia, and this imagery will become explicit in the following events.

Elizabeth leads DeWitt to a *sea of doors* now, a vast ocean region that is composed of an *infinite array of lighthouses* and *docks* that connect them. This panoramic view explains the game's name (INFINITE) and conjures up a beautiful yet terrifying image. For these phallic towers and gateways to the supposed Utopias behind them remind the player of the fragility of human nature, the greed for power and financial gain, be it in the form of the xenophobic Utopia of Columbia or the capitalist one of Rapture. Elizabeth places this inception more clearly when she explains that "[t]here's always a lighthouse [, t]here's always a man, there's always a city ..."[102] and thus addresses a second level of guilt, which revolves around humankind's *failure in losing Utopia again and again*. The player has experienced this loss in both BIOSHOCK and BIOSHOCK INFINITE and compares these events to those of her own empirical world, where troublesome similarities abound.

Given this semantic charge, it is no coincidence that the final scenes of BIOSHOCK INFINITE are heavily discussed by scholars, who mostly describe the ending in a pessimistic manner. For the game to conclude, the player is brought back to the baptism DeWitt initially took to become Comstock. He refuses this

100 Peters, "Global Dystopian," 134.

101 Ascher, "Leuchtturm," 4.

102 BIOSHOCK INFINITE (Irrational Games, 2013), ch. Sea of Doors.

time, and Elizabeth leads him to a revealing moment, where DeWitt gave away Anna. The scenes are changing rapidly now, since Elizabeth leads DeWitt through various periods of time and places, which culminates in the moment of Comstock's birth. Only there can the chain of events be severed and the manifold versions of Comstock averted—most of which created their own Utopia in alternate universes. Several versions of Elizabeth approach DeWitt now, who finally realises that he is both Booker DeWitt and Zachary Comstock. Consequently, DeWitt willingly embraces the women drowning him in the river stream, and with each piano note playing, one Elizabeth disappears after another. It is a selfless act on DeWitt's part to undo the things he has done to his daughter and to attain the forgiveness he was seeking from her. The screen turns black, and the game switches back to DeWitt's office where he awakes from his nightmares and hears the voice of a baby in the adjacent room. When he approaches Anna's cradle, blackness.

Given the events of BIOSHOCK INFINITE's ending, Buinicki describes the game as a "closed anti-utopia."[103] He justifies this claim by arguing that "the apparent death of both protagonist and antagonist do not provide an escape from this vision"[104] and that in such a world, agency is non-existent—for all the choices are predetermined before the player makes them. "Like their avatar, players are not able to free themselves from a past they've both inherited and helped to create."[105] They are simply "fulfilling the role's that's been written for them."[106] Similarly, Peters explains that in BIOSHOCK INFINITE, the player enjoys no agency to counteract the system. He specifically deems interesting the contradiction of the infinite array of lighthouses and the player's triviality in this respect: "In a realm of endless lighthouses—literally, limitless choice—the player-protagonist has none ... the premise of choice is consistently invalidated."[107] Notwithstanding this hopelessness, Peters finds a utopian enclave in Elizabeth. For in contrast to DeWitt, who is only "given the chance to embrace it [his death] willingly,"[108] Elizabeth "has a modicum of control ... The cyborg here is the one who finally does make the active choice to die, not to the aggrandizement of an ego, but to end a cycle of violence and destruction by deleting it entirely."[109]

---

103 Buinicki, "Nostalgia," 734.

104 Ibid.

105 Ibid., 732.

106 Ibid., 733; cf. 732-734.

107 Ibid., 134.

108 Ibid.

109 Ibid., 135; cf. Lange, "Der Freie hat die Wahl," 119-121.

6.5.2 Reuniting the Individual Strings of the BIOSHOCK Multiverse

Although Peters finds a glimmer of hope within a pessimistic ending, I claim that these interpretations fail to regard BIOSHOCK INFINITE in a nuanced enough way. Besides, the events of BIOSHOCK and the DLCs BURIAL AT SEA: EPISODE 1 and 2 have to be taken into account, which will fortify the image of *self-sacrifice* and the *willingness to change* as utopian enclaves. To view BIOSHOCK INFINITE in this context, Elizabeth's actions in BURIAL AT SEA become of importance, since it is here that the now emancipated woman (of whom the player takes control) selflessly struggles for a Little Sister called Sally. The second DLC ends with Elizabeth's death to free the young girl. It is an ambiguous ending with glimmers of hope for Sally, and its events lead to the first BIOSHOCK game. In a vision before her death, Elizabeth sees Jack breaking the circle of greed and violence by saving the Little Sisters and peacefully succumbs to her wounds. Meanwhile, Sally is singing *La Vie en Rose* to her (Édith Piaf, 1945).

The Little Sisters assume a pivotal role in the first BIOSHOCK game, which can be classified as critical dystopia of variant II—because the player may choose the faith of this world in how she deals with the little girls. The events of BIOSHOCK revolve around Andrew Ryan's creation of an objectivist Utopia in Rapture, an underwater Atlantis cut off from libertarian and communist parasites. Ryan chose to leave this world behind and built a city in which *art, science, and industry* would not be restrained by petty moral and governmental institutions. As a Utopia built around these ideals of free market capitalism, BIOSHOCK involves the player in the ramifications of such a thought experiment and virtualises the collapse of this vision. As such, the game targets both the hubris of Ayn Rand's Objectivist Utopia *Atlas Shrugged* (1957), its "*laissez faire politics*,"[110] and similar exuberances of capitalism in contemporary society.[111]

This "rational self-interest" and the "greed"[112] of Rapture's citizens are not only ingrained in the game's audio-visual presentation and plot but also in its mechanics and gameworld processes.[113] Similar to BIOSHOCK INFINITE, the player of BIOSHOCK is "made complicit in the schemes of the game"[114] by participating in

---

110 Packer, "Galt's Gulch," 109.

111 Tulloch, "Ludic Dystopias."

112 Schmeink, "Dystopia."

113 Peters, "Global Dystopian," 209-211, 221.

114 Tulloch, "Ludic Dystopias."

rapturous capitalism in an estranged form. This expresses itself in fighting degenerated objectivists (Splicers) or the player's use of vending machines and plasmids (which are similar to Vigors in that they grant posthuman powers). Yet the most prominent form of complicity is to be found in how the player deals with the Little Sisters.[115]

BIOSHOCK INFINITE mostly followed its predecessor in terms of strategies of estrangement, but chose to deny the player real choice in terms of the plot. This is not the case in the first BIOSHOCK, whose ending depends on how the player treats the Little Sisters. These girls were genetically modified to harvest ADAM from corpses of Splicers—a substance which grants the citizens of Rapture supernatural abilities by altering their DNA—and are protected by Big Daddies.

The first time the player encounters a Little Sister is a vital moment in the game and confronts the player with an important choice. Thereby, the aesthetic staging of the scene and its perspectival arrangements proceed in a meticulous fashion. When the player sees the Little Sister, only few perspectives guide the player's imaginings and the decision she is about to make: to *save* or *harvest* the girl. The scene begins with a helpless child whose Big Daddy has been killed. Her outward appearance resembles that of a monster, but the little girl is afraid when the player approaches. This juxtaposition of perspectives creates *uncertainty* in the player and is fortified by additional positions: 1) Atlas, who the player trusts, tells her that the Little Sister is no child anymore; 2) Dr. Tenenbaum pleads for the girl's protection; 3) the horizon of past perspectives the player has gathered, which evokes an image of insanity and violence permeating Rapture.[116]

These constellations create a *blank* the player will close through imagination and ergodic efforts. However, were it only for these perspectives, the choice would remain bland—and the player would most likely save the little girl for ethical reasons. What fuels the desire to *harvest* the Little Sister, conversely (an indeed semantically charged term), is the player's *greed for power* and *immediate desire for gratification*. This is because before making the choice, she is influenced by Atlas' claim to gain a substantial amount of ADAM for harvesting the girl, while her reward will be less for saving her (or so the player believes). Although Tenenbaum promises to reward the player later on if she spares the girl, the choice speaks to the player's desire for power in a FPS—to level up and complete the game in an effective manner—and, thus, to the gamist and achiever attitude in her. The strategies of estrangement thus work in an implicit manner here and connect the

115 Packer, "Galt's Gulch," 209-211, 215-219, 221.

116 Lange, "Der Freie hat die Wahl," 112-113; Domsch, *Storyplaying*, 156-157; Packer, "Galt's Gulch," 218-219.

gamist/achiever attitude to the modes of behaviour in a capitalist world. To harvest the Little Sisters is therefore to follow "the extreme of the Objectivist ideology of self over others," which fosters the image that "being selfish grants more power" and that this "greed ultimately causes the suffering of innocents."[117]

Given this semantic charge, the outcome of the Little Sister choice is not only connected to the immediate outcome the player witnesses—for harvesting even just one girl in the course of the game will deny the prospect of a utopian horizon to this dystopia. This facet connects the game to literary fictions such as Ursula K. Le Guin's short story "The Ones Who Walk Away from Omelas" (1973), in which the utopian society's welfare is built around *the incarceration and suffering of a child*. Everybody knows about the child's existence and function, and the people of Omelas visit it regularly. Every once in a while, however, some of the visitors cannot bear the sight of the child and their knowledge about its suffering. Consequently, they walk out of its cell and straight out of Omelas. For Sargent, Le Guin's story can thus be classified as a "flawed utopia" which asserts "that neither truth nor harmony (read eutopia) is worth the suffering of a child."[118] It thus raises a "question about our behavior[,] because in modernity people suffer so that others can live in the material eutopia of the world's developed countries."[119] This insight applies to Rapture or Columbia as much as it does to Omelas.

Utopia, so the inevitable conclusion here, is *not worth the suffering of one child*, and so it makes sense that if the player falls victim to the enticing mechanisms of a ludic capitalism, she is denied the utopian enclave at the game's end. Three such endings await the player and are triggered by her actions: 1) save all the Little Sisters, which allows them to lead a happy life outside of Rapture; 2) harvest the Little Sisters, and the Splicer Objectivists escape to the outside world; 3) harvest one or more Little Sister while saving some of them, which triggers the second ending but with a sad undertone of Tenenbaum narrating the events.

The options seem clear at first glimpse; however their ambiguity is expressed in some critics' argument that BIOSHOCK is a closed anti-utopia in all of its endings. This claim they base on several reasons, one of which is the player's apparent lack of agency in the game. After the plot twist, it is revealed that the PC is a clone and that Atlas (alias Fontaine) commanded his actions with the phrase 'would you kindly'. This "revelation … renders any of the 'choices' you thought you had made through the course of gameplay … purely illusory."[120] A second reason can

117 Packer, "Galt's Gulch," 219.

118 Sargent, "Flawed," 227.

119 Ibid., 228.

120 Aldred and Greenspan, "A Man Chooses," 490; cf. 490.

be found in the hopeful ending itself, which, according to Van den Berg, envisions a clichéd prospect for life after the Little Sisters escaped from Rapture, where they graduate from college, find love, marry, and have children. Such a succession of events leads to a "reinstatement of traditional devices of the retainment of capital and a means of production"[121] and, therefore, BIOSHOCK fails to imagine an end to capitalism and a world beyond it.[122]

Despite these pessimistic claims, I come to a more positive conclusion about BIOSHOCK's ending and the series in general. For it is not necessarily the envisioning of a radical end to capitalism but steps in the right direction that describe Utopia's resistance. This route towards the gradual betterment of society can be found in many enclaves BIOSHOCK and BIOSHOCK INFINITE hide within their dystopian systems, which most prominently revolve around the characters' and player's struggles for forgiveness and towards ethical actions. Whereas in BIOSHOCK INFINITE this includes Booker DeWitt on his selfless journey to free Elizabeth, the player has only minor influence on the plot. Her agency mostly revolves around connecting the events to her empirical surroundings on an imaginative level and encountering similarities between both realities. Similarly, Elizabeth needs to glimpse behind the façades of the ideological Utopias before she is driven to action. She thereby assumes a figure exemplary to the player, and once the inception is placed, she continues to struggle for Utopia and for the one child, Sally, she has exploited for selfish reasons.

On the other hand, BIOSHOCK involves the player directly in the decision for Utopia. Choice is no illusion here and lies with the player's treatment of the Little Sisters. To do the "right thing,"[123] the player needs to *delay her desire for immediate gratification* (for power and loot) and save the little girls. Such a choice is a first, crucial step towards Utopia and leads to the downfall of Rapture's extreme form of capitalism. As such, and through a playful trial action in the gameworld, it points the player into the right direction and towards a way of conduct that may attenuate the consequences of her capitalist world.

6.6 CONCLUSION

BIOSHOCK INFINITE represents an anxiety dream in which the dreamer struggles for forgiveness from past deeds and the loss of Utopia. What remains from the

121 Van den Berg, "Playing at Resistance to Capitalism," 26.

122 Ibid., 1, 26.

123 Domsch, *Storyplaying*, 158.

experience once she wakes is the realisation that there is still time to change the inevitable and to strive for Utopia in the real world. Given the complicated nature of social totality, it is not easy to attain a radically different Utopia. The BIOSHOCK games aim to solve this dilemma in a subtle manner without proposing grand solutions to it. They follow the model of the critical dystopia and virtualise gameworlds that are plagued by the issues of theocracy, racism, American exceptionalism, and an extreme form of capitalism that either bewitches the citizens by means of Disneyfication strategies or turns them into rapturous objectivists whose appearance and actions magnify humankind's lust for power and consumption. These are dark worlds against which it is difficult to struggle, but utopian enclaves are hidden within these visions of nightmare.

Both games, thereby, play with the norms of the FPS genre, in which the player is used to gather items and experience points to complete the game in an efficient manner. Such a gamist and achiever attitude is aligned in an estranged way with consumer conduct in a capitalist world: buying goods one does not necessarily need, or not paying attention how these products were produced—be it in the form of child labour or other exploitation of workers. Following Elizabeth's model as she undergoes a process of realisation, the player steadily comes to see behind the shimmering façade of INFINITE's gameworld. Such an insight and aesthetic response is primarily due to her acts of ideation in which the player combines the perspectives that she co-creates. These include the gameworld, its signs, music, and labyrinthine structures, the secondary cast of characters such as Fink, Fitzroy, the Luteces, Lady Comstock, Slate, and the primary cast including DeWitt, Comstock, Songbird, and Elizabeth. Moreover, the tranquil tour around Columbia and its processes are inverted into a battlefield to involve the player in a renegotiation of utopian images. She thereby creates her own perspective on this world by participating in the events and enjoying agency in some instances. Filtered through a first-person perspective, which nonetheless is distanced by the roundness of DeWitt's character, the player conjures up private images in her acts of ideation. These are outlined by the game's strategies of inversion/negation which seduce the player into creating utopian images of Columbia to then negate them in turn. The result is a juxtaposition of perspectives—such as those of the factions of the Founders and Vox Populi—which, through their subtleties, do not result in a clear image of who the antagonists are.

This flux of images and their negations begins with the lighthouse as a symbol of hope, which change to an image of despair in a picturesque environment of infinite lighthouses and possibilities. These stand as a reminder of the glimmering allure of ideological Utopias and how easily humankind succumbs to such promises. In addition, Monument Tower reveals itself as the iron cage of Anti-Utopia, and also DeWitt's character undergoes significant change, from a selfish mercenary to a loving father who will do anything for his child. This struggle for forgiveness and Utopia continues after BIOSHOCK INFINITE's conclusion in Elizabeth's struggle for Sally and the player's potential struggle for the Little Sisters in BIOSHOCK.

The main question, then, is: *how can the unforgivable be forgiven?* Both games take similar routes to answer it, as they describe a selfless struggle for Utopia on the part of some of their characters, but involve the player differently in the resolutions. In this respect, BIOSHOCK INFINITE can be classified as a critical dystopia of variant I in that it explains how the dystopian society came to be and leads the player on a linear journey to an ambiguous ending. This journey is open to the player's imagination but denies any possibility for true agency. On the contrary, BIOSHOCK leaves the choice for Utopia to the player and can be classified as a critical dystopia of variant II. The issue of self-sacrifice and the willingness to agency are expressed in how the player deals with the Little Sisters.

All in all, the player's playful interactions in the estranged gameworld of Columbia and Rapture serve as an example to facilitate her struggle for Utopia in the real world and stand as a reminder of the potential of such trial actions for guiding humankind towards a brighter future. I thereby described the player's acts of ideation for both the individual games and on an intertextual level between them. The resulting images were ambiguous, but they tend towards a counter-narrative as the interpretation of a dream, which suggests that even if the system cannot be changed on a large scale, a selfless and ethical attitude might bring Utopia a little closer. Such a perception of dreams, fiction, and reality views the games not in terms of possible worlds—which, although infinite in number, are closed and finished objects—but as *indeterminate spaces* that may be shaped through player action and desire.

7 THE LAST OF US and the Journey to Nature

THE LAST OF US: REMASTERED (Naughty Dog 2014)[1] is a critical dystopia of variant I that sends the player on an extraordinary journey towards nature, away from the derelict city spaces of a bygone era.[2] It is a route that is extremely dangerous yet well worth pursuing. For the world of 2033 is not one that players know, and a plague by the name of the Cordyceps Brain Infection (CBI) has wiped out large parts of the human population and paved the way for a new order in which nature has astonishingly reclaimed the planet. The post-apocalyptic gameworld the player thus encounters is both terrifying and magnificent. It involves him in a bitter struggle for survival against fellow humans and infected alike and in a torturous journey towards an uncertain future. But it is also a route that promises renewal and re-enchants the player in his perception of the natural world by exposing him to its glimmering allure.

This dialectic between decay and rebirth, between the failures of a past civilisation and the chance for redemption, is a common trope in post-apocalyptic fiction and can be observed in narratives such as Warner Bros. Television's television series REVOLUTION (2012-2014) or Alloy Entertainment's THE 100 (2014-2017). It usually begins with "total breakdown" and "the necessary purgatory for bringing people to their senses" and continues with "a reversion to more primitive

---

1 THE LAST OF US: REMASTERED (PS4, 2014) is the visually and performatively enhanced version of Naughty Dog's original game THE LAST OF US (2013). To simplify things, I will refer to the game as The Last of Us.

2 This chapter is based on the 2016 DiGRA/FDG paper I wrote with Charlotte Ladevèze. (Gerald Farca and Charlotte Ladevèze, "The Journey to Nature: The Last of Us as Critical Dystopia," *Proceedings of the First International Joint Conference of DiGRA and FGD* 13, no. 1 [2016], http://www.digra.org/digital-library/publications/the-journey-to-nature-the-last-of-us-as-critical-dystopia/).

forms of economic and social life."[3] But in THE LAST OF US it differs from H. G. Wells' dystopian tales in which, as Kumar holds, after a certain "period of transition," the "scientific intelligentsia takes power" and gives birth to a renewed and better society.[4] Instead, the game clings to the failures of science and technology and visualises a long-lasting ecocatastrophe as a direct consequence of humankind's missteps. Mismanagement and an utter imbalance with nature are the main culprits here, and THE LAST OF US, consequently, issues a terrifying warning. It is humankind's arrogance, unhindered expansionism, and the unrelenting struggle to (re)claim the top of the food chain which are to blame and that disrupt an otherwise prosperous ecosystem designed for balance.

Figure 21: THE LAST OF US *sends the player on an extraordinary journey towards nature and grasps his attention through astonishing beauty and style. The picture depicts the journey's end.*

THE LAST OF US (Naughty Dog, 2013, 2014), ch. Jackson.

Such an aesthetic response and ethical effect on the player sensitises him to thinking differently about the ecosystem and urges him to revaluate his position in it and the impact the human species has on it. To acquaint the player with such issues, the game refrains from making didactic statements about it but relies on fiction's sensualising powers—by having the player participate in an intimate story

3 Kumar, *Modern*, 222.

4 Ibid.; cf. 222.

that revolves around the relationship of its two main characters, Joel and Ellie. They embark on a perilous yet cathartic journey across the U.S. towards the scientific enclave of the Fireflies who promise a cure for the CBI. Ellie is supposedly immune to the fungus, and it is Joel (and the player controlling him) who is to accompany the fourteen year old girl. Soon, however, their journey turns out to be longer and more tortuous than expected, and on various occasions Joel utters doubt about to the path they took. Consequently, when they finally reach the Fireflies, and when Joel learns that in order to extract a vaccine from Ellie requires surgery that will kill her in the process, his course of action is understandable—and not solely based on the loss of his daughter Sarah during the original outbreak.

The choice for Ellie is certainly a personal choice and implies a decision against a potential cure to the CBI. However, it can also be universalised to the struggle for Utopia: in favouring one utopian proposal over another and cherishing a life in a small community in the woods of Jackson County over a return to modern civilisation. In Tommy's settlement, people have found a way to live in balance with nature, but whether the player agrees with Joel's decision is a different matter and does not concern the level of the ergodic—for the player has no say in it. Yet it is debatable from an interpretive point of view. Although hopeful, the game's ending in Jackson County leaves the player with an ambiguous aftertaste—since there is no certainty about the future of the settlement, nor about the society the Fireflies would have created. Notwithstanding these implications, the road towards the ending has it shine in a clearer light and suggests an ecological, not scientific, solution.

In this respect, the strategies of THE LAST OF US become of particular importance, as they outline a specific insight in the player. Several images arise out of his acts of ideation, and these are themselves in constant flux, guided and at the same time negated by the continual emergence and creation of perspectives. The perspectives, thereby, can be grouped in various *oppositions* and involve the player in a dialectical negotiation of the positions they propose—for none of them is without moral flaws. Consequently, the player's attempts to ideate are first of all reinforced and frustrated by the factions in the game, which adhere to certain ideologies and ways of living. While most of them are flawed (such as the militaristic regime of the Quarantine Zones or the Hunter and Bandit factions), it is the opposing world view of the Fireflies and Tommy's settlement that spur on the player's confusion. The two perspectives evoke differing imaginings about the gameworld's future, and by closing the blanks between them, the player will come to a personal conclusion about the proposed positions.

Were it only for these perspectives, the solution would remain quite ambiguous. However, THE LAST OF US does not content itself with this state of affairs and covertly aims to steer the player's acts of ideation. Especially if viewed from an ecological perspective (which the game promotes), and how the factions adjust to the new situation, THE LAST OF US can be considered a cautionary parable that scratches the surface of the ecotopia. It thereby proceeds by creating further oppositions to address ecological questions—concerning "growth and energy, balance and imbalance, symbiosis and mutuality, and sustainable or unsustainable uses of energy and resources"[5]—but that also speak to the player's emotions. Two of the most prominent of these can be discerned in the juxtaposition of the city's confining indoor spaces and nature's alleviating outdoor spaces and in the relationship between Joel and Ellie.

Hence, THE LAST OF US confronts the player with an *official narrative* that is characterised by an intense struggle for survival and the scavenging for supplies. These procedures and gameplay events are most violent and are primarily to be encountered in the city's confusing mazes and derelict spaces, which aim to evoke in the player feelings of entrapment and suffering. They thus remind him in an estranged way of the precarious confinements of contemporary times: technological and capitalist excess, overpopulation, and a lifestyle marked by stressful interactions and an imbalance with the grander ecosystem. In stark contrast to this are nature spaces, which are depicted in utter beauty. They instigate a subtle *counter-narrative* by tempting the player into beautiful imaginings and they provide a safe space for human dialogue and progression. Nature spaces have a calming and liberating effect on the player and are presented as dynamic enclaves and regenerative solutions to dystopia. The perspective of Ellie furthers this image. For the young girl assumes the role of this story's temptress (which she shares with Mother Nature herself), reminding both Joel and (by extension) the player of nature's preciousness through her youthful enthusiasm and curiosity.

In the following I will illustrate the player's path of realisation towards beauty and emancipation. It is outlined by the genre-specific conventions of the critical dystopia of variant I, which illuminates how the dystopian society came about and maintains possibilities for renewal in suggesting that only a respectful treatment of the ecosystem and a life of balance with it may save humankind from disaster. To place this inception in the player, several oppositions guide his participation and acts of ideation such as fire/water, destruction/rebirth, city/nature, indoor/outdoor, entrapment/liberation, storm/sun, dark/bright, tense/calm, survival/dialogue, regressive/progressive, and so on. Their interdependence and, often, mutual negation contributes to the creation of various images and is reinforced through the

5 Peter Barry, *Beginning Theory: An Introduction to Literary and Cultural Theory. 3rd ed.* (Manchester: Manchester UP, 2009), 254.

context/surroundings in which they occur and the four seasons of summer/autumn/winter/spring. Now, to begin the investigation into the *ecological dystopia* of THE LAST OF US, the image of the CBI becomes of importance, which finds itself in continual negotiation.

7.1 THE CORDYCEPS BRAIN INFECTION AS FICTIONAL NOVUM AND INSTIGATOR TO DYSTOPIA

There are literally thousands of different types of Cordyceps fungi and, remarkably, each specializes on just one species. But these attacks do have a positive effect on the jungle's diversity, since parasites like these stop any one group of animals getting the upper hand. The more numerous a species becomes, the more likely it will be attacked by its nemesis, a Cordyceps fungus.[6]

In the post-apocalyptic storyworld of THE LAST OF US, a mutated version of the virulent and extremely aggressive Cordyceps fungus (Ophiocordyceps unilateralis) has wiped out 60% of Earth's population by 2033 and transformed the once familiar society into the estranged environment the player encounters. The Cordyceps infection progresses in four stages and leads to irrevocable brain damage. It deprives its hosts of vital brain functions and transforms them into degenerate and aggressive versions of their former selves. In addition, the fungus is extremely contagious. It can be transmitted through bites or is spread airborne via spores that are released from the remains of a dead host. This is the premise of THE LAST OF US, whose gameworld is built around the logic of the CBI as fictional novum. In direct extrapolation from contemporary times, players are shown a marvellous place where nature[7] has reclaimed the planet and where *the old order* of a

6 BBC Worldwide, "Cordyceps: attack of the killer fungi - Planet Earth Attenborough BBC wildlife," YouTube, November 3, 2008, https://www.youtube.com/watch?v=XuKjBIBBAL8

7 I will use the concept of nature as something overarching, stretching into the vastness of the universe, and aiming to create harmony between its individual elements in a dialectic between creation and destruction. Such an "intelligent nature" may struggle "against the ecological heresies of humankind" in order to restore balance. (Brian Stableford, "Science Fiction and Ecology," in *A Companion to Science Fiction*, ed. David Seed [Malden: Blackwell Publishing, 2005], 132; cf. 132-133). In this sense, the city is also part of nature's larger ecosystem, but I will nonetheless contrast them, for reasons

bureaucratic consumer capitalism has literally corroded. THE LAST OF US describes the downfall of humanity's hectic lifestyle, its reliance on technology and increasing alienation from nature, which have led to such a catastrophe. In a cleansing apocalypse of brutal renewal, the Cordyceps fungus has cleared the way to achieve a balance lost long ago, in which humankind lives together in harmony with nature. The player, now, is confronted with *a new order*, a return of the natural world and a renewal of Earth's ecosphere, in which he will vicariously participate. Still, in order to compose such a positive image of the CBI, rigorous work is required, and a tedious start shows the fungus in a darker light.

The reasons for the fungus' outbreak are explained relatively early in the game, at the dawn of disaster in a newspaper to be found in Joel Miller's house. Contaminated crops from South America have imported a mysterious infection to the U.S. in 2013, and the player will soon witness its ramifications. The plot of THE LAST OF US consequently begins twenty years in the past in a suburb of Austin, Texas. With the initial cutscene, players are shown a loving father and his daughter Sarah—of whom they take control. She constitutes the first focalizer in the game, and the prologue lets players enact her confusion when she encounters hints that introduce the situation. Television stations broadcast scenes of violence and turmoil, and when Joel arrives, he is agitated and in a hurry. Their neighbours have turned, and Joel has no other option than to shoot them. Eventually his brother Tommy arrives, and the three attempt to make it out of town towards the countryside.

This initial section grants the player an external perspective on Joel. It is an interesting way to commence the game and not only contributes to the Joel's characterisation but also depicts him in a way that anticipates his later treatment of Ellie, which is "paternal, protective, level-headed, resourceful."[8] This foregrounded father-daughter relationship will become an important landmark for the player's acts of ideation and, in the context of Sarah's death, contribute to the creation of various images in the subsequent events.

Right from the start, the gameworld of THE LAST OF US is thus plunged into chaos. The date is September 27th, and the streets are filled with panic as people's neighbours have turned into irrational and violent predators. A rapid and purifying

---

of simplicity and to see the city as an element that regularly disrupt nature's intimate harmony.

8 Daniel Vella, "Modelling the Semiotic Structure of Game Characters," *Proceedings of DiGRA 2014: <Verb that ends in 'ing'> the <noun> of Game <plural noun>* 8 (2014): 15; cf. 15, http://www.digra.org/digital-library/publications/modeling-the-semiotic-structure-of-game-characters/ "Characters"

apocalyptic fire is about to consume humankind. This fire not only manifests itself in the burning chaos and anarchy the player witnesses and enacts but also in the parasite of the CBI, which spreads like wildfire. In a beautiful metaphor of the human condition, the infection relentlessly consumes humankind and turns them into insanity-driven beings who could no longer bear the conditions of their times. It is as Gaston Bachelard pictured the apocalypse. When the symbol of *fire* pervades the scenery and consumes anything it finds on its way: “Le feu est l’ultra-vivant.”[9] “Quand on veut que tout change, on appelle le feu,”[10] “[il] purifie tout.”[11]

It is here that the perspectives players have gathered and co-created are ideated into an initial image of the CBI, which is permeated by the symbol of fire. This is because at this point in the game, the reasons for the outbreak are unknown to the player and will only later be implicitly illustrated. For now, it is only the hint of the crops and the player’s knowledge (of similar post-apocalyptic stories) that fuel his imagination. However, these point in a certain direction. It is known that Cordyceps fungi thrive in humid climates, like the tropical rainforest—and, therefore, humankind’s invasive agricultural expansionism and consequent deforestation can be held responsible for the outbreak and its spread. Such imaginings are underscored by the fact that the idea of *ecology* is based on the relation of organisms with their environment and on the self-regulatory (defensive) forces of nature. Of considerable importance in this respect is the influence of the food chain. For even a minor disturbance to it can have comprehensive effects on the environment[12] and reminds one of the ecosystem’s “complexity and intricacy.”[13]

---

9 Gaston Bachelard, *La Psychanalyse du Feu (Folio Essais)* (Paris: Gallimard, 2015), 23.

10 Ibid., 102.

11 Ibid., 174.

12 An arresting example illustrating the importance of the food chain could be discerned in the Yellowstone National Park in 1995 when wolves were reintroduced into its plains. While absent, a large population of deer were responsible for the park’s meagre vegetation, as they consumed most of it. With the reintroduction of the wolves, however, the deer’s behaviour changed considerably, and they started to avoid certain regions of the park, giving the plants and forests a chance to regenerate. A chain reaction kicked in, and a variety of species like birds, beavers, foxes, and weasels repopulated the area. What was truly extraordinary, though, is that the wolves also changed the flow of the rivers, since the recuperation of the forests resulted in a stabilisation of the river banks. In short, a small number of wolves not only transformed the park’s ecosystem but also had an influence on its geography. (Sustainable Human, “How Wolves Change Rivers,” YouTube, February 13, 2014, https://www.youtube.com/watch?v=ysa5OBhXz-Q).

13 Stableford, “Science Fiction and Ecology,” 127; cf. 127-128.

This "harmony"[14] also applies to the relation between "human beings and their environment," the "intimate bond with the nurturing aspects of Mother Earth"[15] which "modern civilization ha[s] sacrificed on the altar of technology."[16] THE LAST OF US continues to develop this idea of *disbalance* and turns it into an "eco-catastrophe."[17] As a common trope in post-apocalyptic SF, man-made ecocatastrophes are often a direct consequence of issues such as "the 'exhaustion' of the soil crop bearing capacities," "overpopulation,"[18] or the excesses of "capitalism."[19] They depict a "worldwide disaster"[20] and outbreak of dystopia in which the player vicariously participates. He thus engages in a playful trial action that has him scrutinise the aftermath of an ecocatastrophe but also in a world that is purifying itself, characterised by "a redemption of the Earth's ecosphere from the threats posed by human activity."[21]

The player will only later come to appreciate such deliberations, as for now the downfall of human society governs the spectacle. The strategies of the prologue thereby send him on a path that inexorably leads towards *death.* Closed-off alleys, scripted events, lighting, and bright objects point the way—but, most significantly, the level's labyrinthine structure guides the player in a linear fashion. It strips him of agency and leads to Sarah's death when she is shot by a government solider.

This event illustrates a personal tragedy that, on a bigger scale, can be equated with the death of modern society and "the breaking away from order, civility, and the relative comforts of law."[22] The directive strategy of the prologue thus fulfils a dual role: it directs the player towards a certain goal through breadcrumbing and funnelling—strategies to keep the player on the right track or to bring him back to

---

14 Ibid., 133.

15 Ibid., 128.

16 Ibid., 133.

17 Ibid., 137.

18 Ibid.

19 Ibid., 138; cf. 137-138.

20 Ibid., 137.

21 Ibid., 131.

22 Amy M. Green, "The Reconstruction of Morality and the Evolution of Naturalism in The Last of Us," *Games and Culture* 11, no. 7-8 (2016): 4, http://journals.sagepub.com /doi/abs/10.1177/1555412015579489

it[23]—and serves to complement the negative image of the CBI. A strategic placement of events and signs outlines this process. These begin with the first explosions and turmoil, when the player experiences the breakdown of media (telephone lines are dead and television stations have stopped broadcasting), and culminate in Sarah's demise, which is framed by the burning chaos of the CBI. It is these typical things of modern society that have perished in the blink of an eye and whose destruction illustrates the downfall of the capitalist and technology-dependent order, pushing humankind to the brink of extinction. In the vein of the critical dystopia, then, THE LAST OF US explains the circumstances of how dystopia came into being—and this emphasis will continue throughout the experience.

7.2 THE OFFICIAL NARRATIVE: CITY SPACES OF VIOLENCE AND CONFLICT

THE LAST OF US begins in *summer* with the symbol of fire permeating the season and the image of the CBI composing itself in the player's mind. For now, this image does not comprise the positive effects of balancing the grander ecosystem and will be worsened through the accompanying image of the city. Indeed, there is much dispute about city ecosystems in utopian and dystopian fiction, which often foregrounds the opposition between the city and the countryside.[24] The city, thereby, occupies a continuum between Utopia and Anti-Utopia and cannot be allocated to either side.[25] Instead, it enables a potential space for both utopian and dystopian deliberations, and one usually cannot determine whether a recipient would deem a certain city utopian or dystopian (especially considering the thrills of cyberpunk megalopolises).

In THE LAST OF US, conversely, the situation seems clearer. The city spaces of Austin, Boston, and Pittsburgh are marked by combat, violence, and panic; and depict the dilapidated remains of an order that perished on the day of the apocalypse. They hence involve the player in a specific *official narrative* and have him experience an estranged place that on closer inspection seems alarmingly similar to the real world. Such an image begins to foster twenty years after the breakout. Yet when the player again takes control over Joel, he is not the same person as in

---

23 Chris Bateman, "Keeping the Player on Track," in *Game Writing: Narrative Skills for Videogames*, ed. Chris Bateman (Boston, Mass.: Charles River Media, 2007), 89-99.

24 Jameson, *Archaeologies*, 143, 157.

25 Ibid., 161.

the prologue. The post-apocalypse has changed him. Having lost most of his family, he tries survive as a black market smuggler and does not shy away from immoral behaviour. When the game continues, the player thus encounters Joel and his partner Tess on a route through the dystopian microcosm of Boston's Quarantine Zone (QZ) towards another smuggling job. After the Cordyceps outbreak, these were established in urban areas by the U.S. military division FEDRA (Federal Disaster Response Agency) with the intention of safeguarding the remains of humankind from the infected. Initially a noble pursuit, the QZs degenerated into the oppressive regime the player now encounters.

The perspectival structure of the QZ is thereby similar to that of the prologue. Again, the player follows the trajectory of a unicursal labyrinth and is taken on a tour through the new Boston, which presents itself in the worst possible manner. The city's inhabitants live in miserable conditions, the streets are littered with garbage, and the grey shades of the environment foregrounds the despicable character of the place. In the meantime, people are suffering from a shortage of food—and, indeed, the city's currency is now ration cards (with money another element of the old order has disappeared). Moreover, there are the regularly conducted and feared tests of FEDRA. These determine whether citizens show symptoms of the CBI, and in case of a positive result, they are executed right away.

In essence, the QZ evokes the image of a *prison*. It represents a space that is cut off from its surrounding environment and from human rights alike, and when navigating its confines one cannot deny the feeling of claustrophobia and helplessness. The suppression of people's freedom is directly experienced by the player, whose agency is severely reduced and who encounters processes that are both terrifying and resilient. For example, trying to save people from being executed (or coming too close to FEDRA soldiers) will quickly result in failure and in Joel's death. Moreover, there are death sentences for helping supposed criminals, regular curfews, brawls between citizens, and the nonstop surveillance by the FEDRA troops. The image of a prison, then, culminates in two telling examples the player finds on his way: 1) the grey walls separating the QZ from the remains of Boston and 2) the caged up dogs, whose imagery illustrates the atrocious circumstances people live in. Such a journey through dystopia does not leave the player unharmed and involves him on an affective level. The prevalent symbol of the QZ is thus one of *entrapment*, while the perspectives gathered and co-created complement the negative image of the CBI and the establishing one of the city.

7.2.1 The Strategies of the Game and the Opposition Between Indoor and Outdoor Spaces

It has been established that the confines of the QZ are made accessible to the player through visual/spatial demonstration and a reductionist gameplay in which his actions are limited to following a pre-established pathway and disallow further interferences. Yet THE LAST OF US would do no justice to its status as critical dystopia were it not to include hopeful possibilities within the confines of its official narrative. Consequently, *utopian enclaves* are already discernible within this first hour of play. For one, there is the resistance militia group of the Fireflies, who have scattered signs and graffiti across the QZ. Their agenda lies in finding a cure for the CBI and the fight against the FEDRA military in order to secure a return of all government branches. For another, and probably more remarkable, THE LAST OF US confronts the player with the astonishing beauty of the natural world. Plants and animals have steadily reclaimed parts of the city and create an atmosphere that surrounds the player with awe and calamity. In doing so, they bring forth utopian spaces that covertly infiltrate the dystopian integrity of this world and forebode a vital strategy that will subsequently guide the player's acts of ideation and participation in the game.

Already in its beginning, THE LAST OF US juxtaposes the bleak and derelict architecture of the city (and its torturous mazes) with the quaint and tranquil spaces of unclaimed wilderness and the transition between them. This results in a tension between *confining indoor spaces* and *liberating outdoor* spaces that create most interesting blanks to close. The player first encounters such an opposition when Tess and Joel traverse the QZ and come across a section where the natural world has made an astounding return within the confines of the city. Walking towards the exit of a rundown diner, the player is guided by a bright light shining through its door and windows, while dramatic music enhances the feeling of frenzy. Once outside, a dialogue between Tess and Joel underlines this romantic imagery of nature, in which Joel acknowledges that he has not been out for a long time, while Tess connects their stroll to a date. Although brief, the section is revelatory, as it provides a variety of interlocking perspectives that guide the player and initiates *a counter-narrative* that covertly speaks to his emotions. It is as Eugen Pfister explains: the secret lies in a dialectic between architectural ruins that express the decaying body of American civilisation and an extraordinary panorama of the natural world, which has recaptured what is hers and grants the player brief moments of tranquillity.[26]

Consequently, it is no coincidence that not long after this initial contact with nature, the player first comes to meet Ellie Williams. Ellie grew up in the Boston QZ and has never left its bounds, but she is about to set off a series of events that will have a drastic influence on Joel and the player's perception of the gameworld. Shortly, the player will learn that Ellie is immune to the CBI, and Marlene (a further character who is introduced, and head of the Fireflies) tricks Joel into escorting Ellie out of Boston and into the hands of their scientists. What they do not know at this point is that the journey will be longer and more complicated than expected. Step by step, Joel (and also the player) realises that it might be worth the effort. In a cathartic experience, the unexpected journey to nature will not only turn into a second chance for Joel but into one for the entirety of humankind.

As such, THE LAST OF US can be described as a coming-of-age narrative and outlines a father-daughter relationship[27] similar to other post-apocalyptic fiction. In Cormac McCarthy's *The Road* (2006), for example, a loving father takes care of a child without knowledge of the exterior world, and this child becomes the sole "raison d'être" for the father.[28] Coming-of-age narratives thus share a high resemblance to dystopian fictions, as they foreground the moral growth of a character and his quarrel with societal structures. THE LAST OF US continues in this mode and has the player enact a coming-of-age which is brutal for its participants and plastered with hindrances to catharsis.

7.2.2 Defamiliarised City Spaces and the Clutches of a Bureaucratic Consumer Capitalism

The counter-narrative in THE LAST OF US develops once Joel takes care of Ellie. Although reluctantly at first, he will assume a role that is both protective and re-

26 Eugen Pfister, "Jacques Lacan, Caspar David Friedrich und die Zombie-Apokalypse: Eine Erste Annäherung an Mythen in 'The Last of US'," *Spiel-Kultur Wissenschaft Mythen im Digitalen Spiel* (2015): 3-4, accessed February 20, 2016; Green, "The Reconstruction of Morality," 9.

27 Vella, "Characters," 13, 15; Óliver Pérez-Latorre, Mercè Oliva, and Reinald Besalú, "Videogame Analysis: A Social-Semiotic Approach," *Social Semiotics* 27, no. 5 (June 1, 2017): 11, http://www.tandfonline.com/doi/full/10.1080/10350330.2016.1191146

28 Hensel, "Zwischen *Ludus* und *Paidia*," 153; cf. 146, 152-153, 156.

ceptive to the girl's temperament and curiosity of nature's wonders. At first, however, the journey leads them through the dystopian spaces of Boston and Pittsburgh, and from an escape from their suffocating indoor mazes to the scientific enclave of the Fireflies, who promise hope to a world in despair.

Eventually, this route brings Tess, Ellie, and Joel to Boston's financial district and has them pass through the crumbling ruins of two skyscrapers, precariously leaning on one another. It is in the inside of these buildings that the player first has to deal with the infected in a combat situation. Entering the area, Runners (those in the first stage of infection) are standing paralysed in the building's hallways, while Clickers (the third stage) are wandering blindly through the area. One can hear their screams and see them sobbing, as if they were mourning the loss of bureaucracy and the capitalist order that drove them insane in the first place. Throughout the game, the infected are largely encountered in city spaces, their canals, but also in suburban areas. They remind the player, as Pfister claims, of the decaying human body being recaptured by nature and its defamiliarisation[29]—and thus stand for the dying remnants of a past society, paralysed by its system and gone mad in their greed for profit and gain.

The CBI and the ludic experience of the infected thus evoke a frightful image for the player, which is fortified through additional perspectives. In this respect, Joel's ludic capabilities in confronting enemies and the game mechanics that afford them become of interest. These are, as Vella describes, influenced by the genre conventions of the third-person action-adventure and include the ability to walk, run, take cover, pick up and upgrade certain objects, engage in dialogue, and commence combat through melee attacks or weapons. The goal thereby is to ensure Ellie's and the PC's survival and pass a certain area without dying. Whether this is done stealthily or in an upfront manner often (though not always) depends on the player. However, the stealthy approach is fortified by Joel's ability to craft weapons or traps and to hear enemies from long distances by the press of a button—the player will see their silhouettes displayed in the game environment.[30] Consequently, "these affordances emphasize a strong sense of spatial and environmental awareness, privileging careful, studied planning"[31]—which makes

29 Pfister, "Jacques Lacan," 3.

30 Vella, "Characters," 10, 11, 15.

31 Ibid., 10.

sense "in the face of an extremely hostile environment"[32] that involves its characters in a struggle for survival and has Joel's actions appear in the light of "desperate self-defence."[33]

The intensity of the action is thus palpable in every respect. But THE LAST OF US does not content itself with hordes of Runners attacking the player or the treacherous ambushes of Clickers, whose sense of hearing is highly developed and who kill with a single bite. Instead, the game intensifies the player's anxiety through the maze-like structures of the multicursal labyrinth. There is only ever one exit to such a maze, and the combat sections make the player *suffer*—as Victor Navarro-Remesal would say.[34] In doing so, they also serve as perspectives for an important image to develop on the level of significance. Closing the blanks between them (the mazes, processes, city ruins, enemies, and tense combat, etc.) and aligning them with the horizon of past perspectives, the player will soon come to see the city spaces for what they really represent.

In selecting norms and conventions of contemporary society but rearranging them in *analogic* relation in the estranged gameworld, THE LAST OF US evokes imaginings of a terrifying bureaucratic consumer capitalism and the processes that fuel it. The enacted events can thereby be linked to the hectic, consumerist lifestyle of the player's contemporary city surroundings, which now seem surprisingly close to what is occurring in the game. Such an observation also explains conclusions about the game's real-world targets, for example Pérez-Latorre et al.'s observation that it thematises "the current 'great recession'" and contemporary "debates around capitalism, austerity and ways of dealing with economic crises."[35] However, its methods of involving the player and creating such an image go further than relying on the "signs of decay of capitalist civilization" in the environment—"leaning skyscrapers, urban landscapes invaded by wild nature, supermarkets in ruins, traffic jams scenarios with empty dilapidated cars"[36]—and include the labyrinthine structure of this world, its processes, characters, and the events that occur.

Pittsburgh, the second expansive city space in THE LAST OF US, intensifies this terrifying perception of the city. It is the most dangerous city space the player

32 Ibid., 11.

33 Ibid., 10.

34 Victor Navarro-Remesal, "Regarding the (Game) Pain of Others: Suffering and Compassion in Video Games" (lecture, Concerns about Video Games and the Video Games of Concern Conference, IT University of Copenhagen, 2016).

35 Pérez-Latorre et al., "Videogame Analysis," 11.

36 Ibid., 12.

encounters, but it is not the infected that pose the greatest threat, and, instead, gangs of survivors expose Joel and Ellie to severe problems. They call themselves Hunters and prey on their victims to plunder their belongings. Joel and Ellie fall victim to such a trap and only survive through brutal combat and a resourceful scavenging of supplies. Again, the ludic situation is intensified by maze-like structures both within and outside of the city buildings. Various multicursal labyrinths allow for diverse strategies of engagement and offer an advantage to the skilled player. However, they are hard to play on the higher difficulty levels, as they expose the player to constant threat and the need to safeguard Ellie from the city's atrocities. Complicating matters is the need for supplies such ammunition, bindings, alcohol, scissors, rags, and explosive materials, which can be crafted into first-aid kits or Molotov cocktails. While true that on the lower difficulty settings these are abundant, and partially transform the city into a "'supermarket' of free objects,"[37] the higher the difficulty, the scarcer they get. In both cases, the city spaces of THE LAST OF US transform into a horrific scenario where rotten utensils and supplies are essential to the protagonists' survival (whether abundant or scarce) and create an image of a dead city cut off from the grander ecosystem.

Given this context, it is alleviating that Ellie is helpful during combat.[38] There is the potential for her death, of course, in certain situations—for instance, when she is attacked by Hunters or the infected, and when Joel comes to her rescue too late. But she never feels like a burden to the player.[39] This manifests itself in that Ellie searches for supplies/items in the game environment but also in that she attacks enemies with a knife or by throwing bricks on them. She is thus far from the helpless child in *The Road* and represents a "capable daughter" who, although in need of Joel, is about to develop into a potent young woman who will serve as a role model for both Joel and the player.

Consequently, and having in mind the perspectives gathered and co-created, the composed images of Boston and Pittsburgh are outright negative. They fuse

---

37 Ibid., 13.

38 Gerald Voorhees, "Daddy Issues: Constructions of Fatherhood in The Last of Us and BioShock Infinite," *Ada: A Journal of Gender, New Media, and Technology*, no. 9 (2016), http://adanewmedia.org/2016/05/issue9-voorhees/

39 Pérez-Latorre et al. describe Ellie as "a difficulty-creating factor in the game" (Pérez-Latorre et al., "Videogame Analysis," 14), whereas Voorhees clings to the other extreme and misjudges the possibility of Ellie's death—claiming that she is neither "a physic [n]or physical burden to the player." (Vorhees, "Daddy Issues.") The truth lies in between these observations, and Joel (and the player) may fail to protect Ellie during combat sections.

with that of the CBI and evoke the image of *virus* and an "unnatural … toxic and poisonous" environment.[40] The city spaces in *The Last of Us* thus stand as a reminder that although the city is "part of a larger eco-system … it consumes more that it produces,"[41] and that if "mismanagement"[42] causes an imbalance between the availability of resources and human population, the city might easily turn into a "dysfunctional social order."[43]

Figure 22: The spaces of the city are inexorably linked to the old order of a bureaucratic consumer capitalism and the failures of bygone civilisation.

THE LAST OF US (Naughty Dog, 2013, 2014), ch. The Outskirts.

As such, the city spaces add to the player's knowledge of how dystopia came into being in that they craft perspectives that outline a devastating imagery of contemporary society's true nature. Crumbling buildings, including out of order soft drinks vending machines and the crying remnants of its employees, or sites like the Capitol Building in Boston (exhibiting paintings of a proud U.S. history) are reinforced by tense combat encounters. Especially the references to a bygone American history are thereby striking. They are, as Green notices, juxtaposed with bleak, derelict spaces, the infected running madly through the remains of their

40 Jameson, *Archaeologies*, 161.

41 Domsch, "Dystopian," 405.

42 Stableford, "Science Fiction and Ecology," 136.

43 Kim S. Robinson, *Future Primitive: The New Ecotopias* (New York: Tom Doherty Associates, 1994), 10; cf. Stableford, "Science Fiction and Ecology," 133.

offices, and atrocities that abound.[44] This does not imply a nostalgic look backwards, however—even though this could seem so after the experience of homes the player passes. These are filled with objects—such as books, television sets, and children's toys—which conjure up positive imaginings about the old days. What is more difficult to grasp, however, is that this old order has failed—and to do grasp this is to see the truth behind the appearances and see the city for the site of capitalist madness it has always been.

Such an image will be intensified by additional perspectives in the following events—and, consequently, THE LAST OF US creates an official narrative that involves the player in a precarious situation marked by death and destruction. Symbolic for this are the details I described above but also the demise of companions such as Sam and Henry's in a suburban town outside Pittsburgh and Tess' execution in Boston's Capitol Building. Surrounded by incoming FEDRA soldiers and shot to death, the city devours the woman in a symbolism of its destructiveness. Yet even this desperate act to keep its citizens in check and within its perimeters cannot distract from the city's failures and the incapacities of an old order that perished on the day of the apocalypse. The city, as such, resembles a venue of stasis and violence, of entrapment and suffocation, and the player has to learn these rules through games of estrangement. All in all, it is thus not unreasonable to speak about the formation of a certain *chronotope*, defined as an "intrinsic connectedness of temporal and spatial relationships."[45] As such, the *death of the city* is demarcated by the above-mentioned criteria and inherently linked to the season of summer and to the symbol of fire as a metaphor for both world and gameplay.

7.3 The Counter-Narrative as the Journey to Nature

Learning the rules of dystopia is strenuous, experiencing its processes restrictive, and trying to escape its madness dangerous. THE LAST OF US does not surrender to such despair and embarks the player on a renunciation of its official narrative. To do so, it sends him on a journey to nature that leads westwards, from Boston across the country into the wilderness of Jackson County, where Ellie and Joel

44 Green, "The Reconstruction of Morality," 9.

45 Michail M. Bakhtin, "Form of Time and the Chronotope in Novels: Notes towards a Historical Poetics," in *Narrative Dynamics, Essays on Time, Plot, Closure and Frames*, ed. Brian Richardson (Columbus: Ohio State UP, 2002), 15.

settle down in Tommy's community. It is "the classical journey into the west,"[46] but not solely, as Oli Welsh remarks, one designating the death of a nation rather than the establishment of a new one. In the vein of the critical dystopia, THE LAST OF US initiates a counter-narrative that covertly captures the player's attention and guides his coming to awareness: leading from *the old order* (by showing contemporary city spaces in an estranged and shocking manner) to the *establishment of a new* one, which is Utopia, where life has returned to a healthy balance with nature. This transition is brutal. But it serves the purpose of renewal and of renegotiating the image of the CBI while shaping that of the city.

7.3.1 The Guiding Function of Nature as Utopian Enclave

The natural world in THE LAST OF US is presented in an extraordinarily beautiful way and outlines the player's *escape* from the city. The game, thereby, strikes a careful balance between gameplay purpose and that on the level of significance and continues to employ breadcrumbing and funnelling techniques to assure the player stays on track. Consequently, players are guided by bright colours standing out from the environment (often yellow), faded directional signs, and, in general, the beauty of nature itself. This is especially apparent in the escape from the Boston QZ, where the player leads the group through the city's canals, which lead them to safety under the protective shelter of night and rain. Later, the moon's brightness takes over this directive symbolism—and so right from the start, nature assumes a guiding role and outlines a specific image in the player that reminds him of nature's beauty and re-sensitise him to the benefits of a life in balance with her.

A remarkable example of nature's guiding efforts can be found in the element of *water*, flowing steadily towards the countryside. In general, there are three types of water in THE LAST OF US which can be described with Bachelard's observations on the phenomenon. To begin with, there is water that stagnates in the city's sewers. It is dirty and attracts the infected, and often one encounters floating corpses while diving. Using Bachelard's terminology, such *impure water* metaphorically stands for the waste and drainage of the city and for the contaminating character of civilisation itself.[47]

---

46 Oli Welsh, "*The Last of Us* Review: Journey's End: Our Original Review, to mark the Release of *The Last of Us Remastered*," *Eurogamer*, July 31, 2014, http://www.eurogamer.net/articles/2014-07-28-the-last-of-us-review

47 Gaston Bachelard, *L'Eau et les Rêves, Essai sur L'Imagination de la Matière* (Paris: Librairie José Corti, 1987), 13.

Second, there is standing water, which is greenish in colour and can also be encountered in city spaces. It foreshadows danger or death (such as Tess' in the Capitol building) and can be compared to a teardrop and the sorrow or melancholia of contemporary times. Standing water is similar to Bachelard's *dead water* and can additionally be encountered in David's cannibal settlement, where it mourns the hideousness of human nature the in the form of a frozen lake.[48]

Finally, there is flowing water, which is fresh, clear, and steadily streaming towards nature. It helps to produce clear energy and gives people hope in times of discontent (like in Tommy's settlement). This *pure water*, which Bachelard also compares to *springtime water*, shows a childlike character, young and loud, innocent and hopeful.[49]

Flowing water is a steady companion on Joel and Ellie's journey and inexorably leads them towards nature. This symbolism continues in Pittsburgh, where the group follows the lead of a yellow bridge functioning as a landmark and point of orientation. Again, they choose to escape under the cover of night and make their way through an outpost of Hunters. They manage to escape with a big jump into the raging current and almost drown exposed to the forces of nature—which, nonetheless, functions as their saviour. A *storm* is brewing over the city now, underlining the precariousness of the situation, while their route leads towards the sunrise and utopian beauty of the natural world (which connects the journey to Guy Montag's escape in *Fahrenheit 451*, where the protagonist escapes alongside a river stream and will eventually find shelter in the forest).

The descriptions above are revealing, as the ecological strategy of game, in the form of opposition between indoor and outdoor spaces, is again brought into focus. I have already addressed a couple of these, such as fire/water, destruction/rebirth, city/nature, indoor/outdoor, entrapment/liberation, dark /bright, storm/sun, and now wish to discuss additional ones whose effect on the player is intensified through gameplay. THE LAST OF US creates a vital distinction between humankind's confining remains (the derelict city and suburban spaces but also indoor university spaces) and the calming appeal of nature. This opposition is reinforced through gameplay and the antithesis of multicursal city spaces (characterised by intense combat) and linear nature spaces that function as utopian enclaves for *relaxation* and *dialogue*.

This juxtaposition of *the city's agitation* and the *calming appeal of nature* is commonplace and necessarily plays with the player's emotions. It creates a dialectic between interrelated opposites that, as Hensel claims, are in need of each

48 Ibid., 89, 96.

49 Ibid., 47, 193.

other, driven by the alternations between chaos and calamity, the close-up views of combat and the panoramic vistas from mountaintops or elevated buildings.[50] This blindness of proximity and the prudence of utopian visibility speaks to the player on an emotional level. I have discussed his pain and suffering, stemming from the maze-like structures and ludic encounters of the city, which is now "alleviated"[51] through the *calming* and *regenerative* appeal of the environment. Nature sections are linear in structure and represent a space for intimate *dialogue* between Joel and Ellie. They can be compared to Japanese Zen Gardens (Kare-San-Sui), which are also unicursal and aim to trigger deliberations about nature.

One example of such a juxtaposition is when the player escapes the city spaces of Boston and makes his way through a claustrophobic metro system. Having escaped the FEDRA military, a narrow staircase leads towards a bright light. The scene is very stereotypical and results in Joel and Ellie's first encounter with the beauty of the natural world. It enables a utopian space for intimate dialogue in which Ellie says she is sorry about Tess but also shows curiosity about the woods they pass through.

Another example is the campus of the Eastern Colorado University. It is here that Ellie and Joel expect to encounter the Fireflies, but their frustration will be prepared through various strategies of the level. The UEC is probably the most beautiful level in the game from a visual aesthetics point of view. The autumn season is coming to a close, and the scenery is marked by vibrant colours and leaves whistling through the air. Ellie and Joel approach the university by horse and slowly ride through the campus. Unfortunately, Joel has to descend every once in a while. The electricity of the university gates is cut off, and his route leads to the inside of the buildings where, in an allusion to science gone wrong, he encounters a substantial amount of the infected. The terror of these indoor sections evokes the urge to leave their bounds and to escape to the bright and picturesque outdoor spaces of the campus. These are adorned with regenerating grasslands and a return of animal life. Symbolic for this is the scene in which the player encounters lab monkeys roaming the campus. He may follow them quickly by horse, and the scene triggers an intimate dialogue with Ellie. In the meantime, a storm is brewing, and dark clouds are lying over the edifice where the Fireflies are supposed to be.

50 Hensel, "Zwischen *Ludus* und *Paidia*," 147-149, 151-152.

51 Navarro-Remesal, "Regarding the (Game) Pain of Others."

Figure 23: The UEC campus is probably the visually most beautiful level in the game. It juxtaposes frightful indoor university spaces with the outside campus, which has been reclaimed by nature.

THE LAST OF US (Naughty Dog, 2013, 2014), ch. The University.

Consequently, it comes as no surprise that Joel and Ellie must reluctantly accept that the Fireflies have moved on. Several perspectives have prepared this disappointment and harmonised it to the image of *science* and *progress*. Maybe this how things should be. But the player is still undecided in this concern, for Ellie fervently pushes on. The allure of nature, however, has not evaded his attention, and the image of the CBI is about to change. Its dual texture of terror and renewal oscillate in the player's acts of ideation—since he has witnessed the later stages of the infection, which transform the infected into heavily deformed giants called Bloaters. Their fungus grows from several parts of their body, and when death is near, they withdraw to dark, moist areas to emit spores that contaminate anyone who inhales them (as is the case for all the stages of infection). Nature, so it seems, is steadily reclaiming the human body and purifying the land from humankind's virulent aggressions and leftovers.

7.3.2 Ellie as a Figure of Guidance and Temptress Towards Nature

An additional strategy of the game which promotes the natural world is Ellie, who functions as a figure of guidance for both Joel and the player. Ellie grew up in Boston's QZ after the apocalypse reduced humankind to a last few. She is therefore not contaminated by civilisation and the capitalist, bureaucratic order; she can be seen as an emblem of innocence who could show the way to a fresh start. In the course of the game, Ellie discovers the new and to her unfamiliar world. She thus assumes a position similar to the player's, to whom this world is also strange, and constantly reminds Joel (and the player) of nature's beauty.

Her role, thereby, is two sided. On the one hand, she is astounded by the natural world, which she discovers for the first time, while on the other she wishes to find the Fireflies and a cure for CBI. Of course, such a longing is understandable, considering the state of the gameworld, but it might be a short-sighted attempt to change things for the better. This may be partially explained through Ellie's naiveté concerning the old world, which seems fascinating to her. On various occasions she interrogates Joel about it and inquiries about things such as movies, baseball games, and children's toys, or is fascinated by comics they encounter on their way. It creates an intimate bond with Joel and the player alike, and this also expresses itself in instances where the young girl tries to whistle or in optional conversations, for example, when Ellie recites a joke.

Most noteworthy, however, is Ellie's perception of the natural world and her treatment of animals, which is tender and full of juvenile enthusiasm, and that culminates in one of gaming's most memorable moments. Having experienced several setbacks and the torturous encounter with David (to which I will come shortly), Ellie is frustrated and listless. She seems to have lost hope but is awakened by a magnificent event. Leading the way, Ellie discovers a herd of giraffes that now freely roam the land after the breakdown of civilisation—a scene reminiscent of the return of animal life in Richard Matheson's *I am Legend* (1954). The player is now given the chance to express compassion by the option to pet the giraffe. In actualising this possibility and creating the perspective, he takes away Ellie's shyness to approach the animal and becomes affectionately involved in the events. Such an action does not serve any 'winning' purpose but resonates deeply with the player on an affective and aesthetic level.

Ellie, then, assumes the role of the *temptress* towards nature. She makes Joel and the player aware of things they might miss or have failed to recognise and reminds them of nature's beauty. Part of her role is thus didactic and helps the player gain a fresh perspective not only on the gameworld but also on her empirical reality. The other half remains driven by a stubborn, naive resolve to find the Fireflies and a fascination about the old world. It is not very well thought out and unites empirical world norms that blindly lay their trust in science and progress while neglecting the ramifications for the ecosystem. Ellie's perspective is thus twofold and creates an opposition in itself by involving the player in a dialectic between both positions, but it also paves the way to game's darkest moments.

Figure 24: The giraffe scene marks a turning point in the plot and a last, most powerful, reminder of nature's preciousness before the final showdown.

THE LAST OF US (Naughty Dog, 2013, 2014), ch. Bus Depot.

7.3.3 The More Tortu(r)ous Route

Bill: "You know, as bad as those things are [the infected], at least they're predictable. It's the normal people that scare me."[52] *Autumn* in THE LAST OF US is a season of *uncertainty* and the choice between two utopian propositions: Tommy's settlement and the scientific enclave of the Fireflies, whose promise of a cure has impelled Ellie from the beginning. In the course of the journey, however, Joel raises doubts and is only convinced by Ellie to continue the search. This leads them to the disappointment of the Eastern Colorado University and into the abyss of *winter*, where they face the horrors of human nature. It is a fatal choice that, as

52 THE LAST OF US (Naughty Dog, 2013, 2014), ch. Bill's Town.

Lucian Ashworth holds, recalls man's folly for always choosing the "longer and more tortuous" route.[53] But, maybe, this is necessary to see things clearer.

Winter in THE LAST OF US is by far the most torturous season in the game and resembles a maze (a tortuous route) in which Joel and Ellie seem lost and face near death. Until now, they always had a strict goal in mind, but with Joel's severe injury at the UEC (symbolically followed by the first snowflakes), the focus is inevitably laid on *survival*. Like in the city, the spaces of winter expose the player to fierce ludic encounters (including the typical marks of blood on the snow) and multicursal mazes that are made more irritating due to snow storms blinding the player's vision. They are marked by the previously described closeness of sight and metaphorically describe a faction that has utterly failed as community.

The leader of this faction is David, the true antagonist of this story. Under his lead, the community at Silver Lake has devolved into cannibalism, and David himself shows paedophilic traits. It is as Green expresses: "David's actions represent the lowest manifestation of humanity"[54] and come as reminder that humankind is by far the worst predator on Earth. Supporting this argument is the fact that the story is now focalised through Ellie, as the player takes control of her. In a brutal process of coming-of-age, Ellie has to nurse the injured Joel and is confronted by David—which triggers a serious of events that culminate in the playable section of the burning cabin resort where Ellie struggles for her life. It is this change in Ellie that, as Hensel claims, is addressed by the game mechanics and that moves from a *playful, paidic* attitude and external perspective on her to a *disciplined, ludic* one.[55] As the player takes control of Ellie, he therefore enacts her coming-of-age and becomes unfailingly exposed to the darkest parts of the human self. Winter, consequently, evokes the composition of two forceful images that symbolically stand for Ellie's inner life: 1) the *disorienting* snowstorm and 2) the burning cabin resort, which has the player enact Ellie's *vulnerability*, her *fear* and *confusion*.

7.3.4 The Search for Utopian Enclaves

Shortly after the devastating encounter with David is the beginning of spring. It is a period of renewal, where the previously-mentioned giraffe scene revitalises the little girl. However, she is still confident about a scientific solution to the problem, whereas the signs in the environment speak differently—and it is Joel who again

53 Ashworth, "Dystopia and Global Utopias," 69.

54 Green, "The Reconstruction of Morality," 12.

55 Hensel, "Zwischen *Ludus* und *Paidia*," 156.

utters doubt but is persuaded by Ellie to continue the journey. Symbolically, the moment is marked by the herd of giraffes disappearing in the woods of Salt Lake City.

Ellie: Hurry up. C'mon, c'mon. c'mon! Oh man … wow. Look at those things [the giraffes].
Joel: So … this everything you were hoping for?
Ellie: It's got its ups and downs. But … you can't deny the view though.
Joel: We don't have to do this. You know that, right?
Ellie: What's the other option?
Joel: Go back to Tommy's. Just be done with this whole damn thing.
Ellie: After all we've been through. Everything that I've done. It can't be for nothing.[56]

Due to Ellie's perseverance, the faction of the Fireflies appears in a bright light right until the game's end. This hopeful ambiguity of their image stands in contrast to most other factions in the game. All of them were well-intended initially (for example, the QZs tried to safeguard the last of humankind from the infection, and David's community flourished), yet they failed as utopian propositions eventually. This is so for a variety of reasons, but primarily due to their incapacity to live in balance with their surroundings. As Green remarks: in contrast to the animal life, humankind is unable to adapt to the new circumstances, for they cling to the ruins of an old age and "isolate themselves further from the natural world, instead of trying to understand it."[57]

Consequently, it is humankind's relation to their surrounding environment that becomes important for determining utopian enclaves. The city and its inhabitants are thereby easily excluded. They remain caught up in societal ruins[58] and scavenge for supplies/raid passers-by, instead of living in a healthy balance with the surrounding ecosystems. Equally, one can exclude the Bandit factions who regularly attack Tommy's settlement, although their life in the woods provides everything they need, and also David's cannibals. All of these factions recall humankind's potential ugliness, and none of them show progressive aspects, for they are caught up in the static indoor space and chronotope of the city. Such an allocation of norms and conventions to the individual factions (the bureaucratic, capitalist lifestyle of the empirical world) is revealing for the player's acts of ideation—and this applies as well to the Fireflies.

56 THE LAST OF US (Naughty Dog, 2013, 2014), ch. Bus Depot.

57 Green, "The Reconstruction of Morality," 10; cf. 9-12.

58 "[G]esellschaftliche Ruine." Pfister, "Jacques Lacan," 5.

Tommy: "Remember how we thought no one could live like this anymore? Well, we're doing it."[59] In strong contrast to the above-mentioned factions is Tommy's settlement in Jackson County.[60] There, people have achieved a life in balance with nature and established a place for about twenty families. Tommy's settlement lies in the safety of high mountains and is surrounded by an electrocuted fence. They have achieved self-sustainability by growing crops and farming livestock and in their use of natural resources to produce clear energy—with the combined forces of a hydraulic power plant and a river stream flowing close to the settlement. For Green, therefore, Tommy's settlement constitutes a local "of hope"[61] with "no illusions about reclaiming a lost world,"[62] and the player is led to the composition of such an image through a variety of perspectives. From arriving in the woods of Jackson County, he steadily follows a river stream and experiences the sublimity of wilderness. Eventually, he reaches the power plant and, once inside, is given a guided tour of Utopia but additionally experiences its vulnerability, having to defend it against the bandits. The image is thus by no means perfect, but it is the best the player has experienced so far.

Given this description, it comes as no surprise that Tommy's vision of the better life stands in opposition to that of Marlene and the scientific enclave of the Fireflies. With their agenda to bring back humankind to the top of the evolutionary scale, the Fireflies do not shy away from sacrificing Ellie to create a vaccine (the fungus mutated in her brain and has to be surgically extracted). Still, their perspective is not as linear as one could think—for example Green, who suggests that the Fireflies' "light [is] easily extinguished."[63] Like any other faction in the game, the Fireflies have had to endure setbacks in their search for a cure, and they seem desperate. As such, it is an oversimplification to devalue Marlene's "conflict about sacrificing Ellie," which Green considers "little."[64] For the route to Utopia is marked by uncertainty and difficult decisions, and this manifests itself in Joel and Ellie's confusion when, after leaving Tommy's settlement, they are geographically wandering in circles: to the UEC, David's settlement, Salt Lake City, and back to Tommy's.

Right until the end, Ellie and Joel are unsure which route to take, which utopian enclave to follow, and it is finally Joel who takes that decision. For it is not

59 THE LAST OF US (Naughty Dog, 2013, 2014), ch. Tommy's Dam.

60 Green, "The Reconstruction of Morality," 10-11.

61 Ibid., 17.

62 Ibid., 11.

63 Ibid., 13.

64 Ibid.

only Ellie's life he wishes to save (although this is certainly the primary reason), but he implicitly discerns a fatal flaw in the Firefly faction. Their relentless ambition to return to the top of the evolutionary scale and bring back a society that was deeply flawed in the first place has not escaped Joel's (and the player's) view.[65] It is opposed to the game's positive rhetoric about nature, and in this light, Joel's decision to save Ellie from the dangers of surgery and to shoot Marlene point blank is understandable and not *as selfish* as some critics have argued—describing Joel "as an egocentric and egoistic actor with a lack of sense for overarching responsibilities"[66] for choosing Ellie's life over that of the greater good.[67] Such a conclusion is perilous, given the larger context of the gameworld and Joel's reversion to the loving father he was in the prologue. Science and technological progress have failed them once again (in the form of the hospital where they encountered the Fireflies), and it is rather the majority of humankind that "may not be worthy of saving. It seems worthy only of a second chance, over a number of years of hardship and slow evolution, if resistance like Ellie's begins to predominate."[68]

Hope, as Green rightfully concludes, may thus only be found in evolution itself, and in the *posthuman* solution of Ellie—as "logic would dictate that Ellie's value lies in keeping her alive and seeing whether she can pass along immunity as a genetic trait."[69] This enclave is fortified by the many perspectives the player has encountered and co-created during his journey, and their point of convergence culminates in the creation of the *nature chronotope*. Being opposed to that of the city, the chronotope of nature is characterised by the season of *spring* and the symbols of *renewal* and *hope*. It is fortified by the strategies of the game, which include Ellie. She is naturally immune to the CBI and, therefore, shows posthuman characteristics, being genetically adapted to the new environment. Following her model, humankind is maybe offered a true second chance, and one very different from the prospect of the non-actualised Firefly future.

However, it is not only Ellie who is responsible for the emergence of the utopian horizon, and Joel deserves additional merit. For his interference guides Ellie's

---

65 Several perspectives, such as audio logs, notes, medical records, and radiographs have the image of the Fireflies shine in a less positive light and alter the Joel's and the player's perception of the faction.

66 Holger Pötzsch, "Selective Realism: Filtering Experiences of War and Violence in First- and Third-Person Shooters," *Games and Culture* 12, no. 2 (2017): 169, http://journals.sagepub.com/doi/abs/10.1177/1555412015587802

67 Ibid.; Vella, "Characters," 13; Pérez-Latorre et al., "Videogame Analysis," 12.

68 Green, "The Reconstruction of Morality," 15.

69 Ibid., 13; cf. 6-7, 13.

naiveté concerning the old world and directs her playfulness in the right direction.[70] Consequently, this mutual dependence of *ludus* and *paidia* strikes a careful balance for the prospect of human survival and creates a beautiful metaphor for the balanced relation between various ecosystems. Humankind, so the inevitable conclusion, has to be kept in check by the forces of nature and by the saviour of the Cordyceps fungus, which has achieved a harmony lost long ago.

Table 10: Utopian enclaves described through features/symbols and determined by ecological aspects.

Utopian Enclaves?	**Features/Symbols**	**Balance with Nature/Sustainability**	**Yes/ No**
Cities and Quarantine Zones: Austin, Boston, Pittsburgh	violence/conflict/ entrapment/oppression/storm/fire/suffering/old order/absence of hope	decay/power generators/ scavenging gangs/shortage of supplies/stasis/viral	No
Bill's town	personal decay/ isolation/ absence of hope/fire	abandoned/shortage of supplies/stasis	No
Sewer settlement	entrapment/failed	decay/abandoned/death	No
Tommy's settlement	surrounded by wilderness/ safe haven/families/new order/hope/ Ellie (posthuman)	self-sustainable/animals/ clear energy/ progressive/dynamic	Yes
Bandit settlement	raiding/violence	stasis/no progress	No
Fireflies: UEC, Salt Lake City	(false) hope/suffering/ disappointment (UEC)/overcoming the plague	return to old order/ regressive/ would mankind learn?	unclear (No)
David's cannibals	failed/fire/suffering/hate	cannibals/paedophilia	No

70 I disagree with Hensel who claims that Ellie reverts back to a paidic playfulness by leaving Joel's ludic path. (Hensel, "Zwischen *Ludus* und *Paidia*," 156). Both need each other, and this balance is the only way for humankind to succeed.

7.4 Conclusion

Hope has always been the main driving force behind dystopia, and THE LAST OF US represents a magnificent example of the genre. In a mixture of direct extrapolation and analogy with the real world, it suggests that modern society suffers from entrapment and the ravaging violence of a bureaucratic consumer capitalism, and, therefore, issues a terrifying warning. As a solution to dystopia, the game suggests a renunciation of the old order and outlines a route to a new one: a return to nature and the utopian enclave of a life in balance with it. Even though the price may be high, the prospect of Utopia seems to be worth the risk. THE LAST OF US thereby employs the formal characteristics of the critical dystopia (it shows how dystopia came into being and proposes potential solutions to it) and successfully unifies its theme with the dynamic features of the video game medium.

To trigger such an insight and aesthetic response in the player, the game involves him in a meticulous dialectic between various oppositions that centre on the general divergence between *city* and *nature*. The scale, however, tends towards the latter end, for the city is illustrated as viral and infectious. Framed as the game's official narrative, it has the player experience a bureaucratic consumer capitalism gone mad and confronts him with derelict and claustrophobic spaces. These are arranged in confusing mazes and are characterised by intense combat, which makes the player suffer. In stark contrast to this are nature's liberating outdoor spaces, which are linear in structure and provide the player with the opportunity for human dialogue and compassion. They alleviate his suffering and are presented as utopian enclaves that exert a regenerative force on him and lead humankind to the establishment of a new order.

Having gone through this counter-narrative in the act of play and closing the blanks between the interrelated oppositions, the player is confronted with a decision about the future, which the game, however, takes for him. Other games, like PAPERS, PLEASE or FALLOUT 4, allow such a choice (as critical dystopias of variant II). In THE LAST OF US, the plot's logic dictates Joel's decisions, however—which is not to say that the game fails to exert an effect on the player. In this respect, Joel's reversion to the loving father he once was and Ellie's open-minded attitude towards nature are of utmost importance and create an intimate balance. They resensitise those of us who have lost our appreciation of the natural world and pave the way to a future that inspires the player to work towards a similar Utopia in real life. In this way, THE LAST OF US succeeds as critical dystopia and formulates a step in a vigorous utopian direction.

8 Horizons and the Video Game Dystopia

The video game dystopia represents a new strategic enterprise of the utopian philosophy and a genre for cultural reflection. This study has shown that playing dystopia involves the player in a playful trial action in which she may ergodically and imaginatively explore an alternative societal mode through play that is considerably worse than the game designers' empirical present. This experience in the dystopian gameworld warns the player about negative trends within empirical reality and helps her explore emancipatory routes to attenuate the nightmare in both the gameworld and empirical reality. It thus succeeds in triggering an aesthetic response in the player and shows the potential to reach a wide array of people both young and old who are familiar with playing games.

To elicit this kind of reaction in the player requires a tripartite dialectic between dystopian game, player, and culture (world) that can be described in terms of fictionality. In this sense, the fictive permeates the structural concept of the implied player as it reorganises empirical world elements (norms, conventions, troubling tendencies, the state of affairs, and so on) in the dystopian gameworld and its perspectives and partially frees them of their vertical hierarchies. The result is a refracted mirror of the familiar and unfamiliar, which estranges the player and shows her empirical reality in doublings and distortions. In addition, this organised space of play excels through several conflicts that can be lived through by the player (ergodically, imaginatively, psychologically)—that is to say, clashes of perspectives (world/characters, plot/structure, system/agency) she may experience and co-create. It thereby opens up a space for the imaginary to manifest itself in the player's psyche, her ergodic and imaginative actions. This trial space is free from the usual restrictions of the empirical world and reduced in complexity and thus allows for playful explorations. In using her imagination and closing the blanks between the game perspectives, the player experiences an aesthetic effect that is well-known from engaging with representational art but differs from it in that the experience the player lives through is more personal due to her extended involvement in it.

Whether these games of estrangement and fictionality are more subversive than those of non-ergodic art is left undecided, for there are a variety of factors that determine aesthetic response. However, it is indisputable that the VGD is utopian at heart and offers grand potential through virtualising storyworlds into malleable spaces for playful exploration that extend into the past and future and which are organised by a game's perspectival networks. In other words, playing dystopia involves the player in the inner mechanisms of dystopian regimes and functions as a disruption of the present through video game play. By doing so, it indirectly enlightens the player about the true nature of social totality and drives her to catharsis by having experienced a reduced but focused simulation of it in virtuality.

Playing dystopia is thus precarious and gives expression to a utopian impulse which is deeply ingrained in the human psyche. It aims to break free of the confines of *ludus*, to uncover its secrets within a systems of rules that organises the dystopian gameworld and official narrative, and to explore paidic routes and counter-narratives to it. These regenerative solutions affect the player's habitual dispositions, giving her hope where none may be found, and restructure her perception of the empirical world. They function as the opium of Utopia, which drives the player to emancipation and has her ride the steady yet tidal wave of utopianism towards the creation of a brighter future.

To come to such a conclusion required a couple of interrelated steps that gave rise to a variety of results. The most prominent of these is the description of the dystopian genre in video games and how it continues and expands on the legacy of utopian and dystopian fiction in non-ergodic art by involving the player. In this regard, it has been clarified that the utopian impulse manifests itself in a plethora of artwork in overt or disguised manners—and the VGD is a prime example of Utopia's pervasiveness. Thereby, the genre continues the tradition of utopian and dystopian fiction by relying on similar themes and tropes but expands on them through involving the player in a dynamic framework of play that goes beyond the reader's imaginative interaction with a storyworld. Instead, the player uses her imagination to participate ergodically in the gameworld and in dystopia's traditional plot structure—the clash of official narrative and counter-narrative—which has been expanded to a malleable framework in the VGD.

Analysing this framework shows the benefit of approaching the larger structure of dystopia's implied player and, additionally, allows one to distinguish between several subgenres of dystopia in deciphering how they negotiate the question of hope by involving the player in it. As a result, there are four distinct dystopian subtypes that can be situated between the poles of Utopia and Anti-Utopia:

1) the VGD as anti-utopia, which is reluctantly included in the dystopian genre, for it represents a strategy of ideology that aims to coax the player into supporting the status quo by having her fall victim to the game's hypocritical pleasures and criticism; 2) the VGD as classical dystopia, a most powerful subgenre that encloses the player within its nightmarish system and has her follow a counter-narrative that is bound to fail. Hope, as such, is to be found with the player only and in a subversive response to the nightmare; 3) the VGD as critical dystopia of variant I, a variant of the critical dystopia in games that negotiates the poles of Utopia and Anti-Utopia in a more nuanced way. Although these worlds are still despicable, they hold out hope for the diegetic characters and virtualise one or more utopian enclaves to be found by the player—who is sent on a linear route to a hopeful or ambiguous ending; 4) finally, there is the VGD as critical dystopia of variant II, which confronts the player with the choice of either becoming a catalyst or contributing to Utopia's downfall. As such, variant II sets the player in a precarious situation and exposes her to a trial run and a search for utopian enclaves. This task estranges the player and has her ponder similar situations/solutions to empirical world dystopias.

Given these diverse forms of agency to negotiate hope, the VGD represents a genre that sensitises the player to the utopian impulse in different forms. However, whether the game at hand is effective or not in conveying a warning is a related question, and an answer to it may be found in a game's aesthetic complexity and consequent diversity of perspectives. Consequently, it was established that games that involve the player in hectic and pleasurable combat, without giving at least a justification for the spectacle to which the player can attach emotionally, are examples of unsuccessful dystopias. On the other hand, the VGD can usually be endowed with the status of a multi-layered artefact that involves the player in both pleasurable and subversive ventures to an estranged Other. Thereby, several perspectives (character and world situations/dilemmas, diverse ludic interactions, intricate processes, or music pieces) stand in conflict with each other. They structure the plot perspectives of official narrative and counter-narrative in a decisive manner and involve the player in games of estrangement that further aggravate the combination of the perspectives. To attenuate these tensions, the player enters a dialectic between gameworld and empirical world and uses her world knowledge to close the blanks between the perspectives. The result is a meticulous negotiation of both worlds which gives light to the newness of art and to a revelatory effect that has the player discern empirical reality differently.

These insights into the nature of playing dystopia and the player's aesthetic response to it are of additional importance to central concerns in video game studies. They first of all address the nature of the game-narrative hybrids (with the

VGD being a subgenre) and their specific modes of discourse and player involvement. Thereby, a plot framework that is organically integrated into the overall game structure and its system of perspectives is essential and illustrates the prerequisite for VGNs. It structures the loose gamespace elements and endows them with purpose and causality. The player contributes to this creation of plot/meaning and not only enjoys the possibility to explore the gameworld and act within it but also to create diverse plots in doing so. As such, the plot framework formulates a vital part of the implied player and its system of perspectives, which can be held responsible for affording play in the first place and driving the player to an aesthetic response.

In the light of these findings, the discussion of playing dystopia and its structural prerequisites has led to the formulation of the implied player as an intersubjective framework that anticipates play and implies various types of players—for there are many viable ways to play a game. In this respect, the emancipated player's implication in this structure was considered a prerequisite for aesthetic complexity, which unfolds its potential in a tripartite dialectic between (dystopian) game, player, and culture (world). Such a playful trial action in dystopia engages the player with a gameworld on a complex level and drives her towards aesthetic response. What is more, this interaction has to be regarded in terms of fictionality and estrangement, because the specific nature of the dystopian gameworld aggravates the referentiality to the empirical world. To create the connection between both worlds, then, and to experience meaning, the player has to exert efforts of ideation and fill in the blanks between the game perspectives. This facet necessarily construes the concept of fiction I am employing as a semantic phenomenon that involves the player in games of estrangement and holds the function of telling her something about reality that was hitherto inaccessible to direct perception.

As such, a gameworld's incompleteness is the basic premise for the player's participation in the VGD/VGN and integral to unveiling the newness of video game art. This fundamental quality of fictional worlds is refined in video games in how the player approaches the gameworld. Two forms of indeterminacy can thereby be discerned. First of all, there are gaps in the gameworld that can be closed either through imagination or the player's ergodic movements within the diegesis, whereby virtualised potentialities await her ergodic fillings. These create important perspectives on the gameworld that, in combination with the remaining perspectives, are responsible for the emergence of blanks. The player can close these through imagination in order to understand the gameworld, but it also induces her to ergodically act upon the composed images: an integral aspect of video game involvement. In other words, whereas gaps speak to the player's urge to completion, blanks evoke her desire for combination, allowing for an intricate understanding of the gameworld elements and their interrelations.

However, these initial images on the level of the plot give rise to further questions once the player compares her results to facts and situations from the empirical world. As such, the closing of blanks through the formation of images in the player's mind is due to the creation of two interlinked *gestalts*. Whereas the first *gestalt* assembles an image of the occurring plot, the cognitive mapping of the gameworld, its processes and characters, it nonetheless remains partially incomplete. This is because the closing can only come about through the formation of a secondary *gestalt* that unveils something that is not discernible through the mere perception of the object but is imagistic in character.

Hence it follows that fictional communication in VGDs (and VGNs) is similar to and different from how one perceives and interacts with the real world. For while it is based on natural rules of human perception, perceiving the empirical world in perspectival systems, it is only through acts of ideation and the creation of a secondary *gestalt* that unexpected connections between both realities (and their interrelations) are brought to the fore. Playing dystopia thus involves the player in games of fictionality that are structured by the fictive—the ludic organisation of the gameworld that excels with doublings/distortions—and negotiated by the player—in her imagination, ergodic manoeuvres, and resultant emotions. This interaction mitigates the conflicts between the individual perspectives and is propelled by the imaginary as an uncontrollable, paidic force. It creatively engulfs the player and expresses itself in the experience of the aesthetic effect that is so fundamental to conveying the VGD's message.

Given this potential of playing dystopia, the outlook for the genre is positive overall in establishing itself as the dominant genre in the transmedial environment of Utopia. In reaching a wide array of people all around the globe, the VGD is fundamentally important for suggesting emancipatory routes that lead out of the 21st century dystopia and involving the player in a negotiation of possibilities to do so. Thereby, the genre builds on a long tradition of despair and revolution that dates back to the 1980s and continues to arouse the minds of players both young and old in the 21st century. While not all VGDs necessarily fulfil this aesthetic function (as is the case with any medium), they nonetheless embrace the utopian impulse that can be found at their core. In addition, it goes without saying that several creative examples of the genre have emerged. These explore new subversive routes of player involvement and intermingle a variety of perspectives for the player to explore and aid in their creation. As a narrative genre, the VGD is thus permeated by several ludic genres (adventure games, action-adventures, FPSs, role-playing games (RPGs), strategy games, experiential genres, and so on) that afford different player interactions. Moreover, it spices these up with diverse game structures (linear corridors, multicursal labyrinths, one-rooms, or open world structures) and different degrees of estrangement and extrapolation. These possibilities have to be employed in creative ways by the game designers, however, and with the anticipation of more heterogeneous game developer teams (including more women), the VGD may stand at the dawn of a new age of subversiveness that knows few bounds.

In this respect, current VGDs have already contributed their fair share of masterpieces such as the BIOSHOCK games, THE STANLEY PARABLE, FALLOUT 4, METRO 2033, DEUS EX: HUMAN REVOLUTION, MIRROR'S EDGE, THE LAST OF US and PAPERS, PLEASE. Although these games retain a thoroughly pessimistic shell and involve the player in nightmarish gameworlds, their core is fundamentally utopian (to different degrees)—which drives the player towards emancipation and the desire to counteract dystopia in the real world. Such an alignment towards Utopia is not surprising, given the contemporary world situation, which makes it difficult for dystopia to arouse and shock the appreciator/player, for she witnesses injustice and atrocity on a daily basis. Consequently, the VGD has opened its gates to explorations that lead beyond these confinements and give rise to outright utopian solutions (such as in THE LAST OF US or FALLOUT 4) that move the genre in a decisive utopian direction.

List of Abbreviations

AAA games: triple A games (big budget games)
Ai: Artificial Intelligence
ARG: Alternate Reality Game
CBI: Cordyceps Brain Infection
CP: Civil Protection
COD: AW: CALL OF DUTY: ADVANCED WARFARE
DEUS EX : HR: DEUS EX: HUMAN REVOLUTION
DLC: downloadable content
EDTSD: EVERY DAY THE SAME DREAM
ENE: evocative narrative elements
FEDRA: Federal Disaster Response Agency
FPS: first-person shooter
NPC: non-player character
PC: player-character
PP: PAPERS, PLEASE
QTE: quick-time event
QZ: Quarantine Zone
RPG: role-playing game
SF: science fiction
SPC: silent player-character
VGD: video game dystopia
VGN: video game narrative
VG SF: video game science fiction

Glossary

acts/efforts of ideation: refer to the player's implicit comparison of the gameworld to her empirical surroundings through the creation and continual revision of images.

Anti-Utopia: refers to a tradition that seeks to undermine the utopian philosophy by claiming that the present society is the best there is. The agenda behind such a statement is to consolidate dominant ideology.

anti-utopia: refers to real or fictional ways (texts, films, games, etc.) to communicate the anti-utopian philosophy by either depicting a future world or utopian premise that is utterly pessimistic or by seducing the reader/player to believe in shallow and clichéd solutions.

blanks: refer to overall vacancies in the game that allow for interaction to occur and help the player link the fictional events to her/his empirical surroundings. Blanks speak to the player's urge for combination as she/he closes the blanks between the perspectives she has helped create through both imaginative and ergodic interaction.

classical dystopia: refers to a subgenre of the video game dystopia that involves the player in a malignant gameworld and invites her/him to struggle against it. Although these efforts are in vain, it differs from the anti-utopia in that it triggers fictional anger in the player and evokes the urge to counteract such tendencies in the empirical world.

creative dialectic: refers to the empirical player's communication and engagement with a work of art to discover truth—which is influenced and shaped by the specific cultural surroundings.

critical dystopia: refers to a subgenre of the video game dystopia that comes in two variants. The first variant virtualises a negative society that sends the player on a linear route to resistance and the redemption of the flawed world—or, at least, to an ambiguous ending—whereas the second variant lays the prospect of Utopia into the player's hands. Both variants often discuss the origins of dystopia.

empirical world: refers to the reality outside of the game, narrative, film, or work of art. I use the term synonymously with empirical reality, real world, actual world, contemporary society/surroundings, extratextual reality, and so on.

ergodic interaction: refers to the player's physical input on some sort of input device (such as a controller) to effectuate an event or a series of events in the gameworld. The category of the ergodic includes both the player's active impact on the gameworld as well as her/his cognitive readiness to act.

fiction: the term is used in the sense of fictionality and in a functional approach. It refers to the aggravated referentiality between empirical reality and the reality of fiction, which the player has to decipher/construct through her/his acts of ideation.

gamespace: refers to the ontological layer of the gameworld: the virtual gamespace. It is composed of a set of rules and basic buildings blocks (environments, objects, characters, etc.) that create a malleable space of possibility in which the player can manoeuvre and act.

gameworld: refers to how the player perceives the world of the game as a whole: the virtual gamespace and the fictional storyworld in combination. These create a participatory, explorable space in which most diverse narratives may be co-created.

game world: (not to be confused with gameworld), refers to the participatory, private world of the player (or appreciator) and the 'games' she/he is playing with a work world, which include: ergodic, imaginative, psychological, emancipated games, and so on.

gaps: are places of indeterminacy in the game(world) that can either be filled through imaginative interaction (the filling in of indeterminate spaces such as the inside of an inaccessible house or character personalities) or through ergodic interaction, where the player actualises certain possibilities to action ingrained in the game. Gaps speak to the player's urge for completion.

imaginative interaction: refers to the player's imaginative games with the gameworld such as imagining the inside of an inaccessible house, the interrelations between characters, considerations about the gameworld's past or how the plot may proceed, and so on.

Utopia: refers to the philosophy of utopianism at whose core lies humankind's wish to create better social surroundings and the endeavour to lead a happy, fulfilled life.

utopia: refers to a genre that virtualises a future which is better than the designers' and players' empirical present. At its core lies the function of shedding a critical light on the empirical world and suggesting micro or macro solutions to it.

playful trial action: refers to the player's multifarious involvement in the world of game narratives. In the sense I am describing, it revolves around the player's ergodic, imaginative, and psychological involvement in the gameworld and in an intimate comparison to the empirical world through acts of ideation.

representation: in the specific sense I am using the term, representation is synonymous with fiction.

science fiction: refers to a broad genre that encompasses subgenres such as the posthuman, space opera, cyberpunk, steampunk, or post-apocalypse. Although I do not subordinate utopia and dystopia to SF (because of the genres' long traditions), there are many affinities between them.

storyworld: refers to the fictional layer of the gameworld. Through a storyworld, the virtual gamespace is endowed with life, creativity, and meaning. In addition, the concept has significant impact on the creation of gamespaces, for if a plot framework is employed, the shape of the gamespace is determined accordingly.

video games: the term (video) game is used to describe all sorts of digital games. For my specific purposes, it describes the subgenre of the video game narrative (if not otherwise stated).

video game narrative: refers to a subgenre of video games that are considered hybrids between traditional (video) games and a new form of participatory narrative. In this study, most video games I refer to are also VGNs (if not otherwise stated).

video game dystopia: a genre that describes a subset of VGNs. It virtualises a negative society and involves the player in a trial action in which she/he may scrutinise the problems of the gameworld and act to redeem them—or at least attenuate their consequences.

work world: refers to the (story)world of a work of art with which the appreciator, reader, spectator, or player engages.

Ludography

2K Boston. BIOSHOCK. 2K Games, 2007. Played on PC.

3909. PAPER'S PLEASE. 3909, 2014. Played on PC.

4A Games. METRO 2033. THQ, 2010. Played on PC.

Bethesda Game Studios. FALLOUT 4. Bethesda Softworks, 2015. Played on PS4.

—. FALLOUT 3. Bethesda Softworks, 2008. Played on PS3.

Bullfrog Productions. SYNDICATE. Electronic Arts, 1993. Played on PC.

Camouflaj. RÉPUBLIQUE. GungHo Online Entertainment, 2013-2014. Played on iOS.

DICE. MIRROR'S EDGE. Electronic Arts, 2008. Played on PS3.

Dontnod Entertainment. REMEMBER ME. Capcom, 2013. Played on PC.

Eidos Montreal. DEUS EX: HUMAN REVOLUTION. Square Enix, 2011. Played on PC.

Galactic Cafe. THE STANLEY PARABLE. Galactic Cafe, 2013. Played on PC.

Gearbox Software. BORDERLANDS. Take-Two Interactive, 2009. Played on Xbox 360.

—. BORDERLANDS 2. 2K Games, 2012. Played on Xbox 360.

—. BORDERLANDS: THE PRE-SEQUEL. 2K Games, 2014. Played on PC.

Guerrilla Games. HORIZON ZERO DAWN. Sony Interactive Entertainment, 2017. Played on PS4.

Insomniac Games. RESISTANCE: FALL OF MEN. Sony Computer Entertainment, 2006. Played on PS3.

—. RESISTANCE 2. Sony Computer Entertainment, 2008. Played on PS3.

—. RESISTANCE 3. Sony Computer Entertainment, 2011. Played on PS3.

Irrational Games. BIOSHOCK INFINITE. 2K Games, 2013. Played on PC.

—. BIOSHOCK INFINITE: BURIAL AT SEA: EPISODE ONE. 2K games, 2013. Played on PC.

—. BIOSHOCK INFINITE: BURIAL AT SEA: EPISODE TWO. 2K games, 2014. Played on PC.

Kaos Studios. HOMEFRONT. THQ, 2011. Played on Xbox 360.

MachineGames. WOLFENSTEIN: THE NEW ORDER. Bethesda Softworks, 2014. Played on PS4.

Molleindustria. EVERY DAY THE SAME DREAM. Molleindustria, 2009. Played on PC.

Naughty Dog. THE LAST OF US. Sony Computer Entertainment, 2013. Played on PS3.

—. THE LAST OF US: REMASTERED. Sony Computer Entertainment, 2014. Played on PS4.

Ninja Theory. ENSLAVED: ODYSSEY TO THE WEST. Namco Bandai Games, 2010. Played on Xbox 360.

Platinum Games. MAD WORLD. *Sega,* 2009. Played on Nintendo Wii.

—. NIER AUTOMATA. Square Enix, 2017. Played on PS4

Quantic Dream. HEAVY RAIN. Sony Computer Entertainment, 2010. Played on PS3.

Rockstar North. GRAND THEFT AUTO V. Rockstar Games, 2013. Played on PS3.

Sledgehammer Games. CALL OF DUTY: ADVANCED WARFARE. Activision, 2014. Played on PS4.

Telltale Games. TALES FROM THE BORDERLANDS: A TELLTALE GAME SERIES. Telltale games, 2014-2015. Played on PS4.

Thatgamecompany. JOURNEY. Sony Computer Entertainment, 2012 (PS3), 2015 (PS4). Played on PS4.

The Dreamers Guild. I HAVE NO MOUTH, AND I MUST SCREAM. Cyberdreams, 1995. Played on PC.

Ubisoft Montpellier. BEYOND GOOD AND EVIL. Ubisoft, 2003. Played on Nintendo Game Cube.

Ubisoft Montreal. WATCH_DOGS. Ubisoft, 2014. Played on PS4.

—. WATCH_DOGS 2. Ubisoft, 2016. Played on PS4.

Valve. HALF-LIFE 2. Sierra Studios, 2004. Played on PC.

—. HALF-LIFE 2: EPISODE 1. Electronic Arts, 2006. Played on PC.

—. HALF-LIFE 2: EPISODE 2. Valve, 2007. Played on PC.

Bibliography

Aarseth, Espen J. "A Narrative Theory of Games." *Proceedings of the International Conference on the Foundations of Digital Games* (2012): 129-133. http://dl.acm.org/citation.cfm?id=2282365

—. "Allegories of Space – The Question of Spatiality in Computer Games." In *Cybertext Yearbook 2000*, edited by Markku Eskelinen and Raine Koskimaa, 152-171. Jyväskylä: University of Jyväskylä, 2001. http://cybertext.hum.jyu.fi/index.php?browsebook=4

—. *Cybertext: Perspectives on Ergodic Literature*. Baltimore: John Hopkins UP, 1997.

—. "Doors and Perception: Fiction vs Simulation in Games." *Escritas Mutanes* (2006). http://www.luisfilipeteixeira.com/fileManager/file/fiction_Aarseth_jan2006.pdf

—. "I Fought the Law: Transgressive Play and the Implied Player." *Proceedings of the 2007 DiGRA International Conference: Situated Play* 4 (2007): 130-133. http://www.digra.org/wp-content/uploads/digital-library/07313.03489.pdf

—. "Ontology." In *The Routledge Companion to Video Game Studies*, edited by Mark J. P. Wolf and Bernard Perron, 484-492. New York: Routledge, 2014.

Abbott, H. Porter. "Story, Plot, and Narration." In *The Cambridge Companion to Narrative*, edited by David Herman, 39-51. Cambridge: Cambridge UP, 2007.

—. *The Cambridge Introduction to Narrative.* 2nd ed., Cambridge: Cambridge UP, 2008.

Aldred, Jessica and Brian Greenspan. "A Man Chooses, A Slave Obeys: *BioShock* and the Dystopian Logic of Convergence." *Games and Culture* 6, no. 5 (March 2011): 479-496. http://journals.sagepub.com/doi/abs/10.1177/1555412011402674

Amis, Kingsley. *New Maps of Hell*. London: Penguin Books, 2002.

Archibald, Samuel. "Literary Theory (Appendix)." In *The Video Game Theory Reader 2*, edited by Bernard Perron and Mark J. P. Wolf, 361-362. London: Routledge, 2009.

Aristotle. *The Art of Rhetoric*. Translated by W. Rhys Roberts. Megaphone eBooks, 2008. http://www.wendelberger.com/downloads/Aristotle_Rhetoric.pdf

Arsenault, Dominic. "Narratology." In *The Routledge Companion to Video Game Studies*, edited by Mark J. P. Wolf and Bernard Perron, 475-483. New York: Routledge, 2014.

Ascher, Franziska. "Es gibt immer einen Leuchtturm … ." *Paidia: Zeitschrift für Computerspielforschung* (2013): 1-13. http://www.paidia.de/?p=2499.

Ashworth, Lucian M. "Dystopia and Global Utopias: A Necessary Step Towards a Better World." In *Dystopia(n) Matters: On the Page, on Screen, on Stage*, edited by Fátima Vieira, 69-71. Newcastle-upon-Tyne: Cambridge Scholar Publishing, 2013.

Baccolini, Raffaella and Tom Moylan, editors. *Dark Horizons: Science Fiction and the Dystopian Imagination*. New York: Routledge, 2003.

—. "Conclusion. Critical Dystopia and Possibilities." In *Dark Horizons: Science Fiction and the Dystopian Imagination*, edited by Raffaella Baccolini and Tom Moylan, 233-249. New York: Routledge, 2003.

—. "Introduction. Dystopia and Histories." In *Dark Horizons: Science Fiction and the Dystopian Imagination*, edited by Raffaella Baccolini and Tom Moylan, 1-12. New York: Routledge, 2003.

Baccolini, Raffaella. "Living in Dystopia." In *Dystopia(n) Matters: On the Page, on Screen, on Stage*, edited by Fátima Vieira, 44-45. Newcastle-upon-Tyne: Cambridge Scholar Publishing, 2013.

Bachelard, Gaston. *L'Eau et les Rêves, Essai sur L'Imagination de la Matière*. Paris: Librairie José Corti, 1987.

—. *La Psychanalyse du Feu (Folio Essais)*. Paris: Gallimard, 2015.

Bacon, Francis. *The New Atlantis: Or, Voyage to the Land of the Rosicrucians*. London: Forgotten Books, 2008.

Badiou, Alain. "Happiness is a Risk that we Must be Ready to Take." Interview by Mira Davidson. *Verso*, 2015, https://www.versobooks.com/blogs/2032-alain-badiou-happiness-is-a-risk-that-we-must-be-ready-to-take

Bakhtin, Michail M. "Form of Time and the Chronotope in Novels: Notes towards a Historical Poetics." In *Narrative Dynamics, Essays on Time, Plot, Closure and Frames*, edited by Brian Richardson, 15-24. Columbus: Ohio State UP, 2002.

Barry, Peter. *Beginning Theory: An Introduction to Literary and Cultural Theory.* 3rd ed., Manchester: Manchester UP, 2009.

Bartle, Richard. "Hearts, Clubs, Diamonds, Spades: Players who Suit Muds." *Mud*, 1996. Accessed 21 Mar. 2016. http://mud.co.uk/richard/hcds.htm#1.

Bateman, Chris. "Keeping the Player on Track." In *Game Writing: Narrative Skills for Videogames*, edited by Chris Bateman, 85-102. Boston, Mass.: Charles River Media, 2007.

Bateman, Chris and Richard Boon. *21st Century Game Design*. Boston: Course Technology, 2006.

Bateson, Gregory. *Steps to an Ecology of Mind: Collected Essays in Anthropology, Psychiatry, Evolution, and Epistemology*. Lanham, Maryland: Jason Arson, 1987.

BBC Worldwide. "Cordyceps: attack of the killer fungi - Planet Earth Attenborough BBC wildlife." YouTube, November 3, 2008. https://www.youtube.com/watch?v=XuKjBIBBAL8

Berensmeyer, Ingo. *Literary Theory: An Introduction to Approaches, Methods and Terms*. Stuttgart: Klett Lerntraining, 2009.

Blaim, Arthur. "Hell upon a Hill: Reflections of Anti-Utopia and Dystopia." In *Dystopia(n) Matters: On the Page, on Screen, on Stage*, edited by Fátima Vieira, 80-95. Newcastle-upon-Tyne: Cambridge Scholar Publishing, 2013.

Bloch, Ernst. *The Principle of Hope: Volume One.* Cambridge Mass.: MIT Press, 1986.

Bode, Christoph and Rainer Dietrich. F*uture Narratives: Theory, Poetics, and Media-Historical Moment*. Berlin: De Gruyter, 2013.

Bogost, Ian. *Persuasive Games: The Expressive Power of Videogames*. Cambridge, Mass.: MIT Press, 2010.

Bogost, Ian, Simon Ferrari, and Bobby Schweizer. *Newsgames: Journalism at Play*. Cambridge, Mass.: MIT Press, 2010.

Bolotin, Norman and Christine Laing. *The World's Columbian Exposition: The Chicago World's Fair of 1893*. Urbana: University of Illinois Press, 2002.

Booker, M. Keith. "About this Volume." In *Critical Insights: Dystopia*, edited by M. Keith Booker, vii-xii. Ipswich, Mass.: Salem Press, 2013.

—. *Dystopian Literature: A Theory and Research Guide.* Westport, Connecticut: Greenwood Press, 1994.

—. "On Dystopia." In *Critical Insights: Dystopia*, edited by M. Keith Booker, 1-15. Ipswich, Mass.: Salem Press, 2013

—. *The Dystopian Impulse in Modern Literature: Fiction as Social Criticism.* Westport, Connecticut: Greenwood Press, 1994.

Booker, M. Keith and Anne-Marie Thomas. *The Science Fiction Handbook.* Hoboken: Wiley-Blackwell, 2009.

Boon, Richard. "Writing for Games." In *Game Writing: Narrative Skills for Videogames*, edited by Chris Bateman, 43-70. Boston, Mass.: Charles River Media, 2007.

Bosman, Frank G. "'The Lamb of Comstock'. Dystopia and Religion in Video Games." *Online Heidelberg Journal of Religions on the Internet* 5 (2014): 162-182. http://heiup.uni-heidelberg.de/journals/index.php/religions/article/view/12163

Bradbury, Ray. *Fahrenheit 451*. New York: Ballantine Books, 1991.

Brecht, Bertolt. *Bertolt Brecht Poems 1913-1956*, edited by John Willet and Ralph Manheim. New York: Routledge, 1987.

Buinicki, Martin T. "Nostalgia and the Dystopia of History in 2K's *Bioshock Infinite*." *The Journal of Popular Culture* 49, no. 4 (2016): 722-737. http://onlinelibrary.wiley.com/doi/10.1111/jpcu.12440/abstract

Campbell, Colin. "The Big Ideas of *BioShock Infinite*: An in-depth look at Major Themes, with Ken Levine and Shawn Robertson." *IGN*, January 30, 2013. Accessed March 7, 2014. http://www.ign.com/articles/2013/01/30/the-big-ideas-of-bioshock-infinite

Campbell, Joseph. *The Hero with a Thousand Faces*. Bollingen Series XVII, 3rd ed. Novato, California: New World Library, 2008.

Caillois, Roger. *Man, Play and Games*. Translated by Meyer Brash. Urbana: University of Illinois Press, 2001.

Calleja, Gordon. *In-Game: From Immersion to Incorporation*. Cambridge, Mass.: MIT Press, 2011.

Carson, Don. "Environmental Storytelling: Creating Immersive 3D Worlds Using Lessons Learned from the Theme Park Industry." *Gamastura: The Art and Business of Making Games* (2000). https://www.gamasutra.com/view/feature/131594/environmental_storytelling_.php.

Chatman, Seymour. *Story and Discourse in Fiction and Film.* 2nd ed., Ithaca: Cornell UP, 1980.

Ciccoricco, David. "'Play, Memory': *Shadow of the Colossus* and Cognitive Workouts." *Dichtung-Digital: Journal für Kunst und Kultur digitaler Medien* (2007). http://dichtung-digital.de/2007/Ciccoricco/ciccoricco.htm

Claeys, Gregory. *Searching for Utopia: The History of an Idea*. London: Thames & Hudson, 2011.

Clarke, Ian F. "The Pattern of Prediction 1763-1973: Methods of Prediction 1918-1939." *Futures 2*, no. 4 (December 1970): 375-379.

Coleridge, Samuel. T. "Biographia Literaria". *Project Gutenberg,* 2004. Accessed 29 Sep. 2016. http://www.gutenberg.org/files/6081/6081-h/6081-h.htm

Crick, Bernard. Introduction. *Nineteen Eighty-Four*, by George Orwell, 152-153. New York: Clarendon Press, 1984.

Dansky, Richard. "Introduction to Game Narrative." In *Game Writing: Narrative Skills for Videogames*, edited by Chris Bateman, 1-24. Boston, Mass.: Charles River Media, 2007.

Darwin, Charles. *On the Origin of Species*. New York: Bantam Dell, 1999.

Davis, Laurence. "Dystopia, Utopia and Sancho Panza." In *Dystopia(n) Matters: On the Page, on Screen, on Stage*, edited by Fátima Vieira, 23-27. Newcastle-upon-Tyne: Cambridge Scholar Publishing, 2013.

DeMarle, Mary. "Nonlinear Game Narrative." In *Game Writing: Narrative Skills for Videogames*, edited by Chris Bateman, 71-84. Boston, Mass.: Charles River Media, 2007.

DeWildt, Lars. "Precarious Play. To Be or Not to Be Stanley." *Press Start* 1, no. 1 (November 2014): 1-20. http://press-start.gla.ac.uk/index.php/press-start/article/view/10/4

Doležel, Lubomír. *Heterocosmica: Fictional and Possible Worlds*. Baltimore: John Hopkins UP, 1998.

Domsch, Sebastian. "Dystopian Video Games: Fallout in Utopia." In *Dystopia, Science Fiction, Post-Apocalypse: Classics – New Tendencies – Model Interpretations*, edited by Eckart Voigts and Alessandra Boller, 395-410. Trier: Wissenschaftlicher Verlag Trier, 2015.

—. *Storyplaying: Agency and Narrative in Video Games*. Berlin: De Gruyter, 2013.

Donawerth, Jane. "Genre Blending and the Critical Dystopia." In *Dark Horizons: Science Fiction and the Dystopian Imagination*, edited by Raffaella Baccolini and Tom Moylan, 29-46. New York: Routledge, 2003.

Dostoyevsky, Fyodor. *The Grand Inquisitor*. Edited by Helena P. Blavatsky. Auckland: Floating Press, 2009.

Eco, Umberto. *Nachschrift zum >Namen der Rose<*. Translated by Burkhart Kroeber. München: Deutscher Taschenbuch Verlag, 1986.

—. *The Open Work*. Translated by Anna Cancogni. Cambridge, Mass.: Harvard UP, 1989.

Egenfeldt-Nielsen, Simon, Jonas H. Smith, and Susana P. Tosca. *Understanding Video Games: The Essential Introduction.* 3rd ed. London: Routledge, 2015.

Engelns, Markus. *Spielen und Erzählen: Computerspiele und die Ebenen ihrer Realisierung*. Heidelberg: Wissenschaftlicher Verlag der Autoren Synchron Publishers, 2014.

Fahlenbrach, Kathrin and Felix Schröter. "Game Studies und Rezeptionsästhetik." Game Studies: *Aktuelle Ansätze der Computerspielforschung*. Köln: Herbert von Harlem Verlag, 2015.

Farca, Gerald. "The Emancipated Player." *Proceedings of the First International Joint Conference of DiGRA and FGD* 13, no. 1 (2016):1-16. http://www.digra.org/digital-library/publications/the-emancipated-player/

Farca, Gerald and Charlotte Ladevèze. "The Journey to Nature: The Last of Us as Critical Dystopia." *Proceedings of the First International Joint Conference of DiGRA and FGD* 13, no. 1 (2016): 1-16. http://www.digra.org/digital-library/publications/the-journey-to-nature-the-last-of-us-as-critical-dystopia/

Fernández-Vara, Clara. "Game Spaces Speak Volumes: Indexical Storytelling." *Proceedings of the 2011 DiGRA International conference: Think Design Play* 6 (2011): 1-13. http://www.digra.org/digital-library/publications/game-spaces-speak-volumes-indexical-storytelling/

Ferns, Chris. *Narrating Utopia.* Liverpool: Liverpool UP, 1999.

Feil, John. "Writing for Action-Adventure Games." In *Writing for Video Game Genres: From FPS to RPG*, edited by Wendy Despain, 29-36. Boca Raton: A K Peters, 2009.

Ferreira, Aline. "Biodystopias Matter: Signposts of Future Evolution." In *Dystopia(n) Matters: On the Page, on Screen, on Stage*, edited by Fátima Vieira, 49-53. Newcastle-upon-Tyne: Cambridge Scholar Publishing, 2013.

Fitting, Peter. "Unmasking the Real? Critique and Utopia in Recent SF Films." *Dark Horizons: Science Fiction and the Dystopian Imagination,* edited by Raffaella Baccolini and Tom Moylan, 155-166. New York: Routledge, 2003.

Forster, Edward M. *The Machine Stops and Other Stories*. London: Collector's Library, 2012.

Fortunati, Vita. "Why Dystopia Matters." In *Dystopia(n) Matters: On the Page, on Screen, on Stage*, edited by Fátima Vieira, 28-36. Newcastle-upon-Tyne: Cambridge Scholar Publishing, 2013.

Foucault, Michel. *Discipline and Punish: The Birth of the Prison*. 2nd ed. Translated by Alan Sheridan, New York: Random House, 1995.

—. "Of Other Spaces: Utopias and Heterotopias." In *Rethinking Architecture: A Reader in Cultural Theory*, edited by Neil Leach, 350-356. London: Routledge, 1997.

Franklin, H. Bruce. "What is Science Fiction – and How it Grew." In *Reading Science Fiction*, edited by James Gunn, Marleen S. Barr, and Matthew Candelaria, 23-32. Basingstoke: Palgrave MacMillan, 2009.

Frasca, Gonzalo. "Ludologists love Stories too: Notes from a Debate that Never took Place." *Proceedings of the 2003 DiGRA International Conference: Level up* 2 (2003): 91-99. http://www.digra.org/wp-content/uploads/digital-library/05163.01125.pdf

Fraser, Allison. "Whose Mind is the Signal? Focalization in Video Game Narratives." *Proceedings of the 2015 DiGRA International Conference* 12 (2015), 1-16. http://www.digra.org/digital-library/publications/whose-mind-is-the-signal-focalization-in-video-game-narratives/

Freud, Sigmund. "Creative Writers and Day-Dreaming." In *The Standard Edition of the Complete Psychological Works of Sigmund Freud Volume IX*. Translated by The Institute of Psycho-Analysis, 141-154. London: Vintage, 2001.

—. *The Interpretation of Dreams*. Translated by A. A. Brill, Hertfordshire: Wendsworth, 1997.

Frelik, Pawel. "Video Games." In *The Oxford Handbook of Science Fiction*, edited by Rob Latham, 226-238. Oxford: Oxford UP, 2014.

Froschauer, Adrian. "Der Kampf um die Erzählhoheit: Voice-over-Narration im Computerspiel." *Spielzeichen: Theorien, Analysen und Kontexte des zeitgenössischen Computerspiels*, edited by Martin Hennig and Hans Krah, 117-136. Boizenburg: Werner Hülsbusch, 2016.

Frye, Northrop. "Varieties of Literary Utopias." In *Utopias and Utopian Thought*, edited by Frank E. Manuel, 329-80. Boston: Beacon, 1967.

Genette, Gérard. *Die Erzählung.* 2nd ed. Translated by Jane E. Lewin. Paderborn: Fink, 1998.

Geoghegan, Vincent. "Darkness and Light." In *Dystopia(n) Matters: On the Page, on Screen, on Stage*, edited by Fátima Vieira 46-48. Newcastle-upon-Tyne: Cambridge Scholar Publishing, 2013.

Gibbons, William. "Wrap Your Troubles in Dreams: Popular Music, Narrative, and Dystopia in *Bioshock*." *Game Studies* 11, no. 3 (December 2011). http://gamestudies.org/1103/articles/gibbons

Gibson, William. *Neuromancer*. New York: ACE Books, 1984.

Green, Amy M. "The Reconstruction of Morality and the Evolution of Naturalism in *The Last of Us*." *Games and Culture* 11, no. 7-8 (2016): 1-19. http://journals.sagepub.com/doi/abs/10.1177/1555412015579489

Hall, Alexander C. H. "'I am Trying to Believe': Dystopia as Utopia in the *Year Zero* Alternative Reality Game." *Eludamos. Journal for Computer Game Culture* 3, no. 1 (2009): 69-82. http://www.eludamos.org/index.php/eludamos/article/view/vol3no1-8/112

Hardison, O. B. Jr. "A Commentary on Aristotle's Poetics." In *Aristotle's Poetics: a Translation and Commentary for Students of Literature*, edited by O. B. Hardison. New Jersey: Prentice-Hall, 1968.

Heinrich, Maximilian. "The Role of Technology and Idea of Man in *BioShock* and *Half-Life 2*." In *Contact, Conflict, Combat: Zur Tradition des Konflikts in digitalen Spielen,* edited by Rudolf T. Inderst and Peter Just, 183-196. Boizenburg: Werner Hülsbusch, 2011.

Hensel, Thomas. "Zwischen *Ludus* und *Paidia*: *The Last of Us* als Reflexion des Computerspiels." *New Game Plus: Perspektiven der Game Studies. Genres – Künste – Diskurse*, edited by Benjamin Beil, Gundolf S. Freyermuth, and Lisa Gotto, 145-184. Bielefeld: Transcript, 2015.

Herman, David. "Action Theory." In *Routledge Encyclopedia of Narrative Theory*, edited by David Herman, Manfred Jahn, and Marie-Laure Ryan, 2-3. London: Routledge, 2005.

Heron, Michael J. and Pauline H. Belford. "All of Your Co-Workers are Gone: Story, Substance, and the Emphatic Puzzler." *Journal of Games Criticism*2, no. 1 (January 2015). http://gamescriticism.org/articles/heronbelford-2-1

Huizinga, Johan. *Homo Ludens: A Study of the Play Element in Culture*. Boston: The Beacon Press, 1955.

Huxley, Aldous. *Brave New World*. London: Chatto & Windus Vintage, 2004.

—. *Brave New World Revisited*. London: Chatto & Windus Vintage, 2004.

Huxley, Thomas. H. "Evolution and Ethics (the Romanes Lecture, 1893)." In *Evolution and Ethics 1893-1943*, edited by T. H. Huxley and J. S. Huxley. London: Pilot Press, 1947.

Ince, Steve. *Writing for Video Games*. London: A & C Black, 2009.

Inderst, Rudolf. "Stoppt den Endsieg! Das Videospiel *Turning Point* als *Alternative History*: eine digitale Fortsetzung Antinationalsozialistischer Dystopien." *Contact, Conflict, Combat: Zur Tradition des Konflikts in digitalen Spielen,* edited by Rudolf T. Inderst and Peter Just, 155-182. Boizenburg: Werner Hülsbusch, 2011.

—. *Vergemeinschaftung in MMORPGs*. Boizenburg: Werner Hülsbusch, 2009.

Ingarden, Roman. *The Literary Work of Art: An Investigation on the Boundaries of Ontology, Logic, and the Theory of Literature*. Translated by George G. Grabowicz. Evanston: Northwestern UP, 1973.

Iser, Wolfgang. *The Act of Reading: A Theory of Aesthetic Response*. Translated by Wilhelm Fink. Baltimore: John Hopkins UP, 1978.

—. *The Fictive and the Imaginary: Charting Literary Anthropology*. Translated by Frankfurt a.M.: Suhrkamp Verlag. Baltimore: John Hopkins UP, 1993.

Ivănescu, Andra. "The Music of Tomorrow, Yesterday! Music, Time and Technology in *BioShock Infinite*." *Time and Technology in Popular Culture, Media and Communication* 7, no. 2 (July 2014): 51-66. http://ojs.meccsa.org.uk/index.php/netknow/issue/view/34

Jackson, Robert. *BioShock: Decision, Forced Choice and Propaganda*. Winchester: Zero Books, 2014.

Jacobs, Stephen. "The Basics of Narrative." In *Game Writing: Narrative Skills for Videogames*, edited by Chris Bateman, 25-42. Boston, Mass.: Charles River Media, 2007.

Jagoda, Patrick. "Digital Games and Science Fiction." In *The Cambridge Companion to American Science Fiction*, edited by Eric C. Link and Gerry Canavan, 139-152. New York: Cambridge UP, 2015.

Jameson, Fredric. *Archaeologies of the Future: The Desire Called Utopia and Other Science Fictions*. London: Verso, 2005.

—. *The Seeds of Time.* New York: Columbia UP, 1994.

Jenkins, Henry. "Game Design as Narrative Architecture." In *First Person: New Media as Story, Performance, and Game*, edited by Noah Wardrip-Fruin and Pat Harrigan, 118-130. Cambridge, Mass.: MIT Press, 2004.

—. "The Art of Contested Spaces." *Publications Henry Jenkins*, 2002. *MIT.edu*. Accessed 10. Feb. 2014. http://web.mit.edu/21fms/People/henry3/contested spaces.html.

Juul, Jesper. *Half-Real: Video Games between Real rules and Fictional Worlds.* Cambridge, Mass.: MIT Press, 2005.

Keogh, Brendon. "A Play of Bodies: A Phenomenology of Videogame Experience." PhD. diss., RMIT University, 2015. *Researchbank.rmit*. https://re searchbank.rmit.edu.au/eserv/rmit:161442/Keogh.pdf

Kim, John. "The Threefold Model FAQ." *Darkshire*, 1997-2007. Accessed 28 Jun. 2016. http://www.darkshire.net/jhkim/rpg/theory/threefold/faq_v1.html.

Klevjer, Rune. "What is the Avatar? Fiction and Embodiment in Avatar-Based Singleplayer Computer Games." Phd. diss., University of Bergen, 2006. *Semanticscholar*. https://pdfs.semanticscholar.org/b666/b2b75bb7923607a2cb2 02ff6c222707d3703.pdf

Kumar, Krishan. *Utopia and Anti-Utopia in Modern Times*. Oxford: Basil Blackwell, 1987.

—. "Utopia's Shadow." In *Dystopia(n) Matters: On the Page, on Screen, on Stage*, edited by Fátima Vieira, 19-22. Newcastle-upon-Tyne: Cambridge Scholar Publishing, 2013.

Kücklich, Julian. "The Playability of Texts vs. The Readability of Games: Towards a Holistic Theory of Fictionality." *Proceedings of the 2003 DiGRA International Conference: Level Up* 2 (2003): 100-107. http://www.di gra.org/digital-library/publications/the-playability-of-texts-vs-the-readabil ity-of-games-towards-a-holistic-theory-of-fictionality/

Lange, Carsten. “>>Der Freie hat die Wahl, der Sklave gehorcht.<< – Entscheidungsfreiheit und Determinismus in *BioShock* und *BioShock Infinite*.” In *>>I’ll remeber this<< Funktion, Inszenierung und Wandel von Entscheidung im Computerspiel*, 105-124. Boizenburg: Werner Hülsbusch, 2016.

Le Guin, Ursula K. *The Ones Who Walk Away From Omelas: A Story*. New York: Harper Perennial, 2017.

Levitas, Ruth, and Lucy Sargisson. “Utopia in Dark Times: Optimism/Pessimism and Utopia/Dystopia.” In *Dark Horizons: Science Fiction and the Dystopian Imagination,* edited by Raffaella Baccolini and Tom Moylan 13-27. New York: Routledge, 2003.

Lévy, Pierre. *Becoming Virtual: Reality in the Digital Age*. Translated by Robert Bonono. New York: Plenum Trade, 1998.

Lindley, Craig A. “Story and Narrative Structures in Computer Games.” In *Developing Interactive Narrative Content: sagas/sagasnet reader*, edited by Brunhild Bushoff. München: High Text, 2005.

Liwinski, Mateusz. “Nostalgia for Dystopia: Critical Nostalgia in Lucas Pope’s *Papers, Please*.” *(Im)perfection Subverted, Reloaded and Networked: Utopian Discourse across Media.* Peter Lang Edition, edited by Barbara Klonowska, Zofia Kolbuszewska, and Grzegorz Maziarczyk, 223-234. Frankfurt a.M.: Internationaler Verlag der Wissenschaften, 2015.

Lizardi, Ryan. “*Bioshock*: Complex and Alternate Histories.” *Game Studies: The International Journal of Computer Game Research* 14, no. 1 (August 2014). http://gamestudies.org/1401/articles/lizardi

Maziarczyk, Grzegorz. “Playable Dystopia? Interactivity and Narrativity in *BioShock* and *BioShock Infinite*.” *(Im)perfection Subverted, Reloaded and Networked: Utopian Discourse across Media.* Peter Lang Edition, edited by Barbara Klonowska, Zofia Kolbuszewska, and Grzegorz Maziarczyk, 235-254. Frankfurt a.M.: Internationaler Verlag der Wissenschaften, 2015.

McCarthy, John and Peter Wright. *Technology as Experience*. Cambridge, Mass.: The MIT Press, 2004.

McHale, Brian. “Towards a Poetics of Cyberpunk.” *Beyond Cyberpunk: New Critical Perspectives*, edited by Graham J. Murphy and Sherryl Vint, 3-28. London: Routledge, 2010.

McMahan, Alison. “Immersion, Engagement, and Presence: A Method for Analysing 3-D Video Games.” In *The Video Game Theory Reader*, edited by Mark J.P. Wolf and Bernard Perron, 67-86. New York: Routledge, 2003.

Meskin, Aaron and Jon Robeson. “Fiction and Fictional Worlds in Videogames.” *Philosophy of Computer Games* (2009). https://www.academia.edu/244532/Fiction_and_Fictional_Worlds_in_Videogames

Middeke, Martin. "Reception Theory." *English and American Studies: Theory and Practice*, edited by Martin Middeke, Timo Müller, Christina Wald, and Hubert Zapf, 191-196. Stuttgart: J.B. Metzler, 2012.

Milgram, Stanley. *Obedience to Authority: An Experimental View*. New York: Perennial Classics, 2004.

More, Thomas. *Utopia*. Translated by Paul Turner. London: Penguin Group, 2009.

—. "Letter 15: To Peter Giles [41A]." In *For All Seasons – Selected Letters of Thomas More*, edited by Stephen Smith. New York: Sceptre Publishing, 2012.

Moylan, Tom. *Scraps of the Untainted Sky: Science Fiction, Utopia, Dystopia*. Boulder, Colo.: Westview Press, 2000.

—. "Step into the Story." In *Dystopia(n) Matters: On the Page, on Screen, on Stage*, edited by Fátima Vieira, 42-43.Newcastle-upon-Tyne: Cambridge Scholar Publishing, 2013.

Murray, Janet H. *Hamlet on The Holodeck: The Future of Narrative in Cyberspace*. Cambridge, Mass.: MIT Press, 1998.

Mäyrä, Frans. *An Introduction to Game Studies: Games in Culture*. London: SAGE, 2008.

Möring, Sebastian. "Games and Metaphor – A Critical Analysis of the Metaphor Discourse in Game Studies." Phd. diss., IT University of Copenhagen, 2013. *ITU.dk*. https://en.itu.dk/~/media/en/research/phd-programme/phd-defences/2013/20130929-full-thesis-sebastian-moering-itupdf.pdf?la=en

Navarro-Remesal, Victor. "Regarding the (Game) Pain of Others: Suffering and Compassion in Video Games." Lecture presented at Concerns about Video Games and the Video Games of Concern Conference, IT University of Copenhagen, 2016.

Neitzel, Britta. "Medienrezeption und Spiel." In *Game Over!? Perspektiven des Computerspiels,* edited by Jochen Distelmeyer, Christine Hank, and Dieter Mersch, 95-113. Bielefeld: Transcript, 2008.

—. "Narrativity in Computer Games." *Handbook of Computer Game Studies*, edited by Joost Raessens and Jeffrey Goldstein, 227-245. Cambridge, Mass.: MIT Press, 2005.

Neumann, Birgit and Ansgar Nünning. *An Introduction to the Study of Narrative Fiction*. Stuttgart: Klett Lernen und Wissen, 2008.

Newman, James. *Videogames*. London: Routledge, 2004.

Newzoo, "The Global Games Market Reaches $99.6 Billion in 2016, Mobile Generating 37%." Accessed 25 Sep. 2017. https://newzoo.com/insights/articles/global-games-market-reaches-99-6-billion-2016-mobile-generating-37/

Nitsche, Michael. *Video Game Spaces: Image, Play, and Structure in 3D Game Worlds*. Cambridge Mass.: MIT Press, 2008.

Nyman, Elizabeth and Ryan L. Teten. "Lost and Found and Lost Again: Island Utopias and Dystopia in the *BioShock* Series." *Games and Culture* 13, no. 4 (November 2015): 1-15. http://gac.sagepub.com/content/early/2015/11/19/1555412015616510

Origitano, Catlyn. "'The Cage is Somber': A Feminist Understanding of Elizabeth." *BioShock and Philosophy: Irrational Game, Rational Book*. Blackwell Philosophy and Pop Culture Series, edited by Luke Cuddy, 38-48. New York: Wiley Blackwell, 2015.

Orwell, George. *Nineteen Eighty-Four.* London: Penguin Group, 1987.

Packer, Joseph. "The Battle for Galt's Gulch: *Bioshock* as Critique of Objectivism." *Journal of Gaming and Virtual Worlds* 2, no. 3 (December 2010): 209-224. https://www.intellectbooks.co.uk/journals/view-Article,id=10275/

Parker, Felan. "Canonizing *Bioshock*: Cultural Value and the Prestige Game." *Games and Culture* (2015): 1-25. http://gac.sagepub.com/content/early/2015/08/28/1555412015598669

Pérez-Latorre, Óliver, Mercè Oliva, and Reinald Besalú. "Videogame Analysis: A Social-Semiotic Approach." *Social Semiotics* 7, no. 5 (June 1, 2017): 1-18, http://www.tandfonline.com/doi/full/10.1080/10350330.2016.1191146

Perron, Bernard. *Silent Hill: The Terror Engine*. Ann Arbor: University of Michigan Press, 2012.

Peters, Ryan F. "The Global Dystopian: Twenty-First Century Globalization, Terrorism, and Urban Destruction." PhD. diss., Loyola University Chicago, 2015. *eCommons*. http://ecommons.luc.edu/luc_diss/1964/

Pfister, Eugen. "Jacques Lacan, Caspar David Friedrich und die Zombie-Apokalypse: Eine Erste Annäherung an Mythen in 'The Last of US'." *Spiel-Kultur Wissenschaft Mythen im Digitalen Spiel* (2015): 1-6. http://spielkult.hypotheses.org/59

Prince, Gerald. "Narrativity." In *Routledge Encyclopedia of Narrative Theory*, edited by David Herman, Manfred Jahn, and Marie-Laure Ryan, 387-388. London: Routledge, 2005.

Pötzsch, Holger. "Selective Realism: Filtering Experiences of War and Violence in First- and Third-Person Shooters." *Games and Culture* 12, no. 2 (2017): 166-178. http://journals.sagepub.com/doi/abs/10.1177/1555412015587802

Rancière, Jacques. *The Emancipated Spectator*. London: Verso, 2009.

Robinson, Kim S. *Future Primitive: The New Ecotopias*. New York: Tom Doherty Associates, 1994.

Ryan, Marie-Laure. *Avatars of Story*. Minneapolis: University of Minnesota Press, 2006.

—. *Narrative as Virtual Reality: Immersion and Interactivity in Literature and Electronic Media*. Baltimore: John Hopkins UP, 2001.

—. *Narrative as Virtual Reality 2: Revisiting Immersion and Interactivity in Literature and Electronic Media*. Baltimore: John Hopkins UP, 2015.

—. "Towards a Definition of Narrative." In *The Cambridge Companion to Narrative*, edited by David Herman, 22-38. Cambridge: Cambridge UP, 2007.

Saad, Ahmet. "Writing for Sandbox Games." In *Writing for Video Game Genres: From FPS to RPG*, edited by Wendy Despain, 137-148. Boca Raton: A K Peters, 2009.

Sainsbury, Richard M. *Fiction and Fictionalism: New Problems of Philosophy*. London: Routledge, 2010.

Salen, Katie and Eric Zimmerman. *Rules of Play: Game Design Fundamentals*. Cambridge, Mass.: MIT Press, 2004.

Sargent, Lyman T. "Do Dystopias Matter?" In *Dystopia(n) Matters: On the Page, on Screen, on Stage*, edited by Fátima Vieira, 10-13. Newcastle-upon-Tyne: Cambridge Scholar Publishing, 2013.

—. "The Problems of the 'Flawed Utopia'." *Dark Horizons: Science Fiction and the Dystopian Imagination*, edited by Raffaella Baccolini and Tom Moylan. 225-231. New York: Routledge, 2003.

—. "The Three Faces of Utopianism Revisited." *Utopian Studies*5, no. 1, (1994): 1-37. http://www.jstor.org/discover/10.2307/20719246?uid=3737864&uid=2129&uid=2&uid=70&uid=4&sid=21101560295713

—. *Utopianism: A Very Short Introduction*. Oxford: Oxford UP, 2010.

—. "U.S. Eutopias in the 1980s and 1990s: Self-Fashioning in a World of Multiple Identities." In *Utopianism/Literary Utopias and National Culture Identities: A Comparative Perspective*, edited by Paola Spinozzi, 221-232. COTEPRA/University of Bologna, 2001.

Sargisson, Lucy. "Dystopias Do Matter." In *Dystopia(n) Matters: On the Page, on Screen, on Stage*, edited by Fátima Vieira, 40-41. Newcastle-upon-Tyne: Cambridge Scholar Publishing, 2013.

Schmeink, Lars. "Dystopia, Alternate History and the Posthuman in Bioshock." *Current Objectives of Postgraduate American Studies* 10 (2009,). http://copas.uni-regensburg.de/article/view/113

Schnieders, Paul C. *The Books of Enoch*: Complete Edition. Translated by Robert H. Charles. Las Vegas: International Alliance Pro Publishing, 2012.

Schulzke, Marcus. "The Bioethics of Digital Dystopias." *International Journal of Technoethics* 4, no. 2 (2013): 46-57. https://www.igi-global.com/article/the-bioethics-of-digital-dystopias/90488

—. "The Critical Power of Virtual Dystopias." *Games and Culture* 9, no. 5 (July 2014): 315-334. http://gac.sagepub.com/content/9/5/315

Seeber, Hans U. *Die Selbstkritik der Utopie in der Angloamerikanischen Literatur*. Münster: LIT, 2003.

Sharp, John. "Perspective." *The Routledge Companion to Video Game Studies*, edited by Mark J. P. Wolf and Bernard Perron, 107-116. New York: Routledge, 2014.

Sicart, Miguel. "Against Procedurality." *Game Studies* 11, no. 3 (December 2011). http://gamestudies.org/1103/articles/sicart_ap

—. *Beyond Choices: The Design of Ethical Gameplay*. Cambridge, Mass.: MIT Press, 2013.

—. *Play Matters*. Cambridge, Mass.: MIT Press, 2014.

—. *The Ethics of Computer Games*. Cambridge, Mass.: MIT Press, 2009.

Soulban, Lucien and Haris Orkin. "Writing for First-Person Shooters." In *Writing for Video Game Genres: From FPS to RPG*, edited by Wendy Despain, 51-68. Boca Raton: A K Peters, 2009.

Stableford, Brian. "Science Fiction and Ecology." In *A Companion to Science Fiction*, edited by David Seed, 127-141. Malden: Blackwell Publishing, 2005.

Strasen, Sven. *Rezeptionstheorien: Literatur-, sprach- und kulturwissenschaftliche Ansätze und kulturelle Modelle*. Trier: Wissenschaftlicher Verlag Trier, 2008.

Stuart, Keith. "*The Last of Us*, *BioShock: Infinite* and why all Video Game Dystopias Work the Same." *Guardian*, July 1, 2013. Accessed 25 Oct. 2016. https://www.theguardian.com/technology/gamesblog/2013/jul/01/last-of-us-bioshock-infinite-male-view.

Sustainable Human. "How Wolves Change Rivers." YouTube, February 13, 2014. https://www.youtube.com/watch?v=ysa5OBhXz-Q

Suvin, Darko. *Metamorphoses of Science Fiction: On the Poetics and History of a Literary Genre.* New Haven: Yale UP, 1979.

—. "Theses on Dystopia 2001." In *Dark Horizons: Science Fiction and the Dystopian Imagination,* edited by Raffaella Baccolini and Tom Moylan, 187-202. New York: Routledge, 2003.

Swift, Jonathan. *Gulliver's Travels*. London: Wordsworth Editions Limited, 1992.

Tanenbaum, Karen and Joshua Tanenbaum. "Commitment to Meaning: A Reframing of Agency in Games." *Proceedings of Digital Arts and Culture Conference* (2009). http://escholarship.org/uc/item/6f49r74n?query=tanenbaum #page-1

Tavinor, Grant. "Art and Aesthetics." In *The Routledge Companion to Video Game Studies*, edited by Mark J. P. Wolf and Bernard Perron, 59-66. New York: Routledge, 2014.

—. "*BioShock* and the Art of Rapture." *Philosophy and Literature* 33, vol. 1 (2009): 91-106. https://philpapers.org/rec/TAVBAT

—. "Fiction." In *The Routledge Companion to Video Game Studies*, edited by Mark J. P. Wolf and Bernard Perron, 434-421. New York: Routledge, 2014.

—. *The Art of Videogames*. Oxford: Wiley-Blackwell, 2009.

THE MATRIX, directed by Larry Wachowski and Andy Wachowski (Warner Bros, 1999).

Therrien, Carl. "Immersion." In *The Routledge Companion to Video Game Studies*, edited by Mark J. P. Wolf and Bernard Perron, 451-458. New York: Routledge, 2014.

Thoreau, Henry D. *Walden and Civil Disobedience*. New York: Barnes & Noble Classics, 2003.

Tosca, Susana P. "*Amnesia: The Dark Descent:* The Player's Very Own Purgatory." In *Analyzing Digital Fiction*, edited by Alice Bell, Astrid Ensslin, and Hans K. Rustad, 109-123. London: Routledge, 2014.

—. "Transmedial Museum Experiences: the Case of Moesgaard." *NODO: Transmedial Narratives* 18 (2016). https://www.researchgate.net/publication/312024183_nodo_18_3067-13176-1-PB

WESTWORLD. Season 1, episode 7, "Tromp L'Oeil." Directed by Jonathan Nolan and Lisa Joy. Aired November 13, 2016, on HBO. Warner Bros. Television, 2016.

Tulloch, Rowan. "Ludic Dystopias: Power, Politics, and Play." *Proceedings of the Sixth Australasian Conference on Interactive Entertainment*, no. 13 (2009). http://dl.acm.org/citation.cfm?id=1746063.

Van den Berg, Thijs. "Playing at Resistance to Capitalism: *BioShock* as the Reification of Neoliberal Ideals." *Reconstruction* 12, no. 2 (2012). https://reconstruction.eserver.org/122/vandenBerg.shtml

Vella, Daniel. "Modelling the Semiotic Structure of Game Characters." *Proceedings of DiGRA 2014: <Verb that ends in 'ing'> the <noun> of Game <plural noun>* 8 (2014): 1-19. http://www.digra.org/digital-library/publications/modeling-the-semiotic-structure-of-game-characters/

—. "The Ludic Subject and the Ludic Self: Analyzing the 'I-in-the-Gameworld'." PhD Diss., IT University of Copenhagen, 2015. *ITU.dk*. https://en.itu.dk/~/media/en/research/phd-programme/phd-defences/2015/daniel-vella---the-ludic-subject-and-the-ludic-self-final-print-pdf.pdf?la=en

Vieira, Fátima, editor. *Dystopia(n) Matters: On the Page, on Screen, on Stage*. Newcastle-upon-Tyne: Cambridge Scholar Publishing, 2013.

—. "Introduction." In *Dystopia(n) Matters: On the Page, on Screen, on Stage*, edited by Fátima Vieira, 1-7. Newcastle-upon-Tyne: Cambridge Scholar Publishing, 2013.

—. "The Concept of Utopia." *The Cambridge Companion to Utopian Literature*, edited by Gregory Claeys, 3-27. Cambridge: Cambridge UP, 2010.

Voorhees, Gerald. "Daddy Issues: Constructions of Fatherhood in *The Last of Us* and *BioShock Infinite*." *Ada: A Journal of Gender, New Media, and Technology*, no. 9 (2016). http://adanewmedia.org/2016/05/issue9-voorhees/

Walsh, Chad. *From Utopia to Nightmare*. London: Geoffrey Bles, 1962.

Walton, Kendall L. *Mimesis as Make-Believe: On the Foundations of the Representational Arts*. Cambridge, Mass.: Harvard UP, 1990.

Walz, Steffen P. *Toward a Ludic Architecture: The Space of Play and Games*. ETC Press, 2010.

Wardrip-Fruin, Noah and Pat Harrigan, editors. *First Person: New Media as Story, Performance and Game*. Cambridge, Mass.: MIT Press, 2004.

Wardrip-Fruin, Noah, Michael Mateas, Steven Dow, and Serdar Sali. "Agency Reconsidered." *Proceedings of the 2009 DiGRA International Conference: Breaking New Ground: Innovation in Games, Play, Practice and Theory* 5 (2009): 1-9. http://www.digra.org/digital-library/publications/agency-reconsidered/

Wark, McKenzie. *Gamer Theory*. Cambridge, Mass: Harvard UP, 2007.

Weimin, Toh. "The Limits of the Evolution of Female Characters in the *BioShock* Franchise." *Proceedings of DiGRA 2015: Diversity of Play: Games – Cultures – Identities* 12 (2015): 1-15. http://www.digra.org/digital-library/publications/the-limits-of-the-evolution-of-female-characters-in-the-bioshock-franchise/

Wells, Herbert G. *The War of the Worlds*. Mineola: Dover Publications, 1997.

Welsh, Oli. "*The Last of Us* Review: Journey's End: Our Original Review, to mark the Release of *The Last of Us Remastered*." *Eurogamer*, July 31, 2014. Accessed 12 Feb. 2015. http://www.eurogamer.net/articles/2014-07-28-the-last-of-us-review

Wilder, Shannon OB. "A Narrative of Synthetic Fear: Virtualizing Dystopia in a Gaming World." PhD Diss., University of Georgia, 2006. *Open Access Theses and Dissertations*. https://oatd.org/oatd/record?record=oai%5C%3Augakr.libs.uga.edu%5C%3A10724%5C%2F23760

Willumsen, Ea Christina. "Source Code and Formal Analysis: A Hermeneutic Reading of *Passage*." *Proceedings of the First International Joint Conference of DiGRA and FGD* 13, no. 1 (2016): 1-10. http://www.digra.org/digital-library/publications/source-code-and-formal-analysis-a-hermeneutic-reading-of-passage/.

Wysocki, Jan. "Die Präsentation amerikanischer Religion im Spiel *BioShock Infinite* am Beispiel des Charakters des Propheten." Master's thesis, University of Heidelberg, 2014.

Zamyatin, Yevgeny. *We*. Translation by Mirra Ginsburg. New York: Harper Voyager, 1983.

Žižek, Slavoj. *Welcome to the Desert of the Real! Five Essays on September 11 and Related Dates*. London: Verso, 2002.